Common Weeds and their Effects of Agriculture

Common Weeds and their Effects of Agriculture

Nasreen Akhter

RANDOM

RANDOM PUBLICATIONS
NEW DELHI (INDIA)

Common Weeds and their Effects of Agriculture

ISBN 978-93-5111-775-9

Published in 2016 in India by

RANDOM PUBLICATIONS

4376-A/4B, Gali Murari Lal, Ansari Road
New Delhi-110 002
Phone : +9111-43580356, 011-23289044, 011-43142548
e-mail: sales@randompublications.com,
info@randompublications.com, randomexports@gmail.com

Type Setting by : Friends Media, Delhi-110089
Digitally Printed at : Replika Press Pvt. Ltd.

Preface

A weed is a plant considered undesirable in a particular situation, "a plant in the wrong place". Examples commonly are plants unwanted in human-controlled settings, such asfarm fields, gardens, lawns, and parks.

Aquatic insects are one of the most important components of aquatic ecosystems. Providing food for countless vertebrates as well as feeding upon equally countless numbers of lesser animals, are just a few of their important actions. Only a few insects, 3-5 per cent, have adapted successfully to the aquatic environment. In particular, the order Hemiptera stands out as an important group of aquatic insects. Only the order Coleoptera contains more aquatic forms; however, a smaller percentage of the Coleoptera is aquatic. The reason it comprises so many aquatic forms is because it is a larger order overall, not because its members are specifically adapted for the underwater environment. Additionally, the ecological effect of aquatic Coleoptera is probably less than that of the aquatic Hemiptera. This makes Hemiptera arguably the most important group of aquatic insects. The objective of this chapter is to summarize the unique adaptations and ecological importance of the six families of Hemiptera that are fully aquatic.

An alternate definition often used by biologists is any species, not just plants, that can quickly adapt to any environment.Some traits of weedy species are the ability to reproduce quickly, disperse widely, live in a variety of habitats, establish a population in strange places, succeed in disturbed ecosystems and resist eradication once established. Such species often do well in human-dominated environments as other species are not able to adapt. Common examples include the common pigeon,brown rat and the raccoon. Other weedy species have been able to expand their range without actually living in human environments, as human activity has damaged the ecosystems of other species.

The book will provide precious information and serve as a reference material for researchers, practicing engineers, managers, and professionals working on all aspects of aquatic weeds.

– Author

Contents

1

Introduction to Weed

A *weed* is a plant considered undesirable in a particular situation, "a plant in the wrong place". Examples commonly are plants unwanted in human-controlled settings, such asfarm fields, gardens, lawns, and parks. Taxonomically, the term "weed" has no botanical significance, because a plant that is a weed in one context is *not* a weed when growing in a situation where it *is* in fact wanted, and where one species of plant is a valuable crop plant, another species in the same genus might be a serious weed, such as a wildbramble growing among cultivated loganberries. Many plants that people widely regard as weeds also are intentionally grown in gardens and other cultivated settings. The term also is applied to any plant that grows or reproduces aggressively, or is invasive outside its native habitat. More broadly "weed" occasionally is applied pejoratively to species outside the plant kingdom, species that can survive in diverse environments and reproduce quickly; in this sense it has even been applied to humans.

weed: "A herbaceous plant not valued for use or beauty, growing wild and rank, and regarded as cumbering the ground or hindering the growth of superior vegetation... Applied to a shrub or tree, especially to a large tree, on account of its abundance in a district... An unprofitable, troublesome, or noxious growth."

— *The New shorter Oxford English dictionary on historical principles*

ECOLOGICAL SIGNIFICANCE

Certain classes of weeds share adaptations to ruderal environments, that is to say, disturbed environments where soil or natural vegetative cover has been damaged or frequently gets damaged, disturbances that give the weeds advantages over desirable crops, pastures, or ornamental plants. The nature of the habitat and its disturbances will affect or even determine which types of weed communities become dominant.

Examples of such ruderal or pioneer species include plants that are adapted to naturally occurring disturbed environments such as dunesand other windswept areas with shifting soils, alluvial flood plains, river banks and deltas, and areas that are burned repeatedly. Since human agricultural practices often mimic these natural environments where weedy species have evolved, some weeds are effectivelypreadapted to grow and proliferate in human-disturbed areas such as agricultural fields, lawns, roadsides, and construction sites. The weedy nature of these species often gives them an advantage over more desirable crop species because they often grow quickly andreproduce quickly, they commonly have seeds that persist in the soil seed bank for many years, or they may have short lifespans with multiple generations in the same growing season. In contrast, perennial weeds often have underground stems that spread under the soil surface or, like ground ivy (*Glechoma hederacea*), have creeping stems that root and spread out over the ground.

Fig. A dandelion is a common plant all over the world, especially in Europe, Asia, and the Americas. It is a well-known example of a plant that is considered a weed in some contexts (such as lawns) but not a weed in others (such as when it is used as a leaf vegetable or herbal medicine).

Some plants become dominant when introduced into new environments because the animals in their original environment, that compete with them or feed on them are absent; in what is sometimes called the "natural enemies hypothesis", plants freed from these specialist consumers may become dominant. An example is Klamath weed, that threatened millions of hectares of prime grain and grazing land in North America after it was accidentally introduced, but was reduced to a rare roadside weed within several years after some of its natural enemies were imported during World War II. In locations where predation and mutually competitive relationships are absent, weeds have increased resources available for growth and reproduction. The weediness of some species that are introduced into new environments may be caused by their production of allelopathic chemicals which indigenous plants are not yet adapted to, a scenario sometimes called the "novel weapons hypothesis". These chemicals may limit the growth of established plants or the germination and growth of seeds and seedlings.

Another of the ways in which the ecological role of a plant can make it a weed even if it is in itself inoffensive, is if it harbours a pest that is dependent on it for survival; for example, *Berberis* species are intermediate hosts for stem rust fungi, so that they promote serious damage to wheat crops when growing near the fields.

Competition with cultivated and endemic plants

A number of native or non-native plants are unwanted in a specific location for a number of reasons. An important one is that they interfere with food and fiber production in agriculture, wherein they must be controlled in order to prevent lost or diminished crop yields. Other important reasons are that they interfere with other cosmetic, decorative, or recreational goals, such as in lawns, landscape architecture, playing fields, and golf courses. Similarly, they can be of concern for environmental reasons whereby introduced species out-compete for resources or space with desired endemic plants. For all these reasons; horticulture, both functional and cosmetic, andenvironmental, - weeds interfere by:

- competing with the desired plants for the resources that a plant typically needs, namely, direct sunlight, soil nutrients, water, and (to a lesser extent) space for growth;
- providing hosts and vectors for plant pathogens, giving them greater opportunity to infect and degrade the quality of the desired plants;
- providing food or shelter for animal pests such as seed-eating birds and Tephritid fruit flies that otherwise could hardly survive seasonal shortages;
- offering irritation to the skin or digestive tracts of people or animals, either physical irritation via thorns, prickles, or burs, or chemical irritation via natural poisons or irritants in the weed (for example, the poisons found in *Nerium* species);
- causing root damage to engineering works such as drains, road surfaces, and foundations, blocking streams and rivulets.

In weed ecology some authorities speak of the relationship between "the three Ps": plant, place, perception. These have been very variously defined, but the weed traits listed by H.G. Baker are widely cited.

Weeds have long been a concern, perhaps as long as humans have cultivated plants. They are mentioned in various historic texts, such as a Shakespearean sonnet: "To thy fair flower add the rank smell of weeds: / But why thy odour matcheth not thy show, / The soil is this, that thou dost common grow." and the Bible: "Cursed is the ground because of you; through painful toil you will eat of it all the days of your life. It will produce thorns and thistles for you, and you will eat the plants of the field. By the sweat of your brow you will eat your food until you return to the ground,"

BENEFITS OF WEED SPECIES

While the term "weed" generally has a negative connotation, many plants known as weeds can have beneficial properties. A number of weeds, such as the dandelion (*Taraxacum*) and lamb's quarter, are edible, and their leaves or roots may be used for food or herbal medicine. Burdock is common over much of the world, and is sometimes used to make soup and medicine inEast Asia. Some weeds attract beneficial insects, which in turn can protect crops from harmful pests. Weeds can also prevent pest insects from finding a crop, because their presence disrupts the incidence of positive cues which pests use to locate their food. Weeds may also act as a "living mulch", providing ground cover that reduces moisture loss and prevents erosion. Weeds may also improve soil fertility; dandelions, for example, bring up nutrients like calcium and nitrogen from deep in the soil with their tap root, and clover hosts nitrogen-fixing bacteria in its roots, fertilizing the soil directly. The dandelion is also one of several species which break up hardpanin overly cultivated fields, helping crops grow deeper root systems. Some garden flowers originated as weeds in cultivated fields and have been selectively bred for their garden-worthy flowers or foliage. An example of a crop weed that is grown in gardens is the corncockle, (*Agrostemma githago*), which was a common weed in European wheat fields, but is now sometimes grown as a garden plant.

Some people have appreciated weeds for their tenacity, their wildness and even the work and connection to nature they provide. As Christopher Lloyd wrote in *The Well-Tempered Garden*

"Many gardeners will agree that hand-weeding is not the terrible drudgery that it is often made out to be. Some people find in it a kind of soothing monotony. It leaves their minds free to develop the plot for their next novel or to perfect the brilliant repartee with which they should have encountered a relative's latest example of unreasonableness."

Role in mass extinctions

A mass extinction is generally caused by some abrupt disruption to the entire planet's environment. This results in major changes in habitat worldwide, and most endemic species, specially adapted to a single habitat, cannot survive in the new habitats. Thus only weedy species survive, and they dominate the planet in the immediate aftermath.Cockroaches, for example, have survived several mass extinctions. The current Holocene extinction event, then, could lead to a planet inhabited entirely by what are known today as weeds. The fossil record indicates that after mass extinctions, a weed-dominated planet persists for five to ten million years before life re-diversifies.

DISPERSAL

Many weed species have moved out of their natural geographic ranges and spread around the world in tandem with human migrations and commerce. Weed seeds are often collected and transported with crops after the harvesting of grains, so humans are a vector of transport as well as a producer of the disturbed environments to which weed species are well adapted, resulting in many weeds having a close association with human activities.

Some weed species have been classified as noxious weeds by government authorities because, if left unchecked, they often compete with native or crop plants or cause harm to livestock. They are often foreign species accidentally or imprudently imported into a region where there are few natural controls to limit their population and spread.

Weeds as adaptable species

An alternate definition often used by biologists is any species, not just plants, that can quickly adapt to any environment.Some traits of weedy species are the ability to reproduce quickly, disperse widely, live in a variety of habitats, establish a population in strange places, succeed in disturbed ecosystems and resist eradication once established. Such species often do well in human-dominated environments as other species are not able to adapt. Common examples include the common pigeon,brown rat and the raccoon. Other weedy species have been able to expand their range without actually living in human environments, as human activity has damaged the ecosystems of other species. These include the coyote, the white-tailed deer and the brown headed cowbird.

In response to the idea that humans may face extinction due to environmental degradation, paleontologist David Jablonsky counters by arguing that humans are a weed species. Like other weedy species, humans are widely dispersed in a wide variety of environments, and are highly unlikely to go extinct no matter how much damage the environment faces.

PLANTS OFTEN CONSIDERED TO BE WEEDS

White clover is considered by some to be a weed in lawns, but in many other situations is a desirable source of fodder, honey and soil nitrogen. A short list of some plants that often are considered to be weeds follows:

- Amaranth - ("pigweed") annual with copious long-lasting seeds, also a highly edible and resilient food source
- Bermuda grass - perennial, spreading by runners, rhizomes and seeds.
- Bindweed
- Broadleaf plantain – perennial, spreads by seeds that persist in the soil for many years
- Burdock – biennial
- Common lambsquarters - annual
- Creeping charlie – perennial, fast-spreading plants with long creeping stems
- Dandelion – perennial, wind-spread, fast-growing, and drought-tolerant
- Goldenrod – perennial
- Japanese knotweed
- Kudzu – perennial
- Leafy spurge – perennial, with underground stems
- Milk thistle – annual or biennial
- Poison ivy – perennial
- Ragweed – annual
- Sorrel – annual
- *Striga*
- St John's wort - perennial
- Sumac – woody perennial
- Tree of heaven - woody perennial
- Wild carrot – biennial
- Wood sorrel – perennial
- Yellow nutsedge - perennial

WEED CONTROL

Weeds are plants that some people view as undesirable in a particular place. Throughout the long human history of horticulture, people have worked to control weeds for many reasons. Weed control is a highly developed field of knowledge.

Weed control methods vary according to the growth habit of the weeds in questions, as well as the context. For example, different methods of weed control may be used on a food crop versus a fiber crop or a golf course, because there is often more concern about health effects of chemicals used on food crops, because they are ingested.

Weeds can be categorized by their life habit. They can generally either be grouped as annuals or perennials. An annual weed grows from the seeds dropped in the previous growing season. Perennial weeds regrow from previously established roots, dormant stolons, tubers, rhizomes, as well as the seed. If chemical control is chosen as a method of weed control, the grouping of weeds as annual or perennial is important. If chemical control is chosen, then pre-emergent herbicides would reduce annual weeds whereas perennial weeds would require the use of post-emergent herbicides.

Understanding the habit of weeds is also important for non-chemical methods of weed control, such as plowing, surface scuffling, promotion of more beneficial cover crops, and prevention of seed accumulation in fields. For example, amaranth is an edible plant that is considered a weed by mainstream modern agriculture. It produces copious seeds (up to 1 million per plant) that last many years, and is an early-emergent fast grower. Those seeking to control amaranth quote the mantra "This year's seeds become next year's weeds!" However, another view of amaranth values the plant as a resilient food source.

History

It has long been assumed that weeds, in the sense of rapidly-evolving plants taking advantage of human-disturbed environments, evolved in response to the Neolithic agricultural revolution approximately 12,000 years ago. However, researchers have found evidence of "proto-weeds" behaving in similar ways at Ohalo II, a 23,000-year-old archeological site in Israel.

BENEFICIAL WEED

Fig. Clover was once included in grass seed mixes, because it is a legume that fertilizes the soil

A *beneficial weed* is any of various plants not generally considered domesticated, but which nonetheless has some companion planteffect, or else is edible or somehow beneficial. Beneficial weeds include a great many wildflowers, as well as other weeds that are commonly removed or poisoned.

Soil health

Fig. Dandelions benefit neighboring plant health by bringing up nutrients and moisture with its deep tap root

Although erroneously assumed to compete with neighboring plants for food and moisture, some "weeds" provide the soil with nutrients, either directly or indirectly.

- For example, legumes, such as white clover, if they are colonized by the right bacteria (Rhizobium most often) add nitrogen to the soil through the process of nitrogen fixation, where said-bacteria has a symbiotic relationship with its hosts roots, "fixing" atmospheric nitrogen (combining it with oxygen or hydrogen) making the nitrogen plant-available (NH4 or NO3).
- Others use deep tap roots to bring up nutrients and moisture from beyond the range of normal plants so that the soil improves in quality over generations of that plant's presence.
- Weeds with strong, widespread roots also introduce organic matter to the earth in the form of those roots, turning hard, dense clay dirt into richer, more fertile soil.
- In fact, some common plants like tomatoes and corn will "piggyback" on nearby weeds, allowing their relatively weak root systems to go deeper than they could have alone.

Pest prevention

One way they can do this is to repel insects and other pests through their smell [2], as do alliums and wormwood. Another is to entirely mask a companion's scent, or the pheromones of pest insects, as with ground ivy and wild oregano. Some also are unpleasant to small animals, because of their spines or other features, keeping them away from an area to be protected.

Fig. Crow garlic, like any allium, masks scents from pest insects, protecting neighboring plants

Many weeds protect nearby plants from insect pests.

Trap crops

Some weeds act as trap crops, distracting pests away from valued plants. Insects seeking a food plant search by smell, and then land at random on anything green in the area of the scent. If they land on an edible "weed", they will stay there instead of going on to the intended victim. Sometimes, they actively prefer the trap crop.

Host-finding disruption

Recent studies on host-plant finding have shown that flying pests are far less successful if their host-plants are surrounded by any other plant or even "decoy-plants" made of green plastic, cardboard, or any other green material.

- First, they seek plants by scent. Any "weed" that has a scent reduces the odds of them finding crop plants. Obvious examples are Crow Garlic and Ground Ivy, the former being "wild chives" and the latter a form of wild mint, both dramatically masking both plant scent and insect pheromones. They cut down Japanese beetle infestation, and caterpillar infestation, for example cabbage worm, tomato hornworm, and even squash bugs.

- Second, once an insect is near its target, it avoids landing on dirt, but lands on the nearest green thing. Bare earth gardening helps them home in perfectly on the victim crop. But if one is using "green mulch", even grass or clover, the odds are that they will make what's called an "inappropriate landing" on some green thing they don't want. They will then fly a short distance at random, and land on any other green thing. If they fail to accidentally hit the right kind of plant after several tries, they give up.
- If they are there to lay eggs on the crop, weeds provide one more line of defense: Even if they find the right plant, in order to ensure that they didn't hit on a dying plant or falling leaf, they then make short leaf-to-leaf flights before laying eggs. They must land on the "right kind of leaf" enough times in sequence, before they will risk laying their eggs. The more other greenery is nearby, the harder it is for them to remain on target and get enough reinforcement. Enough "inappropriate landings", and they give up, heading elsewhere.

One scientific study said that simply having clover growing nearby cut the odds of cabbage root flies hitting the right plant from 36% to 7%. Statistically, this is enough to dramatically change the productivity of one's crop, all alone.

COMPANION PLANTS

Fig. Queen Anne's Lace provides shelter to nearby plants, as well as attracting predatory insects that eat pests like caterpillars, and appears to boost the productivity of tomato plants

Many plants can grow intercropped in the same space, because they exist on different levels in the same area, providing ground cover or working as a trellis for each other. This healthier style of horticulture is called forest gardening. Among the obvious benefits are providing a wind break or shelter from noonday sun for more delicate plants.

Green mulch

Conversely, some intercropped provide living mulch effect, used by inhibiting the growth of any weeds that are actually harmful, and creating a humid, cooler microclimate around nearby plants, stabilizing soil moisture more than they consume it for themselves.

Plants such as ryegrass, red clover, and white clover are examples of "weeds" that are living mulches, often welcomed in horticulture.

HERBICIDE

Repel plants or fungi, through a chemical means known as allelopathy. Specific other plants can be bothered by a chemical emission through their roots or air, slowing their growth, preventing seed germination, or even killing them.

BENEFICIAL INSECTS

A common companion plant benefit from many weeds is to attract, or be inhabited by, beneficial insects or other organisms which benefit plants.

For example, wild umbellifers attract predatory wasps and flies, that eat nectar, but reproduce by feeding common garden pests to their offspring.

Likewise, some weeds attract ladybugs or the "good" types of nematode, or provide ground cover for predatory beetles.

Human-useful

- Some beneficial weeds, such as lamb's quarters and purslane, are edible. This list of edible flowers includes many wildflowers that are considered weeds when not planted intentionally. Dandelion is an excellent example of an edible weed (see dandelion wine, dandelion coffee).
- A number of weeds have been proposed as natural alternate sources for latex (rubber), including goldenrod, from which the tires were made on the car famously given byHenry Ford to Thomas Edison.
- Cocklebur and stinging nettle have been used for natural dyes.
- Studies have shown that milkweed is actually a more effective insulator than goose down.

Flavor

Some plants seem to subtly change the flavor of other plants around them, in a way humans find desirable, for example stinging nettle, besides being edible if properly cooked, seems to increase essential oil production in nearby herbs.

Examples

- Clover is a legume. Like other beans, it hosts bacteria that fix nitrogen in the soil. Its vining nature covers the ground, sheltering more moisture than it consumes, providing a humid, cooler microclimate for surrounding plants as a "green mulch". It also is preferred by rodents over many garden crops, cutting down on the amount of desired vegetables that get eaten.
- Dandelions possess a deep, strong tap root that breaks up hard soil, benefiting weaker-rooted plants nearby, and draw up nutrients from deeper than shallower-rooted nearby plants can access. They will also excrete minerals and nitrogen through their roots.
- Crow garlic, the wild chives found in sunny parts of a North American yard, has all of the companion plant benefits of other alliums, including repelling japanese beetles,aphids, and rodents, and being believed to benefit the flavor of solanums like tomatoes and peppers. It can be used as a substitute for some alliums in cooking, if its slightly more bitter flavor is masked by sweetness or lipids.
- Bishop's lace (Queen Anne's Lace) works as a nurse plant for nearby crops like lettuce, shading them from overly intense sunlight and keeping more humidity in the air. It attracts predatory wasps and flies that eat vegetable pests. It has a scientifically tested beneficial effect on nearby tomato plants, and when it is young it has an edible root, both of these are true because the domesticated carrot is simply a cultivar of this "weed".

Weed control

Weed control is the botanical component of pest control, which attempts to stop weeds, especially noxious or injurious weeds, from competing with domesticated plants and livestock. Many strategies have been developed in order to contain these plants. The original strategy was manual removal including ploughing, which can cut the roots of weeds. More recent approaches includeherbicides (chemical weed killers) and reducing stocks by burning and/or pulverizing seeds. A plant is often termed a "weed" when it has one or more of the following characteristics:

- Little or no recognized value (as in medicinal, material, nutritional or energy)
- Rapid growth and/or ease of germination
- Competitive with crops for space, light, water and nutrients

The definition of a weed is completely context-dependent. To one person, one plant may be a weed, and to another person it may be a desirable plant. In one place, a plant may be viewed as a weed, whereas in another place, the same plant may be desirable. Weeds compete with productive crops or pasture, ultimately converting productive land into unusable scrub. Weeds can be poisonous, distasteful, produce burrs, thorns or otherwise interfere with the use and management of desirable plants by contaminating harvests or interfering with livestock.

Weeds compete with crops for space, nutrients, water and light. Smaller, slower growing seedlings are more susceptible than those that are larger and more vigorous. Onionsare one of the most vulnerable, because they are slow to germinate and produce slender, upright stems. By contrast broad beans produce large seedlings and suffer far fewer effects other than during periods of water shortage at the crucial time when the pods are filling out. Transplanted crops raised in sterile soil or potting compost gain a head start over germinating weeds. Weeds also vary in their competitive abilities and according to conditions and season. Tall-growing vigorous weeds such as fat hen (*Chenopodium album*) can have the most pronounced effects on adjacent crops, although seedlings of fat hen that appear in late summer produce only small plants. Chickweed (*Stellaria media*), a low growing plant, can happily co-exist with a tall crop during the summer, but plants that have overwintered will grow rapidly in early spring and may swamp crops such as onions or spring greens.

The presence of weeds does not necessarily mean that they are damaging a crop, especially during the early growth stages when both weeds and crops can grow without interference. However, as growth proceeds they each begin to require greater amounts of water and nutrients. Estimates suggest that weed and crop can co-exist harmoniously for around three weeks before competition becomes significant. One study found that after competition had started, the final yield of onion bulbs was reduced at almost 4% per day.

Perennial weeds with bulbils, such as lesser celandine and oxalis, or with persistent underground stems such as couch grass (*Agropyron repens*) or creeping buttercup(*Ranunculus repens*) store reserves of food, and are thus able to grow faster and with more vigour than their annual counterparts. Some perennials such as couch grass exudeallelopathic chemicals that inhibit the growth of other nearby plants. Weeds can also host pests and diseases that can spread to cultivated crops. Charlock and Shepherd's purse may carry clubroot, eelworm can be harboured by chickweed, fat hen and shepherd's purse, while the cucumber mosaic virus, which can devastate the cucurbit family, is carried by a range of different weeds including chickweed and groundsel.

Insect pests often do not attack weeds. However pests such as cutworms may first attack weeds then move on to cultivated crops.

Some plants are considered weeds by some farmers and crops by others. Charlock, a common weed in the southeastern US, are weeds according to row crop growers, but are valued by beekeepers, who seek out places where it blooms all winter, thus providing pollen for honeybees and other pollinators. Its bloom resists all but a very hard freeze, and recovers once the freeze ends.

WEED PROPAGATION

Seeds

Annual and biennial weeds such as chickweed, annual meadow grass, shepherd's purse, groundsel, fat hen, cleaver, speedwell and hairy bittercress propagate themselves byseeding. Many produce huge numbers of seed several times a season, some all year round. Groundsel can produce 1000 seed, and can continue right through a mild winter, whilst Scentless Mayweedproduces over 30,000 seeds per plant. Not all of these will germinate at once, but over several seasons, lying dormant in the soil sometimes for years until exposed to light. Poppy seed can survive 80–100 years, dock 50 or more. There can be many thousands of seeds in a square foot or square metre of ground, thus and soil disturbance will produce a flush of fresh weed seedlings.

Subsurface/surface

The most persistent perennials spread by underground creeping rhizomes that can regrow from a tiny fragment. These include couch grass, bindweed, ground elder, nettles, rosebay willow herb, Japanese knotweed, horsetail and bracken, as well as creeping thistle, whose tap roots can put out lateral roots. Other perennials put out runners that spread along the soil surface. As they creep they set down roots, enabling them to colonise bare ground with great rapidity. These include creeping buttercup and ground ivy. Yet another group of perennials propagate by stolons- stems that arch back into the ground to reroot. The most familiar of these is the bramble.

METHODS

Fig. Pesticide-free thermic weed control with a weed burner on a potato field inDithmarschen

Coverings

In domestic gardens, methods of weed control include covering an area of ground with a material that creates a hostile environment for weed growth, known as a *weed mat*.

Several layers of wet newspaper prevent light from reaching plants beneath, which kills them. Daily saturating the newspaper with water plant decomposition. After several weeks, all germinating weed seeds are dead.

In the case of black plastic, the greenhouse effect kills the plants. Although the black plastic sheet is effective at preventing weeds that it covers, it is difficult to achieve complete coverage. Eradicating persistent perennials may require the sheets to be left in place for at least two seasons. Some plants are said to produce root exudates that suppress herbaceous weeds. *Tagetes minuta* is claimed to be effective against couch and ground elder, whilst a border of comfrey is also said to act as a barrier against the invasion of some weeds including couch. A 5–10 centimetres (2.0–3.9 in)} layer of wood chip mulch prevents most weeds from sprouting. Gravel can serve as an inorganic mulch. Irrigation is sometimes used as a weed control measure such as in the case of paddy fields to kill any plant other than the water-tolerant rice crop.

Bradley

See also Bradley Method of Bush Regeneration, which uses ecological processes to do much of the work. Perennial weeds also propagate by seeding; the airborne seed of thedandelion and the rose-bay willow herb parachute far and wide. Dandelion and dock also put down deep tap roots, which, although they do not spread underground, are able to regrow from any remaining piece left in the ground.

Manual removal

Fig. Weeds are removed manually in large parts of India.

Many gardeners still remove weeds by manually pulling them out of the ground, making sure to include the roots that would otherwise allow them to resprout. Hoeing off weed leaves and stems as soon as they appear can eventually weaken and kill perennials, although this will require persistence in the case of plants such as bindweed. Nettle infestations can be tackled by cutting back at least three times a year, repeated over a three-year period. Bramble can be dealt with in a similar way.

Goat grazing

Companies using goats to control and eradicate leafy spurge, knapweed, and other toxic weeds have sprouted across the American West. Near Red Lodge, Montana, there is Healthy Meadows, a company owned by Ivan Thrane and Chia Chen-Speidel, who as of late summer 2012 manage 250 goats for this express purpose.

"Stale seed bed"

Another manual technique is the 'stale seed bed', which involves cultivating the soil, then leaving it fallow for a week or so. When the initial weeds sprout, the grower lightly hoesthem away before planting the desired crop. However, even a freshly cleared bed is susceptible to airborne seed from elsewhere, as well as seed carried by passing animals on their fur, or from imported manure.

Irrigation

Drip irrigation involves bringing water directly to the roots of the desired plants, thereby limits weed's access to water.

Biological pesticide

Vinegar kills the visible part of the weed. They will wrinkle and die next day, although the root will still be in place to continue growing.

Tilling

Ploughing includes tilling of soil, intercultural ploughing and summer ploughing. Ploughing uproots weeds, causing them to die. In summer ploughing is done during deep summers. Summer ploughing also helps in killing pests.

Mechanical tilling can remove weeds around crop plants at various points in the growing process.

Crop rotation

Rotating crops with ones that kill weeds by choking them out, such as hemp, *Mucuna pruriens*, and other crops, can be a very effective method of weed control. It is a way to avoid the use of herbicides, and to gain the benefits of crop rotation.

Thermal

Several thermal methods can control weeds.

Hot foam (foamstream) causes the cell walls to rupture, killing the plant. Weed burners heat up soil quickly and destroy superficial parts of the plants. Weed seeds are often heat resistant and even react with an increase of growth on dry heat. Since the 19th century soil steam sterilization has been used to clean weeds completely from soil. Several research results confirm the high effectivness of humid heat against weeds and its seeds.

Seed targeting

In 1998, the Australian Herbicide Resistance Initiative (AHRI), debuted. gathered fifteen scientists and technical staff members to conduct field surveys, collect seeds, test for resistance and study the biochemical and genetic mechanisms of resistance. A collaboration with DuPont led to a mandatory herbicide labeling program, in which each mode of action is clearly identified by a letter of the alphabet.

The key innovation of the AHRI approach has been to focus on weed seeds. Ryegrass seeds last only a few years in soil, so if farmers can prevent new seeds from arriving, the number of sprouts will shrink each year. Until the new approach farmers were unintentionally helping the seeds. Their combines loosen ryegrass seeds from their stalks and spread them over the fields. In the mid-1980s, a few farmers hitched covered trailers, called "chaff carts", behind their combines to catch the chaff and weed seeds. The collected material is then burned.

An alternative is to concentrate the seeds into a half-meter-wide strip called a windrow and burn the windrows after the harvest, destroying the seeds. Since 2003, windrow burning has been adopted by about 70% of farmers in Western Australia.

Yet another approach is the Harrington Seed Destructor, which is an adaptation of a coal pulverizing cage mill that uses steel bars whirling at up to 1500 rpm. It keeps all the organic material in the field and does not involve combustion, but kills 95% of seeds.

Hybrid

One method of maintaining the effectiveness of individual strategies is to combine them with others that work in complete different ways. Thus seed targeting has been combined with herbicides. In Australia seed management has been effectively combined with trifluralin and clethodim.

"Organic" approaches

Organic weed control involves anything other than applying manufactured chemicals. Typically a combination of methods are used to achieve satisfactory control. Sulfur in some circumstances is accepted within British Soil Association standards. Soil solarization in some circumstances is very effective at eliminating weeds while maintaining grass. Planted grass tends to have a higher heat/humidity tolerance than unwanted weeds.

Herbicides

The above described methods of weed control use no or very limited chemical inputs. They are preferred by organic gardeners ororganic farmers. However weed control can also be achieved by the use of herbicides. Selective herbicides kill certain targets while leaving the desired crop relatively unharmed. Some of these act by interfering with the growth of the weed and are often based on plant hormones. Herbicides are generally classified as follows:

- **Contact** herbicides destroy only plant tissue that contacts the herbicide. Generally, these are the fastest-acting herbicides. They are ineffective on perennial plants that can re-grow from roots or tubers.
- **Systemic** herbicides are foliar-applied and move through the plant where they destroy a greater amount of tissue. Glyphosate is currently the most used systemic herbicide.
- **Soil-borne** herbicides are applied to the soil and are taken up by the roots of the target plant.
- **Pre-emergent** herbicides are applied to the soil and prevent germination or early growth of weed seeds.

In agriculture large scale and systematic procedures are usually required, often by machines, such as large liquid herbicide 'floater' sprayers, or aerial application.

RESISTANCE

Resistance occurs when a target adapts to circumvent a particular control strategy. It affects not only weed control,but antibiotics, insect control and other domains. In agriculture is mostly considered in reference to pesticides, but can defeat other strategies, e.g., when a target species becomes more drought tolerant via selection pressure.

Farming practices

Herbicide resistance recently became a critical problem as many Australian sheep farmers switched to exclusively growing wheat in their pastures in the 1970s. In wheat fields, introduced varieties of ryegrass, while good for grazing sheep, are intense competitors with wheat. Ryegrasses produce so many seeds that, if left unchecked, they can completely choke a field. Herbicides provided excellent control, while reducing soil disrupting because of less need to plough. Within little more than a decade, ryegrass and other weeds began to develop resistance. Australian farmers evolved again and began diversifying their techniques. In 1983, patches of ryegrass had become immune to Hoegrass, a family of herbicides that inhibit an enzyme called acetyl coenzyme A carboxylase.

Ryegrass populations were large, and had substantial genetic diversity, because farmers had planted many varieties. Ryegrass is cross-pollinated by wind, so genes shuffle frequently. Farmers sprayed inexpensive Hoegrass year after year, creating selection pressure, but were diluting the herbicide in order to save money, increasing plants survival. Hoegrass was mostly replaced by a group of herbicides that block acetolactate synthase, again helped by poor application practices. Ryegrass evolved a kind of "cross-resistance" that allowed it to rapidly break down a variety of herbicides. Australian farmers lost four classes of herbicides in only a few years. As of 2013 only two herbicide classes, called Photosystem II and long-chain fatty acid inhibitors, had become the last hope.

2

Water Hyacinth

DESCRIPTION

Non-Native

Water hyacinth is native to South America but has naturalized much of the Southern U.S.

Water hyacinth is a free-floating perennial plant that can grow to a height of 3 feet. The dark green leave blades are circular to elliptical in shape attached to a spongy, inflated petiole. Underneath the water is a thick, heavily branched, dark fibrous root system. The water hyacinth has striking light blue to violet flowers located on a terminal spike. Water hyacinth is a very aggressive invader and can form thick mats. If these mats cover the entire surface of the pond they can cause oxygen depletions and fish kills. Water hyacinths should be controlled so they do not cover the entire pond.

Submerged portions of all aquatic plants provide habitats for many micro and macro invertebrates. These invertebrates in turn are used as food by fish and other wildlife species (e.g. amphibians, reptiles, ducks, etc.). After aquatic plants die, their decomposition by bacteria and fungi provides food (called "detritus" for many aquatic invertebrates. Water hyacinth has no known direct food value to wildlife and is considered a pest species.

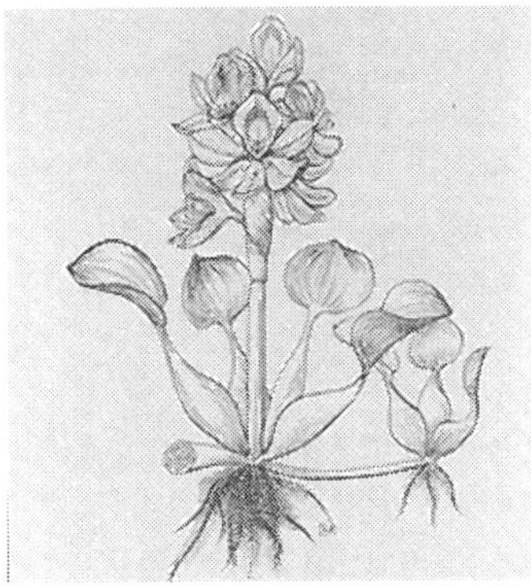

NON-NATIVE INVASIVE FRESHWATER PLANTS

Water Hyacinth (Eichornia crassipes)

General Information

Aquatic plants scientists call water hyacinth (*Eichornia crassipes* and several related species) the worst aquatic plant in the world! It is native to South America, but has been naturalized in most of the southern United States and in many of the world's subtropical and tropical climates. Water hyacinth plants have tremendous growth and reproductive rates and the free-floating mats cause substantial problems. Plant managers and water front residents spend millions of dollars per year in the United States for its management.

Water hyacinth is a popular water garden plant because of the beauty of its large, purple to violet flowers and its interesting floating vegetation. Many plant nurseries and big box stores sell it in Washington as an ornamental pond plant each spring and summer. Although weed scientists believe that water hyacinth cannot survive Washington's winters, its presence as an ornamental makes it possible for escape and growth in the wild. Water hyacinth can survive freezing conditions in other states where it is established and it may be possible for this plant to over winter during a mild western Washington winter.

Growth Habit

Water hyacinth is a free-floating plant that gets its nutrients from the water from dangling roots. The plant reproduces by seeds and vegetatively through daughter plants that form on rhizomes and produce dense plant beds. In one study, two plants produced 1,200 daughter plants in four months. By this mechanism, water hyacinth can form impenetrable mats of floating vegetation. Individual plants break off the mat and can be dispersed by wind and water currents. A single plant can produce as many as 5,000 seeds and waterfowl eat and transport seeds to new locations. Seedlings are common on mud banks exposed by low water levels.

Management

Plant managers control water hyacinth by many methods including harvesting, application of aquatic herbicides, and biological control agents. In Washington, the best way to manage water hyacinth is to prevent it from ever becoming established. If you purchase water hyacinth at a local nursery, do not dispose of excess plants by throwing them into a lake, river, stream, or drainage ditch. A number of water hyacinth plants have been found in drainage ditches in the Longview/Kelso area in southwestern Washington. Drainage ditch managers believe that somebody with a nearby ornamental pond disposes of excess plants by throwing them into the drainage canals. University arboretum staff discovered a large infestation of water hyacinth in a backwater area on Lake Washington where somebody had either deliberately wanted to introduce water hyacinth plants or used the arboretum as a convenient place to dump excess plants. In both instances, managers immediately hand removed all water hyacinth plants from these water bodies.

Help prevent introduction of water hyacinth into Washington waters by:

- Disposing of all excess plants on a compost heap and away from water.
- Only planting water hyacinth into ornamental ponds and not natural water bodies. This is asking for trouble

Identification

Water hyacinth is distinctive.

- They are floating plants with round to oval, shiny green leaves up-to-ten inches in diameter, although smaller leaves are common.
- Leaves are held upright so they act like sails.
- The leaf stalk is thick and spongy and helps to keep the plant buoyant
- A mass of fine roots hang in the water underneath the plant
- The flowers are large (2-3 inches) and attractive. They are blue-purple or lilac-colored with a yellow spot

Another tropical floating-leaved nursery plant, called water lettuce, may be mistaken for water hyacinth However; water lettuce doesn't have showy flowers and has large ribbed leaves. Water lettuce is much less cold-tolerant than water hyacinth and is not thought able to survive in cold climates

Water Hyacinth Factoids

Origin	Amazon River basin -- mostly Brasil
Substrate	Immaterial
Fertilizers	Add if leaves start yellowing
Uses	Beautifies, purifies, provides shade and security
Flowers	Light purple blooms when temp hits 80
Growth rate	Fastest growing plant in the world
Sun or Shade	Full sun best
Threats	Cold weather, salt, herbicides, children
Water	Removes pollutants from water

Craig Hall's pic of his hyacinth ranch. Just kidding, Craig. Water hyacinths are pretty.

Fig. *Really attractive flowers.*

Origins: Originally from what most people call Brazil, water hyacinths came to the U.S. in 1880. They were brought to the Cotton Exposition in

Louisiana, introduced into Florida, and from there quickly spread to all our southern states. Now considered a noxious weed in these states and "the worst aquatic plant," they are illegal to export or import into these and several other states. They grow so fast that they block waterways and impede navigation in many states.

Movie Stars: If you saw the movie "Anaconda," you saw their boat plow thru vast quantities of water hyacinths. It was a sad movie, because the anaconda died at the end.

Floating Plants: You don't plant water hyacinths in special pots; you just toss them in the water. Their air-filled leaves keep them afloat. They float like a green-leaved cork. You don't even need to remember "green side up." Their heavy roots automatically right them no matter what you do.

4

Fig. *Older roots turn black. New roots come in white.*

Root System: Long heavily branched roots looking like they're covered with black whiskers very efficiently remove fish wastes from the water. Older roots turn black. Younger roots are white. Roots can grow as long as 18 inches. If they reach a mud bottom, they grow even faster.

Fig. *Baby plant growing off the bottom left side.*

Starting out: We trim our hyacinths back as far as possible. If you get yours elsewhere, pull off ALL dead or broken leaves. They will not repair themselves. Also pull off the excess roots. Trimming them back (a lot) encourages new growth and makes them look better. Dead or broken leaves will not repair themselves.

Fig. *Even this young plant has three daughter cells growing from its base.*

Reproduction: Water hyacinths can reproduce by seeds in warm climates — not in Iowa. Seeds can survive 15 to 20 years during crummy weather. However, most hyacinths reproduce by stolons (like strawberry runners). In good conditions they double in population every 12 days. In Florida, they can yield 200 tons per acre. All these traits make them a great pond plant for Iowa — but not Florida.

Fig. *Once they look like this, they will not recover. Our fall weather really slows them down.*

Iowa Winters: We see some reports saying hyacinths die at temps below 20°. In Iowa backyards they stop growing at 40 and die out at 32. The first frost wipes them totally. If we had no winters in IOwa, they would be a noxious weed here also.

Substrate: It matters not what you put on the bottom of your pond. Water hyacinths will prosper. If you have a dirt bottom, they will very likely have access to more nutrients.

Fig. *With no fish in here, you can use any fertilizer you want. See photo at bottom for results.*

Fertilizers: You will need to fertilize your fast-growing hyacinths if their leaves turn yellow. Use any fertilizer you know will not hurt your fish. If you have no fish in your pond. use any type of fertilizer you want. Small frequent feedings work better than one massive feeding.

Fig. *At the beginning of the pond season, water hyacinths arrive in bags of 100.*

Excess Plants: Once they get a good start, you wind up with more than you need. At the end of the pond season, disposal of dead hyacinths can present a problem.

We find the best solution involves raking them out of your pond, drying them a couple days, then mulching them with your lawn mower. If you leave them in your pond over the winter, they make a royal mess. Get the dead ones out as soon as possible. In some countries, they compost their hyacinths and use them to grow mushrooms.

Algae Cure: Water hyacinths control algae growth in two main ways. They reduce the amount of sunlight entering the water. They also suck out the nutrients that algae need to grow successfully.

Fig. *About 2 inches across at first, these baby hyacinths will explode in size and number.*

Other Uses: In addition to looking good, water hyacinths also provide the security some fishes need. Their roots also serve as prime spawning areas. The roots also provide a home for tiny insects and other critters that fishes enjoy snacking upon.

Water: Water hyacinths grow in nearly any kind of water without salt in it. Copper will kill them, so will most algae treatments. The only algae treatment that lets them live is AlgaeFix.

Fig. *Trimmed water hyacinths ready for sale in Aqualand's front window.*

Size: Most water hyacinths grow 12 to18-inches tall. They will grow taller — up to three feet tall. Crowded conditions plus fertilizer make them really stretch out.

Fig. *Hyacinths in plastic trays in LA's backyard. They bloom when temperatures hit the eighties.*

Last Word: Water hyacinths will grow for anybody. Lack a green thumb? Try the water hyacinths.

Water hyacinth (Eichhornia crassipes)

Species

Eichhornia crassipes (Mart.) Solms [Pontederiaceae]

Synonyms

Eichhornia speciosa Kunth, *Pontederia crassipes* Mart.

Feed categories

- Aquatic plants
- Other forage plants
- Forage plants

Related feed(s)

Description

The water hyacinth (*Eichhornia crassipes* (Mart.) Solms) is a free floating perennial herb of fresh water ecosystems. It is found at the surface of rivers, lakes, canals and ponds and may root in the mud of shallow waters. It is generally 10-20 cm high but can reach 1 m high when established in dense mats (Ecoport, 2011; Pieterse, 1997). Water hyacinth is a rhizomatous and stoloniferous plant with long, pendant and adventitious roots. The leaves arise from the rhizome nodes and stand above the water. They are dark green, ovate and cordate at the base, borne on swollen bladder-like petioles (Pieterse, 1997). The plant has considerable buoyancy and the leaves act as sails in the wind (Hasan et al., 2009a). The inflorescence bears 8-10 pale violet or blue lily-like flowers. The fruit is a dehiscent capsule containing up to 200 small seeds (Ecoport, 2011; Pieterse, 1997).

Water hyacinth is one of the most noxious weeds in tropical and subtropical regions, and many attempts have been made to eliminate or control it (see *Environmental impact*). Harnessing its considerable productivity is considered as a sustainable and possibly less expensive method of control. Water hyacinth can be used as a vegetable, fodder, green manure, compost and mulch for soil improvement. Much research has been devoted to its use as a feed material for many classes of livestock. In South-East Asia, integrated fish-pig-water hyacinth farming systems have been developed in order to increase global animal production: water hyacinths grown in fish ponds have higher nutritive value (higher protein content) and can be fed to fish and pigs in different forms. Fish and pig manures fertilize fish ponds and provide nutrients to water hyacinth (Yang Huazhu et al., 2001). Animals are usually fed the aerial part (leaves and stems without the roots) or only the leaves, but sometimes the whole plant (root included) is used. Water haycinth can be fed fresh, ensiled or dried but many other processes are used or have been tried, including cooking and fermentation (see *Processes*).

Other products and services derived from water hyacinth have been investigated, including as a source of pigments for poulty nutrition (Lareo et al., 1982), and leaf protein concentrates for food and feed (Abo Bakr et al., 1984; Wu Wenbiao et al., 2011). Water hyacinth is used for making textiles, paper and for camouflaging fish traps (Ecoport, 2011; Pieterse, 1997). It ferments rapidly due to its high water content and can supply biomass for biogas production. Water hyacinth is used as a water-clearing agent, as a substrate for mushroom production and as an ornamental species (Pieterse, 1997).

Note: the term "whole plant" used in the scientific literature is sometimes confusing. For an aquatic plant, it can designate either the whole plant including the roots or the whole aerial part.

Distribution

Water hyacinth originated from Amazonia and spread naturally throughout South America. It was introduced as an ornamental species into the USA, South East Asia and South Africa in the late 19th century and is now naturalized in most tropical and subtropical areas. It can be found between 38°N and 38°S and is referred to as a noxious weed in more than 50 countries on five continents (GISD, 2011).

Water hyacinth thrives in a wide range of fresh water habitats (shallow ponds, marshes, small streams, lakes and rivers) provided they are not saline or do not become saline during drought. The most favourable conditions are average temperatures ranging from 14 to 29°C, water pH about 7, low salinity, N, P, K-rich water, full sunlight and absence of physical disturbance or pests. Under such conditions, water hyacinth readily establishes and forms dense mats due to its efficient vegetative reproduction. The plant cover can double within 6-15 days (Ecocrop, 2011; Hasan et al., 2009a). The growth of water hyacinth is hampered by low temperatures (below 13°C), nutrient scarcity, frost and salinity. It has a low tolerance of flooding, as the physical disturbance caused by the waves breaks the mat and leaves of the plant stranded on the land. Drought and frost also quickly kill the stand (Hasan et al., 2009b).

Processes

Water hyacinth can be used fresh, ensiled, dried, whole, chopped or ground. Roots are usually discarded.

Fresh forage

Water hyacinth is usually chopped and fed directly to animals. However, it is not very palatable due to the presence of oxalate crystals in the leaves that cause mouth irritation in livestock (see *Potential constraints*).

Dried and meal form

The physical structure of the plant makes it unsuitable for the normal methods of making hay and silage. The plants dry rather quickly in the sun, but the neck between the petiole and the lamina is very brittle. The lamina shrinks and breaks off with handling, leaving only the petiole which remains round and full of air. The hay is therefore very bulky. Water hyacinth foliage is often sun-dried, typically for between 2 and 5 days but up to 10 days depending on environmental conditions (Hasan et al., 2009a; Dada, 2002; Biobaku et al., 1991). The process may involve numerous manipulations (Biobaku et al., 1991). The nutritive value per unit of dry matter is too low to warrant the cost of artificial drying (Göhl, 1982). Dried water hyacinth can be ground in a hammermill and turned into a meal, for instance for pigs (Domínguez et al., 1996).

Silage

Ensiling is a good method of perservation for water hyacinth and enhances its palatability. Because of the high moisture content of water hyacinth, it should be wilted in the shade for 48 hours and lacerated or screw-pressed before ensiling (Göhl, 1982). The silage can be mixed with carbohydrate sources such as molasses (up to 20%), dried citrus pulp or maize grain (Nguyen Thanh Van et al., 2010; Bagnall et al., 1974; Aboud et al., 2005). Salt and urea are reported to increase the nutritive value and quality of the silage (Göhl, 1982). For instance, water hyacinth wilted and ensiled with 15% defatted rice bran (25% final DM) or rice bran (22% final DM) was well preserved (Lowilai et al., 1995). Water hyacinth leaves added at between 20 and 35% to a silage based on rice straw, urea, molasses and soybean meal gave a good quality product (Malek et al., 2008). Prior wilting is preferable but even pre-wilted silage may be unsuitable (Aboud et al., 2005). Adding acids gave satisfactory quality silage (Byron et al., 1975). A method for ensiling water hyacinth with by-products (rice straw, chopped sugarcane and molasses), enzymes and lactic acid bacteria has been patented (Simsa et al., 1994).

Cooking

In South-East Asia, the plants are often chopped and mixed with by-products such as banana stems, rice bran, copra meal or fish meal and then boiled to make a suitable feed for pigs, ducks or pond fish (Jafari, 2010). A "recipe" for water hyacinth pig diets has been described as follows: a mixture of water hyacinth and banana stems is boiled slowly for a few hours until the ingredients turn into a paste, to which oil cake, rice bran and sometimes maize and salt are added. The cooked mixture is good for only three days, after which it turns sour. A common formula is 40 kg of water hyacinth, 15 kg of rice bran, 2.5 kg of fish meal and 5 kg of coconut meal (Göhl, 1982).

FORAGE MANAGEMENT

Water hyacinth can be exceptionally productive under favourable conditions. In China, ponds fertilized with sewage nutrients yielded up to 657 t fresh matter/ha/year (Hasan et al., 2009a). In Tanzania, yields of 322 t fresh matter/ha/year (30.5 t DM) have been recorded (Aboud et al., 2005). However, such extreme yields may not be sustainable on a large scale and a production of 200 t/ha/year is a more realistic goal in eutrophic waters in the tropics (Hasan et al., 2009a).

ENVIRONMENTAL IMPACT

Invasive species

Water hyacinth is a major environmental concern in many countries. It quickly invades large water areas, due to a high rate of vegetative propagation, and forms dense mats that obstruct waterways. It affects all water-based economic activities, including fishing, irrigation, traffic, water quality, hydraulic and hydroelectric infrastructures (Ecoport, 2011; Practical action, 2006; Hasan et al., 2009b). Water hyacinth was already causing concern in 1908 in Vietnam, Laos and Cambodia (Pieterse, 1997). In China, where it was introduced as fodder in the 1950s to solve feed shortage, water hyacinth became problematic in the 1980s and is now considered to be responsible for most of the economic damage caused by foreign invasive species (Ding JianQing et al., 2001; Chu JianJun et al., 2006).

Water hyacinth is detrimental to biodiversity, notably by smothering native aquatic plant species. Water hyacinth increases water evaporation and intercepts light: the subsequent reduction of water oxygen content and increase in water acidity can be lethal to fish. The plants provide a breeding habitat for undesirable vectors of human and animal diseases, such as mosquitoes and bilharzia-carrying snails (Ecoport, 2011).

Water hyacinth is difficult to eradicate. The seeds spread through animal faeces and can survive 15-20 years. The most widely used and effective control method is mechanical and consists in removing the plants from the water. Though expensive and slow, this method has several advantages: the excess nutrients are extracted with the plant, which can be used for fodder, green manure or other applications. Chemical control is usually done with the herbicides 2,4-D, glyphosate or Paraquat, which may cause environmental concerns and may not ensure long-term sustainability. The dead plants may also contaminate water when they decay (Chu JianJun et al., 2006). Biological control can be achieved with fungi or insects (*Neochetina eichhorniae* and *Neochetina bruchi*) but only in association with other control methods (Pieterse, 1997).

Phytoremediation

Water hyacinth can be useful for the remediation of water contaminated with heavy metals, as the roots trap and remove from water large amounts of minerals, including heavy metals and radio-active elements (Hasan et al., 2009a). It has been effective in removing cadmium and zinc from moderately contaminated water (Lu XiaoMei et al., 2004).

However, there is still no commercial or large-scale application (Ecoport, 2011; Pieterse, 1997). Combining both water-clearing capacity and biogas production might be profitable: studies have shown that about 1 million L/d/ha of domestic sewage can be treated with water hyacinths, reducing the biochemical and chemical oxygen demand by 89 and 71% respectively (Hasan et al., 2009a).

EICHHORNIA CRASSIPES

Eichhornia crassipes, commonly known as *(common) water hyacinth*, is an aquatic plant native to the Amazon basin, and is often considered a highly problematic invasive species outside its native range.

DESCRIPTION

Water hyacinth is a free-floating perennial aquatic plant (or hydrophyte) native to tropical and sub-tropical South America. With broad, thick, glossy, ovate leaves, water hyacinth may rise above the surface of the water as much as 1 meter in height. The leaves are 10–20 cm across, and float above the water surface. They have long, spongy and bulbous stalks. The feathery, freely hanging roots are purple-black. An erect stalk supports a single spike of 8-15 conspicuously attractive flowers, mostly lavender to pink in colour with six petals. When not in bloom, water hyacinth may be mistaken for frog's-bit (*Limnobium spongia*).

One of the fastest growing plants known, water hyacinth reproduces primarily by way of runners or stolons, which eventually form daughter plants. Each plant can produce thousands of seeds each year, and these seeds can remain viable for more than 28 years. Some water hyacinths were found to grow up to 2 to 5 metres a day in some sites in Southeast Asia. The common water hyacinth (*Eichhornia crassipes*) are vigorous growers known to double their population in two weeks.

HABITAT AND ECOLOGY

Its habitat ranges from tropical desert to subtropical or warm temperate desert to rainforest zones. The temperature tolerance of the water hyacinth is the following; its minimum growth temperature is 12 °C (54 °F); its optimum growth temperature is 25-30 °C (77-86 °F); its maximum growth temperature is 33-35 °C (92-95 °F), and its pH tolerance is estimated at 5.0 to 7.5. It does not tolerate water temperatures >35 °C. Leaves are killed by frost and salt water, the latter trait being used to kill some of it by floating rafts of the cut weed to the sea. Water hyacinths do not grow when the average salinity is greater than 15% that of sea water. In brackish water, its leaves show epinasty and chlorosis, and eventually die.

Because of *E. crassipes* invasiveness, several biological control agents have been released to control it, including two weevils (Coleoptera: Curculionidae), *Neochetina bruchi*Hustache and *Neochetina eichhorniae* Warner, and the moth *Niphograpta albiguttalis* (Warren) (Lepidoptera: Pyralidae). *Neochetina eichhorniae* causes "a substantial reduction in water hyacinth production" (in Louisiana); it reduces plant height, weight, root length, and makes the plant produce fewer daughter plants. *N. eichhorniae* was introduced from Argentina to Florida in 1972.

Azotobacter chroococcum, an N-fixing bacteria, is probably concentrated around the bases of the petioles. But the bacteria do not fix nitrogen unless the plant is suffering extreme N-deficiency.

Fresh plants contain prickly crystals. This plant is reported to contain HCN, alkaloid, and triterpenoid, and may induce itching. Plants sprayed with 2,4-D may accumulate lethal doses of nitrates, as well as various other nocive elements in polluted environments. See further down.

INVASIVE SPECIES

Water hyacinth has been widely introduced in North America, Asia, Australia, Africa and New Zealand. In many areas it is has become an important and pernicious invasive species. In New Zealand it is listed on the National Pest Plant Accord which prevents it from being propagated, distributed or sold. In large water areas such as Louisiana, theKerala Backwaters in India, Tonlé Sap in Cambodia and Lake Victoria it has become a serious pest.

When not controlled, water hyacinth will cover lakes and ponds entirely; this dramatically impacts water flow, blocks sunlight from reaching native aquatic plants, and starves the water of oxygen, often killing fish (or turtles). The plants also create a prime habitat for mosquitos, the classic vectors of disease, and a species of snail known to host a parasitic flatworm which causes schistosomiasis (snail fever). Directly blamed for starving subsistence farmers in Papua New Guinea, water hyacinth remains a major problem where effective control programs are not in place. Water hyacinth is often problematic in man-made ponds if uncontrolled, but can also provide a food source for goldfish, keep water clean and help to provide oxygen to man-made ponds. Water hyacinth often invades bodies of water that have been impacted by human activities. For example, the plants can unbalance natural lifecycles in artificial reservoirs or in eutrophied lakes that receive large amounts of nutrients.

Eichhornia crassipes, the Common water hyacinth, has become an invasive plant species on Lake Victoria in Africa after it was introduced into the area in the 1980s.

United States

The water hyacinth was introduced in 1884 at the World's Fair in New Orleans, also known as the World Cotton Centennial. The plants had been given away as a gift by a group of visiting Japanese. Soon after, the water hyacinth was choking rivers, killing fish and stopping shipping in Louisiana, and an estimated 50 kilograms per square meter choked Florida's waterways. There were many attempts to eradicate the flower, including one by the U.S. War Department to pour oil over many of the flowers, but none worked. In 1910, a bold solution was put forth by the New Foods Society.

Their plan was to import and release hippopotamus from Africa into the rivers and bayous of Louisiana. The hippopotamus would then eat the water hyacinth and also produce meat to solve another serious problem at the time, the American meat crisis.

Known as the American Hippo bill, H.R. 23621 was introduced by Louisiana Congressman Robert Broussard and debated by the Agricultural Committee of the U.S. House of Representatives.

The chief collaborators in the New Foods Society and proponents of Broussard's bill were Major Frederick Russell Burnham, the celebrated American scout, and Captain Fritz Duquesne, a South African scout who later became a notorious spy for Germany.

Presenting before the Agricultural Committee, Burnham made the point that none of the animals that Americans ate, chickens, pigs, cows, sheep, lambs, were native to the U.S., all had been imported by European settlers centuries before, so why should Americans hesitate to introduce hippopotamus and other large animals into the American diet? Duquesne, who was born and raised in South Africa, further noted that European settlers on that continent commonly included hippopotamus, ostrich, antelope, and other African wildlife in their diets and suffered no ill effects. The American Hippo bill nearly passed, but fell one vote short.

Africa

The plant was introduced by Belgian colonists to Rwanda to beautify their holdings and then advanced by natural means to Lake Victoriawhere it was first sighted in 1988. There, without any natural enemies, it has become an ecological plague, suffocating the lake, diminishing the fish reservoir, and hurting the local economies. It impedes access to Kisumu and other harbors. The water hyacinth has also appeared to the north in Ethiopia, where it was first reported in 1965 at the Koka Reservoir and in the Awash River, where the Ethiopian Electric Light and Power Authority has managed to bring it under moderate control at the considerable cost of human labor. Other infestations in Ethiopia include many bodies of water in the Gambela Region, the Blue Nile from just below Lake Tana into Sudan, and Lake Ellen near Alem Tena.

Asia

Water hyacinth has also invaded the Tonlé Sap lake in Cambodia. One project in Cambodia (Osmose) is trying to fight it by having local people make baskets from it.

Control

There are three commonly used control efforts used to suppress water hyacinth infestations. They are physical, chemical, and biological controls. However, no one control is better than the other because each has its advantages and disadvantages. The choice of control is dependent on the specific conditions of each affected location such as the extent of water hyacinth infestation, regional climate, and proximity to human and wildlife.

Chemical control

The use of chemical controls is the least used out of the three controls of water hyacinth, because of its long-term effects on the environment and human health. The use of herbicides requires strict approval from governmental protection agencies of skilled technician to handle and spray the affected areas. The use of chemical herbicides is only used in case of severe infiltration of water hyacinth.

However, the most successful use of herbicides is when it is used for smaller areas of infestation of water hyacinth. This is because in larger areas, more mats of water hyacinths are likely to survive the herbicides and can fragment to further propagate a large area of water hyacinth mats. In addition, it is more cost-effective and less laborious than mechanical control. Yet, it can lead to environmental effects as it can penetrate into the ground water system and can affect not only the hydrological cycle within an ecosystem but also negatively affect the local water system and human health.

It is also notable that the use of herbicides is not strictly selective of water hyacinths; keystone species and vital organisms such microalgae can perish from the toxins and can disrupt fragile food webs. The chemical regulation of water hyacinths can be done using common herbicides such as 2,4-D, glyphosate, and diquat. The herbicides are sprayed on the water hyacinth leaves and leads to direct changes to the physiology of the plant. The use of the herbicide known as 2,4-D leads to the death of water hyacinth through inhibition of cell growth of new tissue and cellular apoptosis (Jimenez, 2005). It can take almost a two-week period before mats of water hyacinth are destroyed with 2,4-D. It has been found that up to 150,000 acres of water hyacinth are treated annually in Louisiana.

The herbicide known as diquat is a liquid bromide salt that can rapidly penetrate the leaves of the water hyacinth and lead to immediate inactivity of plant cells and cellular processes. For the herbicide glyphosate, it has a low toxicity than the other herbicides; therefore, it takes longer for the water hyacinth mats to be destroyed (about three weeks). The symptoms include steady wilting of the plants and a yellow discoloration of the plant leaves that eventually leads to plant decay.

Physical control

Physical control is performed by land based machines such as bucket cranes, draglines, or boorm or by water based machinery such as aquatic weed harvester, dredges, or vegetation shredder. Mechanical removal is seen as the best short-term solution to the proliferation of the plant.

A project on Lake Victoria in Africa used various pieces of equipment to chop, collect, and dispose of 1500 hectares of water hyacinth in a 12-month period.

It is, however, costly and requires the use of both land and water vehicles, but it took many years for the lake to become in poor condition and reclamation will be a continual process. It can have an annual cost from $6 million to $20 million and is only considered a short-term solution to a long-term problem. Another disadvantage with mechanical harvesting is that it can lead to further fragmentation of water hyacinths when the plants are broken up by spinning cutters of the plant-harvesting machinery.

The fragments of water hyacinth that are left behind in the water can easily reproduce asexually and cause another infestation. However, transportation and disposal of the harvested water hyacinth is a challenge because the vegetation is heavy in weight.

The harvested water hyacinth can pose a health risk to humans because of the plant's propensity for absorbing contaminants, and it be considered toxic to humans. However, the practice of mechanical harvesting is not effective in large-scale infestations of the water hyacinth, because this aquatic invasive species grows much more rapidly than it can be eliminated. In addition, only one or two acres of water hyacinth can mechanically harvested daily because of the vast amounts of water hyacinths in the environment. Therefore, the process is very time-intensive.

Biological control

Fig. In 2010 the insect Megamelus scutellaris was released by the Agricultural Research Service as a biological control for the invasive species *Eichhornia crassipes,* more commonly known as waterhyacinth.

As chemical and mechanical removal is often too expensive and ineffective, researchers have turned to biological control agents to deal with water hyacinth. The effort began in the 1970s when USDA researchers released three species of weevil known to feed on water hyacinth into the United States, *Neochetina bruchi, N. eichhorniae*, and the water hyacinth borer *Sameodes albiguttalis*. The weevil species were introduced into the Gulf Coast states, such as Louisiana, Texas, and Florida, where there was thousands of acres of infestation from water hyacinth.

It was found that a decade later in the 1980s that there was a decrease in water hyacinth mats by as much as 33%. However, because the life cycle of the weevils is only ninety days, it puts a limitation on the use of biological predation to efficiently suppress water hyacinth growth. These organisms regulate water hyacinth by limiting water hyacinth size, its vegetative propagation, and seed production.

They also carry microorganisms that can be pathological to the water hyacinth. These weevils eat stem tissue, which results in a loss of buoyancy for the plant, which will eventually sink. Although meeting with limited success, the weevils have since been released in more than 20 other countries. However, the most effective control method remains the control of excessive nutrients and prevention of the spread of this species.

May 2010 the USDA's Agricultural Research Service released *Megamelus scutellaris* as a biological control insect for the invasive waterhyacinth species. *Megamelus scutellaris* is a small planthopper insect native to Argentina. Researchers have been studying the effects of the biological control agent in extensive host-range studies since 2006 and concluded that the insect is highly host-specific and will not pose a threat to any other plant population other than the targeted water hyacinth. Researchers also hope that the biological control will be more resilient than existing biological controls to the herbicides that are already in place to combat the invasive water hyacinth.

USES

Bioenergy

Because of its extremely high rate of development, *Eichhornia crassipes* is an excellent source of biomass. One hectare of standing crop thus produce more than 70,000 m^3 ofbiogas. According to Curtis and Duke, one kg of dry matter can yield 370 liters of biogas, giving a heating value of 22,000 kJ/m^3 (580 Btu/ft^3) compared to pure methane (895 Btu/ft^3)

Wolverton and McDonald report only 0.2 m^3 methane per kg, indicating requirements of 6000 MT biomass/ha to attain the 70,000 m^3 yield projected by the National Academy of Sciences (Washington). Ueki and Kobayashi mention more than 200 MT/ha/yr. Reddy and Tucker found an experimental maximum of more than a half ton per day.Bengali farmers collect and pile up these plants to dry at the onset of the cold season; they then use the dry water hyacinths as fuel.

They then use the ashes as fertilizer. In India, a ton of dried water hyacinth yield *circa* 50 liters ethanol and 200 kg residual fiber (7,700 Btu). Bacterial fermentation of one ton yields 26,500 cu ft gas (600 Btu) with 51.6% methane, 25.4% hydrogen, 22.1% CO_2, and 1.2% oxygen. Gasification of one ton dry matter by air and steam at high temperatures (800°) gives *circa* 40,000 ft^3 (*circa*1,100 m^3) natural gas (143 Btu/cu ft) containing 16.6% hydrogen, 4.8% methane, 21.7% CO, 4.1% CO_2, and 52.8% N. The high moisture content of water hyacinth, adding so much to handling costs, tends to limit commercial ventures., A continuous, hydraulic production system could be designed, which would provide a better utilization of capital investments than in conventional agriculture, which is essentially a batch operation.,

The labour involved in harvesting water hyacinth can be greatly reduced by locating collection sites and processors on impoundments that take advantage of prevailing winds.Wastewater treatment systems could also favourably be added to this operation. The harvested biomass would then be converted to ethanol, natural gas, hydrogen and/or gaseous nitrogen, and fertilizer. The resulting byproducts of water and fertilizer can both be used to irrigate nearby cropland.

Phytoremediation, waste water treatment

The roots of *Eichhornia crassipes* naturally absorb pollutants, including lead, mercury, and strontium-90, as well as some organic compounds believed to be carcinogenic, in concentrations 10,000 times that in the surrounding water. Water hyacinths can be cultivated for waste water treatment.

Water hyacinth is reported for its efficiency to remove about 60–80 % nitrogen (Fox et al. 2008) and about 69% of potassium from water (Zhou et al. 2007). The roots of water hyacinth were found to remove particulate matter and nitrogen in a natural shallow eutrophicated wetland (Billore et al. 1998).

Edibility

The plant is used as a carotene-rich table vegetable in Taiwan. Javanese sometimes cook and eat the green parts and inflorescence.

Medicinal use

In Kedah (Java), the flowers are used for medicating the skin of horses. The species is a "tonic.",

Potential as bioherbicidal agent

Water hyacinth leaf extract has been shown to exhibit phytotoxicity against another invasive weed *Mimosa pigra*. The extract inhibited the germination of *Mimosa pigra* seeds in addition to suppressing the root growth of the seedlings. Biochemical data suggested that the inhibitiory effects may be mediated by enhanced hydrogen peroxide production, inhibition of soluble peroxidase activity, and stimulation of cell wall-bound peroxidase activity in the root tissues of *Mimosa pigra*.

Other uses

In East Africa, water hyacinths from Lake Victoria are used to make furniture, handbags and rope. The plant is also used as animal feed and organic fertilizer although there is controversy stemming from the high alkaline pH value of the fertilizer. Though a study found water hyacinths of very limited use for paper production, they are nonetheless being used for paper production on a small scale.

3

Crop Weeds

India produces 201.6 million mt of food in 124 million ha with a productivity of 1,625 kg/ha. The oilseed production is about 22.3 million mt grown in 25.26 million ha with a productivity of 873 kg/ha. Pests and diseases cause huge damage in the production of various crops. It is estimated that 53.1 million mt of food grains and 56.8 million mt of sugarcane worth Rs.292.4 billion were lost to pests and diseases attack. Normally the farmers depend on chemical pesticide for the control of pests and diseases. The use of chemical pesticides has caused a lot of environmental hazards apart from resistance developed by the pests to the chemical. In view of the above reasons, nonpesticide methods of controlling pests and diseases have gained considerable importance.

STATUS OF DIFFERENT NONPESTICIDE METHODS

MECHANICAL METHODS

Collecting and Killing

The moths of several harmful insects emerge during *Kharif* (June-September) season as well as in subsequent months. After the emergence, male and female moths mate and gravid female lay eggs from 500 to 2,000 in soil or on crop plants. These moths normally rest under clods, in cracks or under dry vegetation matter in soil. Such moths can be physically collected and burnt. This helps in reducing population of pests. Collection and killing or trapping of moths of red-headed hairy caterpillar, *Amasacta albistriga* Walker (Lepidoptera: Arctiidae), cotton bollworms, *Earias insulana* Boisd. (Lepidoptera:Noctuidae) and *Earias vitella* Fabr. (Lepidoptera:Noctuidae), pigeon pea and chickpea pod borer, *Helicoverpa armigera* Hübner, (Lepidoptera:Noctuidae) have been reported from the endemic areas of Karnataka, Tamil Nadu, Kerala and Andhra Pradesh in South India. For instance, in Pavagada taluk of Tumkur (13°18' N, 77°12' E) and in Karimnagar (18°26' N, 79°10' E) and Mahbubnagar (16°45' N, 77°59' E) districts of Andhra Pradesh in South India, villagers collect the moths and caterpillars in large numbers, deploy light-traps for attracting moths and mass-kill the insects. Apart from the adults, the larvae or grubs in soil can be collected large numbers and destroyed. The root grub comes out of the soil during summer (February-April) or pre-monsoon (May-June) showers and feeds on the leaves or fresh twigs of neem (*Azadirachta indica)*, tamarind (*Tamarindus indica*) and *Ficus* species. Adults of root grubs are attracted to light and bonfire. Such adults can be collected in containers containing kerosene and water and destroyed. It is an important, proven integrated pest management (IPM) tool, practiced in many South-East Asian countries. Mass collection and destruction of root grubs has been reported from endemic areas of Karnataka, Gujarat, Maharashtra, Rajasthan, etc. For instance, in Karnataka, sugarcane (*Saccharum officinarum*) root grubs are collected in large numbers and destroyed in Bhadravathi taluk of Shimoga (13°57' N, 75°32' E), Gundlupet in Mysore district (13°15' N, 77°O' E), of coffee (*Coffea arabica* and *Coffea robusta*) in parts of Chickamagalur (13°15' N, 75°45' E) district, of areca (*Areca catechu*) in Thirthahalli, Sagar and Soraba taluks of Shimoga, and of groundnut (*Arachis hypogaea*) in Chintamani taluk of Kolar district (13°12' N, 78°15' E).

The practice of collecting and destroying in huge numbers the eggs, grown-up larvae and adults of pests like grasshoppers, locusts, hairy caterpillars on bunds, sticks, weeds, stretches of wastelands, is common. Such insects can be collected and destroyed or put in a tin pasted with grease on the top portion. Such larvae are killed in tin. *H. armigera* is a pest which is difficult to suppress by methods other than mechanical methods. The larvae can be collected by stirring the plants and by sweeping the larvae fallen underneath the plants. Such larvae are concentrated in pits and can be destroyed/burnt.

Hot Water Treatment

Sugarcane setts treated in hot water for 10 minutes controls whip smut (*Ustilago seitaminea*) in Gujarat. It is a recommended practice in all sugarcane areas in India.

Soil Solarization

Soil solarization in the vegetable nursery with the help of polythene mulch (25 microns) for 40 days after flood irrigation kills *Pythium* spp. and *Fusarium* spp. up to 30 cm soil depth. Raising vegetable nursery later in such solarized beds will have lower incidence of pest emergence, damping off, higher seed germination and better seedling vigor in Himachal Pradesh, Northwest India.

Detopping

Clipping of terminal shoots that serve as oviposition sites for 110 days in cotton (*Gossypium hirsutum* and *G. arboreum*) and 40-50 days in chickpea (*Cicer arietinum*) reduces the *H. armigera* infestation in egg stage, which is being practiced in cotton, as well as in chickpea.

5. Bird Perches [SSH]

Wooden or metallic 'T' or 'Y' perches are placed randomly in the cropped area to attract predatory birds to devour larvae/adults of pest insects. In India especially insectivorous birds like drongo (*Dicrurus adsimilis*) and mynah (*Acrido theres*) have been reported to be effective agents of suppressing populations of *H. armigera, Achaea janata* Linn. (Lepidoptera:Noctuidae) and hairy caterpillars in pigeon pea, chickpea, cotton, castor (*Ricinus communis*) ecosystems.

Birds feed on the larvae and reduce their population. For instance, about 80 percent of *H. armigera*. Spreading boiled rice balls or balls of Bengal gram preparation in semi-looper (*A. janata*) affected castor fields help in attracting birds, which eat larvae of semi-looper and reduce the populations. Spreading fried rice (puri) in *H. armigera* affected fields of chickpea also attract birds which predate on larvae of *H. armigera* and other pests.

PHYSICAL METHODS

Light traps in the form of gas lights or mercury lamps placed in large contiguous areas help in trapping moths of *A. albistriga*, wherever the pest is endemic. This will reduce the moth population and multiplication. This has been reported from endemic areas like Pavagada taluk in Tumkur, Kadur in Chickamagalur and Hosadurga in Chitradurga district (14°15' N, 76°28' E) Karnataka, South India.

CULTURAL METHODS

TILLAGE PRACTICES

Fall ploughing with a furrow turning plough by the end of the cropping season is advocated to expose the diapausing insect stages in several agro-ecosystems. The tillage practices expose the larvae and pupae of pest insects like cutworms (*Agrotis* spp.), hairy caterpillars (*Amsacta* spp.), grasshoppers (*Hieroglyphus* spp.), root grubs (*Holotrichia* spp.) and locusts (*Locusta* spp.) either to inimical conditions or to predators like insectivorous birds.

Crop Rotation

Monocropping, ratooning and growing the same crop year after year favors the pest and disease buildup. Therefore, it should be avoided. The population of the fungus, *Fusarium* sp. can be reduced when soil is kept fallow, compared to growing successive pigeon pea in the same soil, pigeon pea-fallow-chilies and pigeon pea-fallow-tobacco cropping pattern appreciably reduces *Fusarium* spp. Growing maize, (*Zea mays*) chickpea crop sequence decreases root-knot nematode population in chickpea (All India Coordinated Research Project [AICRP]).

Planting Time

Planting time has been exploited by many entomologists and pathologists to check insect-pests and disease buildup. For instance, sowing black gram (*Phaseolus mungo*) in the second fortnight of June reduces the foliar diseases in the hill region of Uttar Pradesh. The pigeon pea fusarium wilt is reduced when the crop is sown after July and also by sowing under rainfed conditions in Kanpur, Uttar Pradesh, North India. Sowing early and synchronizing preferably with short duration genotypes lowers bollworms (*Earias* spp.) incidence and damage in cotton in Southern India. Planting rice during second fortnight of July reduced rice neck-blast in Haryana, India. Sowing mustard in October reduces white rust incidence in Karnataka.

Trap Cropping

Trap cropping is yet another technique to successfully suppress pests or pathogens in cultivated ecosystems. Planting highly preferred crops in and around the cotton field serves as trap crops for bollworms. Variation in attractiveness of crops for egg laying or larval feeding is exploited as an effective tool to concentrate early pest populations on selected plants. The major benefit of this is that either the insect can be collected manually or control measures with bioagents or chemicals can be effected, to a limited area, thus avoiding chemical applications to the entire area.

The overall application will be less than in conventional farming. Okra (*Ebelmoschus esculentus*), marigold (*Tagetus* spp.), pigeon pea, chickpea, tomato (*Solanum esculentus*), tobacco (*Necotiana tuberosum*), maize (*Z mays*), sorghum (*Sorghum vulgare*), and beans (*Phaseolus* spp.) have been used as trap crops in India and proved effective to lower the incidence of *H. armigera* on cotton. Several studies have been made in India to suggest that okra and marigold have the potential to work as effective trap crops for *H. armigera* on cotton. Trap crops may also attract natural enemy by the presence of semio-chemicals, thereby enhancing effectiveness of natural control. Suppression of diamond back moth (DBM) (*Plutella xylostella* Linn.) (Lepidoptera:Yponomeutidae) by trap cropping is popular among cabbage (*Brassica oleracea*) growers in and around Bangalore, Karnataka. For suppressing DBM, paired rows of mustard (*Brassica juncea*) for every 25 rows of cabbage planting and second row sown 25 days after planting exhibited 90 percent attack to mustard than cabbage. Similarly planting marigold one row for every 16 rows of tomato attracts *H. armigera* moths to oviposit on marigold (*Tagetes* spp.).

Plant Geometry

Inter- and intra-row spacing has direct and indirect influence on the population of bollworm (*Earias* spp.) because of variation in plant habit and in architecture. Closer spacing (75 × 30 cm) attracted higher numbers of bollworms than wider spacing (90 × 30 cm). Wider spacing alters the microclimatic conditions to make it unfavorable for pest multiplication and also make easy to reach pest with bioagents and chemicals. Closer spacing of 25 × 10 cm reduces mung bean yellow mosaic virus compared to 40 × 10 cm.

Intercropping

Intercropping with cowpea, (*Vigna unguiculata*), maize, onion, finger millet (*Eleusine coracana*), mung bean, soybean (*Glycine max*) groundnut in south and central zones of India to colonize biological control agents have been advocated. Growing cotton with groundnut/green gram at 1:3 or with onion 1:4 or five proportions have been agronomically accepted practice for remunerative yield in Karnataka. Intercropping barley (*Hordeum vulgare*) with chickpea reduced blight disease (*Ascochyta rabiei*) in Rajasthan, India, marigold or cowpea planted around cotton crops as border crop reduces incidence of bollworms (*Earias* spp. and *H. armigera*) in cotton.

Water/Moisture Management

Managing or altering the moisture levels or extent of moisture in cultivated tracts is important for the management of pests and diseases as illustrated in instances given below. Early termination of irrigation avoids crop prolongation and this reduces bollworm succession. However, this is mainly for pink bollworm, *Pectinophora gossypiella* (Lepidoptera: Gelechiidae). Increased soil moisture due to mulching, farm yard manure and low plant population alone or in combination in cluster beans (*Cyamposis tetragonoloba*) helps to increase the population of resident bacteria, which reduces the dry rot caused by *Macrophomina phasiolina*.

Nutrient Management

Manuring and fertilization are practices adopted to promote rapid growth in plants and shorten the period of susceptible stage of attack to the herbivores. Nutrients provide greater tolerance and compensates for insect damage. However, such practices may also change the physiology of plants in favor of insects to make the plant most desirable for survival and multiplication.

Selected examples are provided below for the role of nutrients in conferring resistance/susceptibility to insect attack or pathogen invasion:

i. Increased nitrogen level lead to increased oviposition and increased infestation and damage by *H. armigera* in cotton.
ii. Use of 100 kg nitrogen/ha induces moderate incidence of false smut in paddy compared to higher doses of nitrogen in Gujarat.
iii. One hundred kg potash/ha decreases stem rot disease (*Macrophomina phasiolina*) and increased fiber yield in jute in West Bengal.
iv. Boron application at 2.5 ppm in pearl millet (*Pennisetum typhoides*)

reduced the primary incidence of downy mildew to the extent of 76.4 percent at 35 days after sowing in Rajasthan.

PLANT RESISTANCE

Host plant resistance is deployed as unique approach for pest management with predictable problems. It helps in suppressing the pest population at low cost with least disturbance to the ecosystem. Unequivocal advantage of plant resistance is its compatibility with all other methods of pest control. Even low levels of field tolerance/resistance are desirable in cultivated plant varieties.

BIOLOGICAL CONTROL METHODS

Several species of natural enemies including parasitoids, predators and pathogens have been recorded in paddy, cotton, pigeon pea, chickpea, tomato and cabbage ecosystem. These natural enemies help in killing the eggs, larvae of different pest. These natural enemies play an important role in the agro-ecosystem.

BIORATIONAL TOOLS

Biotechnology

The important contribution of biotechnology is the capacity to express pesticidal proteins within transgenic plants. Bt cotton could provide effective control of target pests. Transgenic bollworm (*H. armigera*)-resistant cotton is seen as an improvement over traditional cotton and has achieved significant social and economic benefits. The use of insecticides in Bt (*Bacillus thuringiensis*) cotton decreased by 60-80 percent and the density of predators increased by 24 percent. The Government of India has given provisional approval for commercial cultivation of Bt cotton in March 2002.

Pheromones

Pheromones are semio-chemicals secreted by insects to complete normal life cycle activities. Entomologists have exploited these chemicals to interfere with the normal activities such as mating and reproduction and thus reduce their populations in the cultivated patches.

Pheromones are also key component of cotton, pigeon pea, chickpea IPM and are mostly utilized for monitoring the pest populations so that timely interventions are made. Monitoring helps in early detection of bollworm populations and its buildup. Use of pheromones has helped in monitoring *H. armigera* population on tomato. Two sex pheromone traps per acre were fixed and lures have to be changed in 15 days for monitoring *H. armigera* which helps in assessing attack of pests in advance and intensity of pest which helps to take control measures.

SUCCESS STORIES

Red-headed Hairy Caterpillar

In Tumkur district of Karnataka during 2001-02, the Department of Agriculture organized mass collection of eggs, larvae and adults. An estimated 341 kg of moths were collected and burnt.

Three thousand moths weighed 1 kg. Similarly eggs were collected in 2,000 ha. The eggs were collected and put in a tin coated with grease. Such eggs have been destroyed. During 2000-01, in the same area 65,728 mt of larvae were collected and burnt.

Reasons for Success[SSH]

- Government sanctioned special package for the purpose.
- Action plan developed involving farmers, people's representatives, the press, scientists, NGOs, Rotary Club members, merchants, media persons, etc.
- Awareness was created through street plays, video cassettes were shown, colorful wall posters were displayed, wall paintings were done, meetings were conducted in villages, banners were put at popular junctions, publicity was made through mike-systems, visits were made to areas where operations concerning hairy caterpillars were successfully carried out.
- Active groups were formed in each village.
- Distributed 2,300 gaslights under subsidy program for using in light traps.
- Set up 30 monitoring centers.
- Set up 200 mercury lamps for light traps.
- Clear-cut responsibilities were fixed and on works to be done.
- With rainfall, follow-up action were taken.
- Seeds of pulse crops were distributed to be grown as intercrops under subsidy schemes.

Root Grub *Holotrichea serrata* Linn. (Coleoptera: Scarabaeidae) [SH]

In Bellary, Dharwad district of Karnataka during 1986-87 and 1987-88 collected 20 mt of root grubs in each place after the receipt of summer showers in May 1986 and May 1987.

The farmers were briefed through meetings about the pest, identified the endemic area with the help of farmers, conducted night campaigns during summer showers rainfall, light traps, bonfires with crop stubbles and bicycle waste tires were set up at night.

The adult root grubs were attracted to light traps and fire. The beetles were collected and put in water mixed with kerosene.

Farmers Field School (FFS) [SH]

1. FFS in Cotton[SSH]

During 2000-01, with the help of Food and Agriculture Organization (FAO) technical assistance, 22 agricultural officers were trained as IPM facilitators. The training of facilitators was held at the Regional Research Station, Raichur, in Karnataka, South India. During 2000-01, 20 FFSs were conducted with the help of above IPM facilitators. The school was conducted in the farmer's field only for 20 weeks. Every week, the farmers gathered for the field school and learned to analyze the cotton agro-ecosystem, determine the population levels of parasitoids and herbivorous insects. They carried out experiments and understood the function of useful insects. Based on their understanding of the ecosystem, farmers were able to make informed decisions on growing a healthy crop. The following suggestions were evaluated by farmers to determine if they are useful:

i. Okra, marigold, pigeon pea and cowpea were grown as trap crop to control sucking insects, *H. armigera* and *Erias* bollworm.
ii. Seed treatment with *Trichoderma* at 4 g/kg and with imidachloprid 7.5 g/kg.
iii. Stem pasting with imidachloprid @ 1:20 ratio to control sucking insects.
iv. *Trichogramma* released 100,000/acre to parasitize eggs of *H. armigera*.
v. NPV spray @ 200 LE/acre.
vi. Set up 25 bird perches/acre for control of small larvae of *H. armigera*.
vii. Caterpillars were collected and destroyed.
viii. In 40 days old cotton crop, pheromone traps at 2/acre were set up to monitor the intensity of *H. armigera* and *Earias* bollworms.
ix. Twenty-five yellow sticky tins were set up to control whiteflies.

Farmer would usually apply 19 chemical sprays to control pests and diseases in their fields and spend Rs.9,000 whereas the cost of protection was only Rs.2,097 in IPM plots. The yield obtained in farmer's plot where IPM was not practiced was 600 kg/acre, earning Rs.9,000 at the cost of Rs.12,936 whereas in IPM plots, the farmers were able to get 1,100 kg of yield by investing Rs.6,043 and fetch the net returns of Rs.10,468 as profit.

2. FFS in Paddy [SSH]

In Tumkur district, Karnataka during 1995, one officer participated in a season-long training to become an IPM facilitator with the technical assistance of the FAO. Four FFSs were conducted for 10 weeks in which 20 officers and 120 farmers participated. Every week the farmers would assemble for half a day in the study field. They researched on parasitoids and other useful insects including pests using insect zoos. From the research conducted by farmers, they learned more about useful insects and this helped them to conserve the beneficials by not using insecticides. The farmers after this exercise did not use even a single chemical spray whereas in control plot the farmers had sprayed four times. The FFS concept is being continued in different areas every year.

Nonpesticide Method of Controlling Pests and Diseases in Chickpea [SH]

In 2001-02, 50 ha demonstrations were conducted at Chitradurga district, Karnataka, involving 16 farmers.

The farmers were shown the following practices:

i. Seeds treated with 4 g/kg *Trichoderma*.
ii. Sex pheromone traps were put @ 2/acre.
iii. Nipping of tips as soon as eggs were noticed.
iv. Churumuri (roasted rice) spreading at 6 kg/acre to attract birds, which predates on *Helicoverpa* larvae.
v. Erection of 10 bird perches/acre to attract birds that feed on caterpillars in the crop.

This has resulted in 750 kg/ha yield against 500 kg/ha in the control. In addition to regular cost of production, an additional cost of Rs.470 was spent on nonpesticide method for controlling pests and diseases. In farmer practice, Rs.1,200 was the additional cost for purchase of chemicals to control the pests and diseases. Hence, use of nonpesticide methods result in 250 kg of additional yield, which fetched about Rs.4,000 per net profit.

GOVERNMENT POLICIES AND PROGRAMS

INTEGRATED PEST MANAGEMENT

Indiscriminate and injudicious use of chemical pesticides in agriculture have resulted in adoption of IPM as cardinal principle and main plank of plant protection in the overall Crop Production Program. IPM is an eco-friendly approach which focused on farmer education to make farmers more efficient in their production activities.

The IPM concept is being promoted through 26 Central Integrated Pest Management Centers (CIPMC) located in 22 States and one Union Territory aiming at human resource development, these centers are imparting training to extension functionaries and farmers in IPM skill by conducting FFSs. In addition, these centers are engaged in mass-production and field releases of biological control agents. With a view to provide technical knowledge to the extension functionaries and farmers in the States, IPM approach has been tried out for 51 crops like rice, cotton, vegetables, pulses, plantation crops, spices, fruit crops and oilseeds. A participatory IPM approach based on farmer education is being implemented throughout the State by the extension service. Towards human resource development, facilitation skills have been developed through continuous education of Agriculture Extension Officers (AEO) and farmers. IPM FFSs are conducted in the States to enhance the skills of farmers. Aiming at organic farming production technology for biocontrol agents, biopesticides have been evolved which are being used by the private as well as public sector units for production of biocontrol agents to be used by farmers to save their crops from the ravages of pests and diseases.

This concept prevents dependence on chemical/pesticides for plant protection. Since the inception of IPM program, 7,113 FFSs have so far been conducted thereby training 32,786 AEOs and 212,870 farmers in IPM skills. Eleven hundred and thirty-two IPM facilitators have been trained by conducting 35 season-long training of facilitators courses on cotton, rice, vegetables, pulses, oilseeds, etc. With the background of IPM knowledge farmers are practicing farming with minimum use of pesticides as agrochemicals. Instead they are depending on biological control agents. As a result of implementation of IPM program, there is a significant reduction in the consumption of pesticides from 61,357 million mt (technical grade) during 1994-95 to 43,584 million mt (technical grade) during 2000-01. In the case of red-headed hairy caterpillar the government has given sanction to take up effective awareness campaign, providing gaslights under subsidy program, providing funds for catching larvae at Rs.35/kg and moth at Rs.50/kg. Farmers contact centers have been started at hobli level to deliver the different nonpesticide/IPM methods to farmers. Government has given permission to grow commercial Bt cotton in farmer fields in March 2002.

ISSUES AND PROBLEMS [H]

a) *Trichogramma* parasitoids and NPV are not available on a large scale.
b) Quality monitoring and testing facilities for bioagents. Parameters and checkup facilities are not adequate.
c) Awareness of people on nonpesticide methods is inadequate.
d) People participation in control of moths, grub on community basis is not up to the expected level.
e) Monitoring of pests and their populations is inadequate.
f) Prevailing method of giving plant protection chemicals under credit directly from the commission agents to the farmers hinders the use of nonpesticide methods.

SUGGESTION FOR IMPROVEMENTS [H]

a) Large-scale training to agriculture departmental staff/farmers is required.
b) Block demonstrations are to be laid out exclusively on nonpesticide methods to build more confidence in farmers.
c) Availability of parasitoids/NPV on large scale should be ensured.

d) More testing facilities should be created.
e) Large-scale monitoring of pests is required.
f) Farmers should be encouraged to adopt more nonpesticide methods.

CHEMICAL WEED CONTROL

The proper application of herbicides can dramatically reduce turfgrass weed populations in a short period of time. A combination of their cost, effectiveness and simplicity of use often makes herbicides the primary means of weed control -- with little attention given to other weed reduction measures. Remember, herbicides are only one facet of the total weed control puzzle. Effective long-term weed control and a reduction of pesticides in the environment will be achieved only when an effective cultural weed control programme has been implemented.

CULTURAL WEED CONTROL

The basic principle involved in cultural weed control is to grow a stand of grass that is dense and competitive enough to prevent weed encroachment. Weeds are not the cause of poor turf, but rather the result

Weeds require light, water and nutrients to grow. As turf loses density, light will penetrate the canopy and cause weed seeds at the thatch and soil surface to germinate. Once germinated, weeds can develop rapidly, especially if turf continues to decline.

Weeds probably will receive sufficient light to develop if you can see bare soil or thatch when you look straight down on a stand of grass. Management practices that discourage turf weeds include proper mowing, watering, fertilizing, thatch control and cultivation. The chance for weed encroachment can be reduced by using locally adapted turfgrasses and establishing them during the correct season of the year.

- *Mowing:* To prevent weed germination, mow frequently at the tallest recommended mowing height. Weeds germinate rapidly when turf is scalped by mowing too short or when it is not mowed frequently enough. Both mistakes decrease turf density and cause an open canopy that favours weeds. Experts recommend a range of mowing heights to meet specific turf activities. Lower mowing heights require more frequent mowing. Each mowing should not remove more than one-third of the total leaf height. Annual grassy weeds -- such as crabgrass -- are especially a problem on turfs that lack density as the result of poor mowing.
- *Fertility:* Apply nitrogen at least once a year. Areas that receive chemical weed control but no nitrogen are probably being mismanaged. Even in low- maintenance situations, the first defence against weeds is dense turf. For cool-season grasses, apply nitrogen

in the fall. For warm-season grasses, apply it in the summer. Additional nitrogen should be supplied during the growing season to provide adequate growth and density with less attention given to turfgrass colour. On cool-season grasses avoid heavy applications of nitrogen in the spring. That will encourage excessive vertical leaf growth, bringing with it a greater chance for scalped turf and an eventual decrease in turf density. Don't apply nitrogen to dormant warm-season grasses, since it will encourage winter annual weeds, such as chickweed, henbit and speedwell.

- *Establishment:* Weeds can prevent the establishment of grasses at certain times of the year. Spring seeding of cool-season grasses is seldom successful -- because the fertility requirements and frequent watering necessary for turf establishment will also create intense weed pressure from crabgrass and summer annual broadleaf weeds. Consequently, late summer and early fall are preferred for establishment of cool-season grasses. Cool-season turfgrasses that are germinating by early September will not be affected by summer weeds and will usually establish a competitive stand of grass before winter annuals, such as chickweed and henbit, become a problem.
- *Thatch and Cultivation:* These have a dual effect in terms of weed control. Weed seeds below a thatch layer lack the necessary light for germination, so weeds that germinate in the thatch layer usually die because of the poor environment for seedling development in thatch. In this respect, thatch provides some degree of weed suppression. On the other hand, thatch can cause turf decline by harbouring disease-causing pathogens, reducing water infiltration and tying up pesticides and fertilizers. Power-raking and verticutting are typically used to remove thatch.
- *Cultivation:* Soil compaction from traffic and tight soils will often decrease turf competition and favour some weeds, especially goosegrass and knotweed. Two ways to fix the problem are coring and slicing. These procedures improve the soil surface and the growing environment for turfgrass.

DAMAGED CORN FOR SILAGE

Corn that has been damaged by drought, high temperatures, blight, frost or hail can be salvaged for silage. Quality will not be as high as where corn has reached the dent stage.

Feeding value will depend upon the state of development and how it is handled after the damage occurred. Silage from immature corn usually is higher in moisture, does not ferment in the same manner, frequently has a sour odour and is more laxative when fed in large quantities.

Frosted corn has a low carotene content and should be cut as soon as possible. It will dry out quickly and lose leaves. It may be necessary to add water to corn that has frosted and become too dry to pack well. Drought corn also may need added water. When the corn forage is dry, keep the chopper knives sharp and chop as fine as possible.

Immature corn that has been damaged by extremely high temperatures should not be immediately ensiled. Although these plants may never produce an ear, some additional stalk growth and consequently some additional feed may be produced by delaying harvest.

If the plants are harvested for silage soon after they have been extensively damaged by heat, the stalk will have so much moisture that a very low quality silage will result. Nutrients also will decline significantly through seepage. Corn that has been damaged by leaf diseases such as the southern corn leaf blight often are made into silage. The blight organism is not believed to be toxic to ruminants.

It also has been shown that it does not survive the ensiling process. In severe cases, a secondary infection of molds on the damaged areas of the plant may produce a harmful toxin. However, limited research indicates that this is unlikely.

The greatest problem with this type of silage stems from its lack of energy due to reduced grain formation and improper fermentation due to the excessive dryness of the damaged plant. In severe cases where large areas of the corn plant are dead, fermentation problems caused by the lack of plant moisture could arise.

Nitrates

Under certain soil and environmental conditions, the corn plant may store an excessive amount of nitrogen compounds.

Nitrate poisoning and drought conditions often are associated, but the condition is difficult to define. The degree of drought and nitrogen availability generally confuses the issue. As the soil moisture level becomes acute, nitrates move towards the soil surface above the corn roots. Some drought-stricken corn may be short of nitrate instead of oversupplied. If heavy rains occur at this point, excess nitrogen is leached downward and may be taken up by the corn root system. This could cause abnormally high nitrate levels in the plant. Under these conditions, avoid harvesting silage from the stricken fields for a few days.

One measure that may be helpful in reducing the amount of nitrate in the silage is, to allow the corn plant to grow beyond the period of drought damage. If the plant is capable of making some regrowth following a drought, it may be able to use much of its excessive stores of nitrogen for additional regrowth. The highest concentrations of excessive nitrogen usually are in the lower portion of the stalk. Raising the chopper cutter blade so that the lower 18 or 20 inches of the stalk remain in the field also may reduce the concentration in the silage. Testing for nitrates before silage harvest is frequently unreliable as a feeding guide. Nitrate level in the plant changes rapidly from day to day and usually is reduced about 1/3 in the ensiling process. If silage is suspected to contain excessive nitrates, it can be detected by test prior to feeding. This analysis should be taken as near to the time when the silage will be fed as practical.

Silo Gases

Lethal gases may occur at any time during silo filling. The greatest danger is 12 to 72 hours after filling, but gas may occur up to 10 days after the last silage is put in the silo.

Silo gases may appear in any ensiled material, grown on any type of soil and under any level of fertilization.

If the gases are present in higher concentrations, two of them, nitrogen dioxide (NO_2) and nitrogen tetroxide (N_2O_4) may be recognized by their irritating odour and colour. Nitrogen dioxide is reddish brown and nitrogen tetroxide is yellow. Nitric oxide (NO) is a colorless gas that may be present in deadly concentrations without being visible.

The gases are heavier than air and remain beneath the air mass over the silage. They often will layer on top of the silage just below the upper edge of the top door. They may settle down the chute to accumulate at the bottom of the silo, in unloading chutes, in adjacent feed rooms and may even move into the barn, loading areas or milk houses. They usually leave a yellow stain on silage, wood or other materials they contact. The presence of dead birds and small animals around these areas is one indication that the gases are at dangerous levels.

A few simple rules will prevent tragedy and injury:

- Run the blower 15 to 20 minutes before going into a partly filled silo. Keep the blower running while anyone is inside.
- Stay out of the silo for at least a week or preferably two after the silo is filled.
- If you experience the slightest throat irritation or coughing, get into fresh air quickly. Immediate treatment by a doctor is an absolute must.
- Ventilate the silo room for at least two weeks after filling by opening outside doors and windows to carry away fumes. Removing the chute doors on the silo down to the level of the settled silage will permit natural ventilation where gas tends to be concentrated.
- Keep the doors between the silo room and the barn closed to protect livestock.

Length of Cut

Corn silage should be cut into particles 1/2 to 3/4 inches in length. Particles of this size will pack more firmly in the silo and are more palatable to cattle. Very finely cut silage may be made with a recutter. This will increase the amount of dry matter that can be stored in a silo, but very finely cut silage is less palatable and has resulted in lower butterfat tests when this feed was the primary source of roughage for dairy cattle.

Adding Water to Dry Silage

If silage is too dry, it may be necessary to add water in order to establish airtight conditions. As a rule of thumb, add four gallons of water per ton of silage for each 1 percent desired rise in moisture content. Add this water as the silo is being filled. If water is added after the silo is filled, it tends to seep down the silo walls and does not permeate the silage mass. This may cause leaching of silage nutrients, seepage that may break the air seal and improper fermentation. Frozen silage: Frozen silage is sometimes a problem, especially with trench or bunker silos. Freezing does not impair the keeping quality of silage so long as the silage is not disturbed, but frozen silage may cause digestive disturbances when eaten by cattle. It is best to thaw silage before feeding it.

Silage additives: Top quality silage can be made without the addition of any additives or preservatives. There is no reliable evidence that adding enzymes, yeast cultures, antibiotics or acid-forming bacteria will increase the feeding value of corn silage.

The two additives most often used in corn silage are limestone and non-protein nitrogen compounds. Molasses and grain may be added to corn forage at the time of ensiling. However, this practice only tends to enrich the resulting corn silage as a feed rather than to improve the quality of the corn forage itself. There is also some fermentation loss with the added molasses and grains. *Molds:* Molds are common in silage, especially around silo doors and edges of bunker silos.

White and gray molds, which are caused by exposing silage to air, are seldom toxic to livestock, but intake is sometimes reduced. A mold known as monascus causes silage to form lumps or small clumps, which are white on the outside but have a reddish centre. Cattle usually will eat silage containing this mold. There have been no reported cases of toxicity.

In the spring as warm temperatures occur, a rapid-forming red mold may appear on the face of the silage between morning and evening feedings. This is commonly known as bakery mold, monilla sitophilia.

The mold spores are activated by the heat produced in the silage process and grow quickly when exposed to air. It has not produced any known cases of toxicity.

Silo Capacities and Filling

Tables giving estimated silo capacities often vary due to the crop ensiled, length of cut and effectiveness of distribution and packing at filling time. Most tables are based on silage harvested at 30 to 35 percent dry matter.

Silo capacities actually vary little in terms of dry matter storage, despite differences in moisture content at filling time. An exception to this is material that is 50 percent dry matter or more at the time it is placed in the silo. With this type of material, dry matter capacity may be reduced as much as 10 percent for estimates based on 30 to 35 percent dry matter.

With large-capacity silos and high-speed filling methods, the distribution and packing of silage in silos has been sacrificed. Improper distribution and packing may cause excessive seepage, poor fermentation and losses in storage capacity. Half the capacity of the silo, which is 14 feet in diameter, is in the outside 2 feet. If an overabundance of fluffy, light material is in this outside area, silo capacity may be reduced as much as 20 percent. Most large tower silos should be equipped with a distributor to aid in proper silage distribution and packing. Length of cut of corn should be 1/2 to 3/4 inches to ensure proper placing.

Covering Filled Silos: A loss of nutrients occurs in all silage during the ensiling process. Microorganisms that carry out the fermentation process require energy. This loss depends on the effort made to exclude air at time of filling and to prevent loss of carbon dioxide, which is necessary to arrest respiration of the ensiled plant cells, as well as prevent seepage losses and undesirable fermentation or spoilage due to surface exposure. There is probably a greater range of losses within a particular type or design of silo than occurs in different types of silos where equally good ensiling practices are followed. So, regardless of silo structure, good ensiling practices save feed.

A plastic cover on a trench or bunker silo or a large-diameter tower silo can materially cut feed losses. For best results, the cover must be applied immediately after the last load is packed in the silo.

On trench or bunker silos, it is important to mound or crown the forage so that the rain water will drain off the silo. Plastic covers that are not weighted to hold them firmly on the surface of the silage will be only partly effective. Covering the plastic with old tires plus limestone, sawdust or similar material can be of great benefit.

HARVESTING

Corn should be harvested for silage after the ear is well dented but before the leaves turn brown and dry. The quantity and quality of corn silage are at their peak in this stage of development. The ear has accumulated most of its potential feeding value, but there has been little loss from the leaves and stalks. After the dent stage, feeding value of corn stalks and leaves decreases while field losses increase. Fall temperatures influence the maturity rate of the grain. Maturity usually refers to the time when the ear has accumulated 100 percent of its dry matter production potential. In many years this potential is not achieved because of cool temperatures and cloudy weather. Values listed may be used as a guide to determine when maximum dry matter production has occurred, but variety and weather interactions will exhibit some influence on the result. Ears usually will be well dented somewhere between the 32- to 35-percent moisture stage.

Table : Relationship of Kernel Moisture to Yield Potential

Water in kernels	*Yield of grain as percent of maximum*
40 percent	93.5
38 percent	94.8
36 percent	96.3
34 percent	98.0
26 percent	100.0
21 percent	98.0

Corn harvested for silage in the milk or dough stage will yield less feed nutrients per acre than if harvested later. Corn also may ferment improperly in the silo if harvested too soon. Corn silage often is made too early because it is believed that feed is lost if undigested corn kernels appear in the manure. This is not true. Digestibility is as high for well-dented kernels as for immature corn grain.

Corn silage that is cut late and has brown and dead leaves and stalks usually will make fair- to good-quality silage, but total production per acre may be sharply reduced. Field losses as high as 30 percent have been found when silage is made late into the fall or early winter. A 10 percent reduction in the amount of dry matter stored in the silo also has been noted with late-cut silage.

Results of feeding trials with late-cut silage tend to vary, but in most cases the quality of late-cut corn silage has been slightly lower than silage made from corn cut soon after the dent stage of the ear.

Present research does not support recommending late-cut or mature corn silage as a standard farm practice. However, it does indicate that in emergency conditions, corn may be harvested over a wide period of time and still make a satisfactory feed.

4

Aquatic Abitats (Insects)

Less than 3 per cent of the world's total water occurs on land, and most of this is frozen in polar ice caps. Streams and rivers are one of the more conspicuous features of the landscape; however, their total area is about 0.1 per cent of the land surface, whereas lakes represent about 1.8 per cent of the total land surface. Some authors have questioned whether insects have been successful in water because aquatic species represent only a small portion of the total hexapod fauna. However, 13 orders of insects contain species with aquatic or semiaquatic stages, and in 5 of these (Ephemeroptera, Odonata, Plecoptera, Megaloptera, and Trichoptera) all species are aquatic with few exceptions. Few aquatic insects spend all of their life in water; generally any insect that lives in water during a portion of its development is considered to be " aquatic." Usually, but not always, for most "aquatic" species, it is the larval stage that develops in aquatic habitats, and the adults are terrestrial. The pupae of some taxa undergoing complete metamorphosis (*i.e.*, holometabolous) remain within the aquatic habitat; in others the last larval instar moves onto land to pupate, providing the transition stage from the aquatic larva to the terrestrial adult.

The success of insects in freshwater environments is demonstrated by their diversity and abundance, broad distribution, and their ability to exploit most types of aquatic habitat. Some species have adapted to very restricted habitats and often have life cycles, morphological, and physiological adaptations that allow them to cope with the challenges presented by aquatic habitats. One aquatic environment in which insects have not been very successful is saltwater habitats, although some 14 orders and 1400 species of insects occur in brackish and marine habitats; only one group occurs in the open ocean.

Table. Occurrence of Life Stages in Major Habitat Types for Aquatic and Semiaquatic Representatives of Insect Orders

Order	Terrestrial	Freshwater
Collembola	A, L	A, L
Ephemeroptera	A	L
Odonata	A	L

Heteroptera	A	A, L
Orthoptera	A, L	A, L
Plecoptera	A	L
Coleoptera	A, L, P	A, L
Diptera	A, P	L, P
Hymenoptera	A	A, L, P
Lepidoptera	A	L, P
Megaloptera	A, P	L
Neuroptera	A, P	L
Trichoptera	A	L, P

One of the most widely accepted attempts to explain why more insects do not live in marine environments is that successful resident marine invertebrates evolved long before aquatic insects and occupy many of the same niches inhabited by freshwater insects. Thus, marine invertebrates, such as crustaceans, have barred many insects from marine habitats by competitive exclusion. Problems with osmoregulation have been given as another reason for the paucity of saltwater species; however, one of the two multicellular animals found in the Great Salt Lake is a member of the order Diptera, providing evidence that some insects display a strong ability to osmoregulate. The first aquatic insects are believed to have inhabited flowing water as early as the Permian and Triassic. It was not until the late Triassic and early Jurassic that evidence of abundant lentic, or still-water, fauna arose, accompanied by rapid diversification of water beetles, aquatic bugs (Heteroptera), and primitive Diptera.

On the basis of several lines of evidence including osmoregulation, fossil evidence, secondary invasions to water of many taxa, and great variation in gill structure among and within orders, some authors have suggested that the first insects may have lived in water rather than in terrestrial habitats. However, the general consensus is that an aquatic origin for insects seems unlikely and that aquatic insects may not have shown up until 60- 70 million years later than their terrestrial counterparts.

Freshwater systems are often divided into standing (lentic) and flowing (lotic) waters. Although such a division is useful for indicating physical and biological differences, habitat diversity can vary tremendously within these two broad categories, and some of the same taxa may be found in both lentic and lotic habitats, depending on the physiological constraints of a given habit. Many factors influence successful colonisation of aquatic insects to a given habitat;

However, most of these would fall under four broad categories:

1. Physiological constraints (*e.g.*, oxygen demands, respiration, osmoregulation, temperature effects),
2. Trophic considerations (*e.g.*, food acquisition),
3. Physical constraints (*e.g.*, coping with harsh habitats), and
4. Biotic interactions (*e.g.*, predation, competition). However, these

categories are so interrelated that detailed analysis of each factor separately is very difficult.

CLASSIFICATION SYSTEM

The classification system used here for lotic and lentic habitats stresses the basic distinction between flowing water (*i.e.*, streams, rivers) and standing-water (*i.e.*, ponds, lakes, swamps, marshes) habitats. This separation is generally useful in describing the specific microhabitats (*e.g.*, sediments, vascular hydrophytes, detritus) in which aquatic insects may be found. Both stream/river currents and lake shoreline waves often create erosional (riffle-type) habitats and may resemble each other in their physical characteristics, whereas river floodplain pools and stream/river backwaters create depositional (pool-type) habitats that may resemble lake habitats as well.

Within a given habitat, the modes by which individuals maintain their location (*e.g.*, clingers on surfaces in fast-flowing water, sprawlers on sand or on surfaces of floating leaves, climbers on stem-type surfaces, burrowers in soft sediments) or move about (*e.g.*, swimmers, divers, surface skaters) have been categorised. The distribution pattern resulting from habitat selection by a given aquatic insect species reflects the optimal overlap between habit and physical environmental conditions that comprise the habitat, such as bottom type, flow, and turbulence.

Food in aquatic habitats is almost always distributed in a patchy fashion, the match between habitat and habit is maximized in certain locations. This combination will often result in the maximum occurrence of a particular species. In view of the complex physical environment of streams, it is not surprising that benthic invertebrates have evolved a diverse array of morphological adaptations and behavioral mechanisms for exploiting foods. Throughout this article we will follow the functional classification system originally described by K. W. Cummins in 1973, which is based on the mechanisms used by invertebrates to acquire foods.

These functional groups are as follows:

- Shredders, which are insects and other animals that feed directly on large pieces of organic matter (*e.g.*, decomposing leaves and fragments of wood >1mm in size) and their associated fungi and bacteria, and convert them into fine particulate organic matter (FPOM) through maceration, defecation, and physical degradation.
- Collector-filterers, which have specialised anatomical structures (*e.g.*, setae, mouth brushes, fans, etc.) or silk and silklike secretions that act as sieves to remove fine particulate matter less than 1 mm in diameter from the water column.
- Collector-gatherers, which gather food, primarily FPOM, that is deposited within streams or lakes.

- Scrapers, which have mouthparts adapted to graze or scrape materials (*e.g.*, periphyton, or attached algae, and the associated microbes) from rock surfaces and organic substrates.
- Predators, which feed primarily on other animals by either engulfing their prey or piercing prey and sucking body contents.

These functional feeding groups refer primarily to modes of feeding or the means by which the food is acquired, and the food type *per se*. For example, shredders may select leaves that have been colonized by fungi and bacteria; however, they also ingest attached algal cells, protozoans, and various other components of the fauna along with the leaves.

HABITATS

Streams vary greatly in gradient, current velocity, width, depth, flow, sinuosity, cross-sectional area, and substrate type, depending on their position in the landscape with respect to geology, climate, and the basin area they drain. Anyone who has spent much time around or wading in streams is aware that these can be extremely diverse habitats, often manifesting great differences over short distances.

In the upper reaches of a catchment or drainage basin, small streams often display a range of habitats characterised by areas that are shallow, with fast flow over pebbles, cobbles, and boulders. There are also areas with steep gradients, cascades, or waterfalls when the underlying substrate is bedrock. There also may be areas of slow velocity in pools of deeper water.

In many streams draining forested watersheds, pools are found. Pools are depositional areas during normal flow as organic and inorganic particles settle to the substrate, and a similar settling process often occurs in side channels or backwater areas of streams. Pools are also created upstream of large instream pieces of wood, which may form obstructions known as debris dams.

Because pools are generally characterised by reduced water velocity, many of the small particles normally suspended in fast flows settle to the bottom. In many low-gradient streams, including large rivers, bottom substrate often consists of silt, sand, and gravel-sized particles that are frequently moved by the force of the flowing water. In such systems, large pieces of woody debris entering the river from bank erosion. Substratum characteristics are often perceived as a major contributor to the distribution of many invertebrates; however, many other factors, including water velocity, food, feeding habits, refuge, and respiratory requirements, can be associated with specific substrates. Substratum particle size is influenced by several items, including geology, physical characteristics of the rock, past and present geo-morphic processes (flowing water, glaciation, slope, etc.), climate and precipitation, and length of time over which the processes occur. These in turn influence landform, which exerts a major influence on various hydrological characteristics of aquatic habitats. Unlike many lentic environments, in lotic systems the velocity of moving water is sufficient to pass the water around the body of an insect and turbulence provides reaeration; thus, dissolved oxygen is rarely limiting to stream inhabitants. Local transport and storage of inorganic and organic materials by the current may be either detrimental (*e.g.*, scouring action) or beneficial (as a food source). For example, most aquatic insects in flowing waters are passive filter feeders and depend on the water current for delivery of their food. Scouring flows may therefore remove from the streambed large organic particles (*e.g.*, leaves) as well as smaller ones, creating temporary reductions in food supplies. In contrast, moderately rapid flows may facilitate feeding of some scraper or grazer insects by preventing excessive sedimentation buildup on the surfaces on which they feed.

EROSIONAL HABITATS

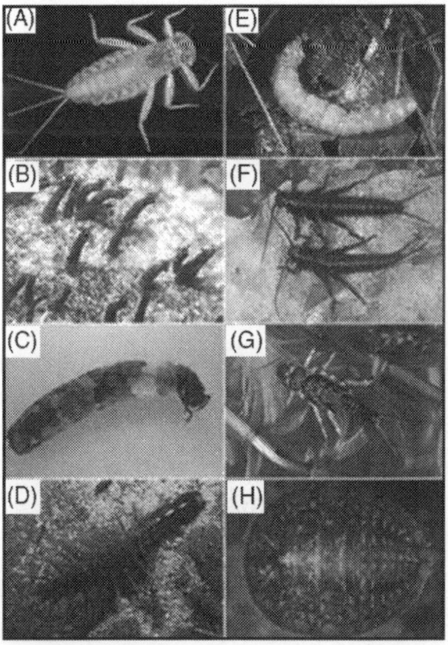

Fig. Typical Insects Inhabiting Lotic Environments. (A) Ephemeroptera: Heptageniidae (Rhithrogena). (B) Diptera: Simuliidae (Simulium), (C) Trichoptera: Limnephilidae (Dicosmoecus), (D) Megaloptera: Corydalidae (Corydalus), (E) Diptera: Tipulidae (Tipula), (F) Plecoptera: Pteronarcyidae (Pteronarcys), (G) Plecoptera: Perlidae, Coleoptera: Psephenidae (Psephenus).

Adaptations of aquatic insects to torrential or "rapid flow" habitats include the dorsoventral flattening of the body, which serves two purposes: it increases the organism's area of contact with the surface substratum, and it offers a mechanism by which animals can remain in the boundary layer when water velocity diminishes, thereby reducing drag under subsequent exposure to high velocities. However, this second idea may be an oversimplification. Indeed, some authors have suggested that the dorsoventrally flattened shape may actually generate lift in the insect. Examples of animals inhabiting stones in torrential habitats include a number of mayflies (Ephemeroptera) belonging to the families Heptageniidae and Ephemerellidae; some Plecoptera, such as Perlidae; some Megaloptera (*i.e.*, Corydalidae); and caddisflies (Trichoptera), such as Leptoceridae (Ceraclea).

An addition to body shape, many mayflies and stoneflies have legs that project laterally from the body, thereby reducing drag and simultaneously increasing friction with the substrate. Most of these taxa are either scrapers or gatherers on surfaces of stones or predators on other aquatic insects. Undoubtedly, the diverse physical forces encountered in aquatic environments, especially streams, influence the array of morphologies found among aquatic insects. In some caddisflies (*e.g.*, Glossosomatidae), the shape of the case rather than the insect is modified.

The larvae of Glossosomatidae in their tortoiselike cases are frequently seen grazing on the upper surfaces of stones in riffle areas. Another curious caddisfly grazer on stone surfaces is Helicopsyche, whose larvae construct coiled cases of sand grains shaped like snail shells. Both glossosomatids and helicopsychids reach their greatest abundances in sunny cobble riffles, where they feed on attached periphyton or algae. Another lotic insect that relies on a rather streamlined case is the limnephilid cad-disfly Dicosmoecus. Larvae of the dipteran family Blephariceridae are unusual in that they possess hydraulic suckers.

A V-shaped notch at the anterior edge of each of the six ventral suckers works as a valve out of which water is forced when the sucker is pressed to the substrate. The sucker operates as a piston with the aid of specialised muscles. In addition, a series of small hooks and glands that secrete a sticky substance aid sucker attachment as the larvae move in a zigzag fashion, releasing the anterior three suckers, lifting the front portion of the body to a new position, and then reattaching the anterior suckers before releasing and moving the posterior ones to a new position. These larvae are commonly found on smooth stones in very rapid velocities and are usually absent from stones covered with moss and from roughened stones that interfere with normal sucker function. Several other aquatic insects have structures that simulate the action of suckers.

The enlarged gills of some mayflies function as a friction pad, and Drunella doddsi has a specialised abdominal structure for the same purpose. Some chironomids have "pushing prolegs" represented by circlet of small spines that function as a false sucker when pressed to the substrate. Mountain midge larvae (Deuterophlebiidae) possibly use a similar mechanism to attach their suckerlike prolegs. Most of these animals are primarily grazers on thin films of epilithon (algae, associated fine organic matter, and microbes) found on the surface of stones. Flowing water usually carries many organic (and inorganic) particles and a number of insects exploit these suspended particles. Filter-feeding collectors exploit the current for gathering food with minimal energy expenditure. For example, certain filtering collectors exploit locations where flows converge over and around substrates, thus allowing the animals to occupy sites of greater food delivery. Examples include caddisfly larvae belonging to the families Hydropsychidae and Brachycentridae. Silk is used for attachment by a number of caddisflies (*e.g.*, Hydropsychidae, Philopotamidae, and Psychomyiidae), which build fixed nets and retreats.

Although the Philopotamidae are found in riffle habitats, their fine-meshed, tubelike nets are usually found in crevices or undersides of stones in low velocity microhabitats. The nets of the caddisfly, Neureclipsis, are limited to moderately slow (<25cms_1) velocities and the large (up to 20 cm long), trumpet-shaped nets are used to capture small animals drifting downstream. Neureclipsis larvae are often very abundant in some lake outflow streams where drifting zooplankton are abundant. Some case-making caddisflies (*e.g.*, Brachycentrus sp.) also use silk for attaching their cases to the substrate in regions of moderately

Larval blackflies use a combination of hooks and silk for attachment. The thoracic proleg resembles that of chironomids and deuterophlebiids, described earlier, and the last abdominal segment bears a circlet of hooks, which it uses to anchor itself to substrates. The larva moves forward, inchwormlike, spins silk over the substrate, and attaches the proleg and then the posterior circlet of hooks to the silken web. Most blackfly larvae possess well-developed cephalic fans, which are used to filter small particles from suspension. These attached larvae twist their bodies longitudinally from 90° to 180° with the ventral surface of the head and fans facing into the current.

The fusiform body shape of blackfly larvae reduces turbulence and drag around their bodies, which are often located in regions of relatively rapid flow. Blackfly pupae are housed in silken cases that are attached to the substrate. Although unidirectional current is the basic feature of streams, most lotic insects have not adapted to strong currents, but instead have developed behaviour patterns to avoid current. Very few lotic insects are strong swimmers, probably because of the energy expenditure required to swim against a current. Downstream transport or drift requires only a movement off the substrate to enter the current. Streamlined forms, such as the mayflies Baetis spp., Centroptilum, Isonychia spp., and Ameletus spp., are capable of short, rapid bursts of swimming, but most lotic insects move by crawling or passive displacement. One characteristic of these latter mayflies is the possession of a fusiform, or streamlined, body shape: examples include several Ephemeroptera such as Baetis, Centroptilum, and Isonychia, as well as a number of beetle (Coleoptera) larvae. A fusiform body shape reduces resistance in fluids, and within the mayflies the shape is often associated with excellent swimming abilities.

The benthic fauna in streams often can be found in cracks and crevices, between or under rocks and gravel, within the boundary layer on surfaces, or in other slack-water regions. Another method of avoiding fast currents is living in debris accumulations consisting of leaf packs and small woody debris. This debris offers both a food resource and a refuge for insects and contains a diverse array of aquatic insects including stoneflies such as Peltoperlidae and Pteronarcyidae, caddisflies such as Lepidostomatidae and some Limnephilidae, as well as dipterans such as chironomids and tipulid crane flies.

In some streams with unstable sandy or silt substrates, woody debris can represent a "hot spot" of invertebrate activity. Wood debris provides a significant portion of the stable habitat for insects in streams when the power of the flowing water is insufficient to transport the wood out of the channel. In addition to the insect component using wood primarily as a substrate, there is often a characteristic xylophilus fauna associated with particular stages of wood degradation. These include chironomid midges and scraping mayflies as early colonizers, and larvae and adults of elmid beetles.

In western North America, an elmid (Lara avara) and a caddisfly (Heteroplectron) are gougers of firm waterlogged wood, chironomids are tunnelers, and the tipulids, Lipsothrix spp., are found in wood in the latest stages of decomposition. Woody debris is most abundant in small, forested watersheds, but it is also an important habitat in larger streams with unstable beds. In the southeastern coastal plain of the United States and in low-gradient mid- and southwestern streams and rivers with unstable bottom substrate, woody debris or "snags" often represent the major habitat for aquatic insect abundance and biomass.

High populations and biomass of filter-feeding animals such as net-spinning caddisflies (Hydro-psyche spp., Cheumatopsyche spp., and Macrostenum), and blackflies occur in these streams and rivers. In addition to filter feeders, other groups such as odonates, mayflies, stone-flies, elmid beetles, non-filtering caddisflies, and dipteran larvae can be locally abundant on large pieces of woody debris. Invertebrate shredders and scrapers promote decomposition of outer wood surfaces by scraping, gouging, and tunnelling through wood. In fact, wood gouging habits of some net-spinning caddisflies have been blamed for the failure of submerged timber pilings that had been supporting a bridge!

Sand and silt substrates of rivers and streams are generally considered to be poor habitats because the shifting streambed affords unsuitable attachment sites and poor food conditions. An extreme example of this instability is the Amazon River, where strong currents movc the bedload downstream, resulting in dunes of coarse sand up to 8 m in height and 180 m in length, thus largely preventing the establishment of a riverbed fauna. However, sandy substrates do not always result in poor habitat for all aquatic insects: some sandy streams are quite productive. Blackwater (*i.e.*, high tannic acid concentrations from leaf decomposition) streams of the southeastern United States have extensive areas of sand, with some of insects, such as small Chironomidae (<3mm in length), exceeding 18,000m-2 in abundance. Their food is derived from fine organic matter, microbes, and algae trapped in the sandy substrate. Numerically, the inhabitants of sandy or silty areas are mostly sprawlers or burrowers, with morphological adaptations to maintain position and to keep respiratory surfaces in contact with oxygenated water. At least one insect, the mayfly Ametropus, is adapted for filter feeding in sand and silt substrates of large rivers. Ametropus uses the head, mouthparts, and forelegs to create a shallow pit in the substrate, which initiates a unique vortex (flow field in which fluid particles move in concentric paths) in front of the head and results in resuspension of fine organic matter as well as occasional sand grains. Some of these resuspended fine particles are then trapped by fine setae on the mouthparts and forelegs.

Many predaceous gomphid (Odonata) larvae actually burrow into the sediments by using the flattened, wedge-shaped head and fossorial (adapted for digging) legs. The predaceous mayflies Pseudiron spp. and Analetris spp. have long, posterior-projecting legs and claws that aid in anchoring the larvae as they face upstream. Some mayflies (*e.g.*, Caenidae and Baetiscidae) have various structures for covering and protecting gills, and others (*e.g.*, Ephemeridae, Behningiidae) have legs and mouthparts adapted for digging. The predaceous mayfly Dolania spp. burrow rapidly in sandy substrates and have dense setae located at the anterior-lateral corners of the body as well as several other locations. The larva uses its hairy body and legs to form a cavity underneath the body where the ventral abdominal gills are in contact with oxygenated water.

Dense setae also are found in burrowing mayflies belonging to the family Ephemeridae that are common inhabitants of sand and silt substrates. They construct shallow U-shaped burrows and use their dorsal gills to generate water currents through the burrow, while using their hairy mouthparts and legs to filter particles from the moving water. Hairy bodies seem to be a characteristic of many animals dwelling on silt substrates, which include other collector mayflies such as Caenis, Anepeorus, and some Ephemerellidae.

Many dragonflies (*e.g.*, Cordulegaster spp., Hagenius spp., and Macromiidae) have flattened bodies and long legs for sprawling on sandy and silty substrates. Some caddisflies, such as Molanna, have elongate slender bodies but have adapted to sand and silt substrates by constructing a flanged, flat case. They are camouflaged by dull colour patterns and hairy integuments that accumulate a coating of silt. The eyes, which cap the anterolateral corners of the head, are elevated over the surrounding debris.

The genus Aphylla (Gomphidae) is somewhat unusual in that the last abdominal segment is upturned and elongate, allowing the larvae to respire through rectal gills while buried fairly deep in mucky substrate. Some insects burrow within the upper few centimetres of the substratum in depositional areas of streams. This practice is found among some dragonflies and a number of caddisflies, including Molanna, and various genera of the families Sericostomatidae and Odontoceridae.

SPECIALISED FLOWING WATER HABITATS

The hyporheic region is the area below the bed of a stream where interstitial water moves by percolation. In gravelly substrates or glacial outwash areas, it may also extend laterally from the banks. In some situations, an extensive fauna occurs down to 1 m in such substrates. Most orders are represented, especially taxa with slender flexible bodies or small organisms with hard protective exoskeletons. Some stoneflies in the Flathead River of Montana spend most of their larval period in this extensive subterranean region of flow adjacent to the river. Stonefly larvae have been collected in wells over 4 m deep, located many meters from the river. Rivers draining glaciated regions where there are large boulders and cobble appear to have an exceptionally well-developed hyphoreic fauna.

Other specialised flowing water habitats include the madi-colous (or hygropetric) habitats, which are areas in which thin sheets of water flow over rock. These often approach vertical conditions (*e.g.*, in waterfalls) and have a characteristic fauna. Among common animals in these habitats are caddisflies, including several microcaddisflies (Hydroptilidae), Lepidostomatidae, beetles such as Psephenidae, and a number of Diptera larvae belonging to the Chironomidae, Ceratopogonidae, Thaumaleidae, Tipulidae, Psychodidae, and some Stratiomyidae. Thermal (hot) springs often have a characteristic fauna, which is fueled by algae and bacteria adapted to high temperatures. The common inhabitants include a number of dipteran families such as Chironomidae, Stratiomyidae, Dolichopodidae, and Ephydridae, as well as some coleopterans. A number of these survive within rather narrow zones between the thermal spring and cooler downstream areas.

LENTIC HABITATS

Lentic or standing-water habitats range from temporary pools to large deep lakes and include marshes and swamps, as well as natural (*i.e.*, tree holes, pitcher plants) and artificial (*i.e.*, old tires, rain barrels) containers. These habitats include the littoral zone, which comprises the shallow areas along the shore with light penetration to the bottom and normally contains macro-phytes (rooted vascular plants). The limnetic zone is the open-water area devoid of rooted plants, whereas the deeper profundal zone is the area below which light penetration is inadequate for plant growth, water movement is minimal, and temperature may vary only slightly between summer and winter. The aquatic and semiaquatic insect communities inhabiting these zones are known as the pleuston (organisms associated with the surface film), plankton and nekton (organisms that reside in the open water), and benthos (organisms associated with the bottom, or solid-water interface). Nektonic forms are distinguished from plankton by their directional mobility, and the latter are poorly represented in lentic waters by insects; the majority of insects found in standing-water habitats belong to the benthos. Their composition and relative abundance is dependent on a variety of factors, some of which are integrated along depth profiles. The overall taxonomic richness of benthic insect communities generally declines with increasing depth.

Among the aquatic communities of lentic habitats, the following orders of aquatic and semiaquatic insects are commonly found within the littoral, limnetic, and profundal zones: the springtails (Collembola), mayflies (Ephemeroptera), true bugs (Heteroptera), caddisflies (Trichoptera), dragonflies (Anisoptera) and dam-selflies (Zygoptera), true flies (flies, gnats, mosquitoes, and midges) (Diptera), moths (Lepidoptera), alderflies (Megaloptera), and beetles (Coleoptera). Not all these groups occur in lakes, and many are associated with ponds or marshes.

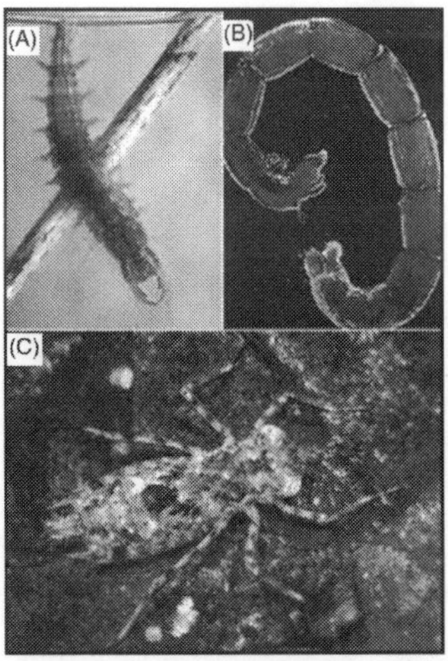

Fig. Typical Insects Inhabiting Lentic Environments. (A) Coleoptera: Hydrophilidae (Hydrochara). (B) Diptera: Chironomidae (Chironomus), (C) Odonata: Libellulidae (Pantala).

UNIQUE PROPERTIES OF THE WATER SURFACE

The unique properties of the water surface or air-water interface constitute the environment of the pleuston community. The Collembola, or springtails, are small in size, have a springing organ (furcula), and a water-repelling cuticle that enables them to be supported by and move across water surfaces. Among the true bugs, the Gerridae (water striders) and related families, the Veliidae (broad-shouldered water striders) and Hydrometridae (water measurers), are able to skate across the water. Adaptations for this habit include retractable preapical claws to assist in swimming, elongate legs and body to distribute the insect's weight over a large area of the surface film, and hydrofuge (non-wettable) hairpiles for support on the surface.

Some gerrids also are capable of detecting surface vibrations caused by potential prey. Adult whirligig beetles (Gyrinidae) live half in and half out of water with each eye divided into upper and lower halves, permitting vision simultaneously in both the air and the water; glands keep the upper portion of the body greased to repel water. The middle and hind legs of adult gyrinids are paddle shaped, enabling them to be one of the most effective swimming invertebrates.

Among the Diptera, only the mosquitoes (Culicidae) may be considered to be permanent members of the pleuston of lentic waters. The larvae and pupae of most species use the underside of the surface film for support. Larval Anopheles lie horizontally immediately beneath the air-water interface, supported by tufts of float hairs on each. Larvae of most other genera (Aedes, Culex, Culiseta) hang upside down, with an elongated terminal respiratory siphon penetrating the surface film. Feeding adaptations associated with pleuston specialisation include predation by the Hemiptera and Coleoptera, to practically all functional feeding modes by different mosquito larvae, including coUecting-filtering and gathering, scraping, and shredding.

THE NEKTON AND PLANKTON COMMUNITIES

The nekton are swimmers able to navigate at will (*e.g.*, Coleoptera, Hemiptera, some Ephemeroptera), whereas plankton are floating organisms whose horizontal movements are largely dependent on water currents. The phantom midge Chaoborus sp. (Chaoboridae) is normally regarded as the only planktonic insect and is abundant in many eutrophic (nutrient-rich) ponds and lakes.

The tracheal system in these larvae is reduced to kidney-shaped air sacs that function solely as hydrostatic organs, and the larvae slowly descend or rise by adjusting the volume of the air sacs. Chaoborus remains in ben-thic regions during the day but moves vertically into the water column at night. These journeys are dependent on light and oxygen concentrations of the water. The larvae avoid predation by being almost transparent, and they have prehensile antennae that are used as accessory mouthparts to impale zooplankton. The only other group of insects that may be considered to be planktonic are the early chirono-mid instars, which have been reported in the open-water column. Among the Heteroptera, nektonic species are in the Notonectidae (back swimmers), Corixidae (water boatman), and Belostomatidae (giant water bugs), all of which are strong swimmers.

Many of these rise to the water surface unless continuously swimming or clinging to underwater plants. Notonectids have backs formed like the bottom of a boat and navigate upside down. They hang head downward from the surface or dive swiftly, using their long hind legs as oars.

On the underside of the body, they carry a silvery film of air, which can be renewed at regular intervals, for breathing while submerged. Two genera of back-swimmers (Anisops and Buenoa) use hemoglobin for buoyancy control, and this adaptation has enabled these insects to exploit the limnetic zone of fishless lentic waters, where they prey on small arthropods.

They have been considered for use as biological control agents for mosquito larvae in some areas of North America. In contrast to notonectids, corixids always swim with the back up, using their elongate, flattened oarlike legs. Although some water boatmen are predators, they are the only group of semiaquatic Heteroptera that have members that are collectors, feeding on detritus and associated small plant material. The Belostomatidae are strong swimmers, but probably spend most of their time clinging to vegetation while awaiting prey, rather than actively pursuing their food in the open water. They are masters of their environment and capture and feed on a variety of insects, tadpoles, fish, and even small birds.

The eggs of many belostomatids are glued to the backs of the males by the females and carried in this position until nymphs emerge, a remarkable adaptation for protection of the eggs. Although most aquatic beetles (Coleoptera) are associated with the substrate, members of the Dytiscidae (predaceous diving beetles) and the Hydrophilidae (water scavenger beetles) are often found swimming in the water column and together constitute the majority of all species of water beetles. The dytiscids are mainly predators in both the adult and larval stage, while adult hydrophilids are omnivorous, consuming both living and dead materials.

The larvae of hydrophilids are predaceous. To respire, hydrophilid adults, having their largest spiracles on the thorax, break the surface film with their antennae; dytiscids, having their largest spiracles on the abdomen, come up tail end first, as do the larvae of both families.

Overall, there are actually few truly nektonic insects, and most of them pass through the limnetic zone when surfacing for emergence. This may be, partly, because with no resting supports in the limnetic zone, maintaining position requires continuous swimming or neutral buoyancy. The vast majority of lentic insects occur in shallow water with emergent plants and are considered to be part of the benthos.

THE BENTHOS COMMUNITY

Benthos, derived from the Greek word for bottom, refers to the fauna associated with the solid-water interface and includes insects residing on the bottom or associated with plant surfaces, logs, rocks, and other solid substrates. In lentic habitats, many insects fall into this category, particularly the

Chironomidae, which often represent over 90 per cent of the fauna in the profundal (deep-water) zone of lakes and ponds. These inhabitants are mostly bur-rowers that feed on suspended or sedimented organic materials and are capable of tolerating low dissolved oxygen or even anaerobic conditions. Chironomid larvae build U- or J-shaped tubes with both openings at the mud-water interface. Body undulations cause a current of water, providing conditions under which oxygen and particulate food can be drawn through the tube.

Some midge larvae found in sediments (mainly Chironomus sp.) are bright red and are known as bloodworms. The red colour is caused by the respiratory pigment hemoglobin, which enables a larva to recover rapidly from anaerobic periods because the pigment takes up oxygen and passes it to the tissues more quickly than is possible by diffusion alone. Other members of the benthos of deeper waters include the mayfly, Hexagenia (Ephemeridae), which inhabits the silt and mud of nearshore lake bottoms and has legs modified for digging to construct U-shaped burrows .

Mayfly numbers have been increasing because of improved water quality standards for lakes and streams. Exceptions to the main constituents of the profundal zone are some immature mayflies, stoneflies, and caddisflies that have been collected at depths from 30 to 100 m in Lake Superior, Michigan. Also, a stonefly, Utacapnia lacustra (Capniidae), occurs at depths of 80 m in Lake Tahoe, California-Nevada, and completes its entire life cycle at this depth, never needing to surface. Several orders of aquatic insects reach their greatest abundance and diversity in the shallow littoral zone of ponds and lakes as benthos typically associated with macrophytes (macroalgae and rooted vascular plants). The occupants are burrowers, climbers, sprawlers, clingers, swimmers, and divers and include the Ephemeroptera, Heteroptera, Odonata, Trichoptera, Megaloptera, Lepidoptera, Coleoptera, and Diptera.

The same groups occupy marshes and some swamps, which generally tend to be shallow, with an associated plant zone across the entire surface. Mayflies belonging to the families Baetidae and Siphlonuridae are generally swimmers, clingers, and climbers in vegetated ponds and marshes and mainly feed by means of collecting-filtering or -gathering. Heteroptera include the water scorpions (Nepidae), which have long slender respiratory filaments and are well concealed by detritus and tangled plant growth because of their sticklike appearance.

These sit-and-wait predators capture organisms that frequent their place of concealment. Other families of Heteroptera adapted for moving through vegetation in ponds are the Pleidae or pygmy backswimmers and creeping water bugs, the Naucoridae. The Odonata, particularly the Gomphidae, are all predators and usually conceal themselves by either burrowing in substrate, sprawling among fine sediment and detritus, or climbing on vascular plants. Sprawlers are more active hunters and include the Libellulidae and Corduliidae. Numerous setae give them a hairy appearance to help camouflage the larvae, and colour is protective in patterns of mottled greens and browns. Most Zygoptera (damselflies) and the dragonfly (Anisoptera) family Aeshnidae are mainly climbers or clingers, lurking in vegetation or resting on stems of aquatic plants. The larvae stalk their prey, and both dragonfly and damselfly larvae have a unique lower lip (the labium) armed with hooks, spines, teeth, and raptorial setae that can extend to seize prey and then bring it back into the mouth, holding the food while it is being eaten. The food of larval odonates consists of other aquatic insects such as midges, semiaquatic bugs, and beetles, as well as small fish.

Predators of larval odonates include aquatic birds, fish, and large predaceous insects. In the order Megaloptera, which includes the hellgrammites or dobsonfly larvae of streams, only the predaceous larvae of the alder-fly (Sialis) is common in ponds and lakes. They are generally found in sand or mud along the margins, but occasionally in deeper water, and they prey on insect larvae and other small animals.

The only aquatic family in the related order Neuroptera is the Sisyridae (the spongilla flies), and these are found feeding on freshwater sponges that occur in some streams and the littoral zones of lakes and ponds. The larvae, which occur on the surface or in the cavities of the host, pierce the sponge cells and suck the fluids with their elongated mouthparts.

Although most caddisflies are observed living in lotic waters, several families of caddisflies are either associated with temporary ponds in the spring, aquatic vegetation in permanent ponds, lakes and marshes, or wave-swept shorelines of lakes. The Hydropsychidae (net spinners), Helicopsychidae (snail case makers), Molannidae, and Leptoceridae are often found along wave-swept shorelines of lakes, and their feeding habits range from those of scrapers and collector-filterers to predators. The Phryganeidae and several genera within the Limnephilidae are climbers, clingers, and/or sprawlers among vegetation in temporary and permanent ponds and marshes; generally, they are shredders of vascular hydrophytes and other decaying plants.

The cases of lentic caddisfly families vary with the environment they are found in. Some cases consist of narrow strips of leaves put together in spiral form around a cylinder (Phryganeidae: Phryganea sp.), others consist of plant materials such as leaves and bark arranged transversely to produce a bulky cylindrical case (Limnephilidae: Limnephilus). Both aquatic and semiaquatic moths (Lepidoptera) occur in len-tic habitats, and several genera form close associations with vascular hydrophytes. Larvae of the family Pyralidae (Parapoynx sp.) spend the first two instars on the bottom and feed on submerged leaves of water lilies, whereas older larvae generally become surface feeders. Silk spun by the caterpillars is often used to build protective retreats, and pupation usually takes place in silken cocoons or silk-lined retreats.

Larval habits of aquatic and semiaquatic moths include leaf mining, stem or root boring, foliage feeding, and feeding on flower or seed structures. One semiaquatic lepidopteran called the yellow water lily borer (the noctuid Bellura gortynoides), mines the leaves as a young caterpillar and then bores into the petioles of lilies as an older caterpillar. Within the petiole, larvae are submerged in water and must periodically back out to expose the posterior spiracles to the air before submerging again. The larvae swim to shore by undulating their bodies and overwinter under leaf litter in protected areas. In addition to the water scavenger and predaceous diving beetles that may occur as nekton swimming through the water column, larvae and adults of other beetles are considered to be part of the benthos of ponds and marshes. These include the Haliplidae (crawling water beetles), which are clingers and climbers in vegetation, and the Staphylinidae (rove beetles), which are generally found along shorelines and beaches, as well as in the marine intertidal zone.

The Scirtidae (marsh beetles) are generally found associated with vascular hydrophytes but also are a prominent inhabitant of tree holes. The aquatic Chrysomelidae (leaf beetles) occur commonly on emergent vegetation in ponds, especially floating water lily leaves. The larvae of one genus, Donacia, obtain air from their host plant by inserting the sharp terminal modified spiracles into the plant tissue at the base of the plant. Water lilies can be heavily consumed by larvae and adults of the chrysomelid beetle, Galerucella sp., and some of the aquatic herbivorous beetles belonging to the family Curculionidae (weevils) include pests of economic importance such as the rice water weevil (the curculionid Lissorhoptrus).

The Diptera is clearly one of the most diverse aquatic insect orders, inhabiting nearly all lentic habitats and representing all functional feeding groups and modes of existence. Although the benthic Chironomidae may reach their highest densities in the profundal zone of eutrophic lakes and ponds, they also are largely represented in the littoral zone associated with submergent and emergent plants, where they often graze on the algae attached to leaf surfaces or are vascular plant miners. Among these, a few are of particular interest because of their high diversity and/or abundance in these habitats, namely the crane flies (Tipulidae), the shore and brine flies (Ephydridae), and the marsh flies (Sciomyzidae). The Tipulidae, the largest family of Diptera, are found along the margins of ponds and lakes, freshwater and brackish marshes, and standing waters in tree holes.

A few littoral species inhabit the marine intertidal zone. To these are added the large numbers of species that are semiaquatic, spending their larval life in saturated plant debris, mud, or sand near the water's edge or in wet to saturated mosses and submerged, decayed wood. Ephydridae larvae have aquatic and semiaquatic members and occupy several different. Lentic habitats ranging from saltwater or alka line pools, springs, and lakes to burrowers and miners of a variety of aquatic plants in the littoral margins of these freshwater lentic habitats. All larvae utilise a variety of food, but algae and diatoms are of particular importance in their diet. The Sciomyzidae share some of the same habitat with the shore and brine flies, particularly fresh- and saltwater marshes, and along margins of ponds and lakes among vegetation and debris.

The unique aspect of their larval life is that they are predators on snails, snail eggs, slugs, and fingernail clams. The aquatic predators float below the surface film and maintain buoyancy by frequently surfacing and swallowing an air bubble. Prey may be killed immediately or over a few days.

MARINE HABITATS

As noted earlier, insects have been largely unsuccessful in colonizing the open ocean, except for some members of the heterop-teran family Gerridae. Most marine insects live in the intertidal zone (*i.e.*, between high and low tide marks), especially on rocky shores or associated with decaying seaweed on sandy beaches . Although several orders have representatives in the intertidal zone, only a few orders, notably the Diptera, Coleoptera, and Collembola, have colonized these habitats in any numbers. The harsh physical environment of this area has forced these groups to occur buried in sand or mud and to hide in rock crevices or under seaweed.

UNUSUAL HABITATS

Because of adaptive radiation over evolutionary time, insects have colonized virtually every aquatic habitat on earth. Therefore, it is not surprising that these organisms are found in the most unusual of aquatic habitats. The title of most versatile aquatic insect must be shared among members of the dipteran family Ephydridae, or shore flies. Shore flies can breed in pools of crude petroleum and waste oil, where the larva feed on insects that become trapped on the surface film. Other species of this family (Ephydra cinera), known as brine flies, occur in the Great Salt Lake, Utah, which has a salinity six times greater than that of seawater. Larvae maintain water and salt balance by drinking the saline medium and excreting rectal fluid that is more than 20 per cent salt.

Another related family of flies, the Syrphidae, or "rat-tailed maggots," occur in sewage treatment lagoons and on moist substrates of trickling filter treatment facilities. Both families have larvae with breathing tubes on the terminal end, which permits the larvae to maintain contact with the air while in their environment. Some Stratiomyidae, or soldier flies, live in the thermal hot springs of Yellowstone National Park with temperatures as high as 47°C! Other members of this family inhabit the semiaquatic medium of cow dung and dead corpses. A few species of insects have invaded caves and associated subterranean habitats. Another unusual aquatic habitat that several insect orders occupy is referred to phytotelmata or natural container habitats and include tree holes, pitcher plants, bromeliads, inflorescences, and bamboo stems. Synthetic container habitats, such as old tires, cemetery urns, rain gutters, and similar natural habitats such as hoofprints also harbour similar insects. Some of these habitats are extremely small and hold water only temporarily, but nevertheless can be quite diverse.

The most common order found in these habitats is the Diptera with more than 20 families reported. Over 400 species of mosquitoes in 15 genera alone inhabit these bodies of water and some of these species are important vectors of disease agents.

Insect communities inhabiting pitcher plants (Sarracenia purpurea) in North America are exemplified by a sarcophagid or flesh fly (Blaesoxipha fletcheri), a mosquito (Wyeomyia smithii), and a midge (Metriocnemus knabi). The relative abundance of these pitcher plant inhabitants is related to the age, inasmuch as each of the three species consumes insect remains that are in different stages of decomposition.

Specifically, the larvae of the flesh fly feed on freshly caught prey floating on the pitcher fluid surface. The mosquito larvae filter feed on the decomposed material in the water column, and the midge larvae feed on the remains that collect on the bottom of the pitcher chamber. Temporary habitats are important because they are populated by a variety of species, often with unique morphological, behavioral, and physiological properties.

MOVEMENT OF AQUATIC INSECTS

Aquatic entomology is one of the most important aspects of fly fishing that both lake and stream anglers will be exposed to and learn from. This is a huge subject that one could literally study throughout their entire angling life. However, it is essential that we are at least able to identify those insects that are important trout food and that we also have a basic understanding of the insect's life cycle and habitat preferences. An integral part of the aquatic entomology subject is recognising how the various insect life stages move through the water and then applying that information to the way in which we present and/or retrieve our flies.

In streams, much of the aquatic insect life are able to utilise the current to move or drift them as they complete emergence cycles or simply move from one type of habitat to another. In lakes, the larval, nymphal and pupal stages of insects must propel themselves to locate food, avoid predators or reach the surface in order to emerge into the adult form. One of the ways fish recognise prey items is by the unique swimming or movement actions displayed by the food source. Let's take a closer look at how the various life stages of the important aquatic insect orders move. Keep in mind that these are staple food sources for trout and other fish species found throughout the stillwater world.

DAMSELFLIES AND DRAGONFLIES

Damselflies and dragonflies of the Order Odonata are active swimmers while in the nymphal stage. Damselfly nymphs utilise a 3 lobed set of caudal lamellae or abdominal gills to propel them in a fairly slow, sinusoidal or snake-like motion through the water.

They will swim for 3 to 8 inches; pause for up to a couple of seconds then resume moving. Fully developed nymphs swim more continuously as they urgently search out emergent plant stems to crawl up onto and complete the transformation to the adult stage. Dragonfly nymphs swim by venting water through the tip of their abdomen. This action gives the nymph a short, 3 to 5 inch quick burst of speed. Nymphs can sustain several of these fast movements in succession prior to taking a brief rest. In most situations dragonfly nymphs are happy to be crawling along the vegetation or lake bottom in search of food and usually only employ the "jet propulsion" swimming system to avoid predators.

CADDISFLIES

Caddisflies (Order Trichoptera) undergo a complete metamorphosis or life cycle. Most species inhabiting stillwaters belong to the case maker group. These larvae live within a case which is built from bits of vegetation, particles of sand or other debris. Larvae move by using their 3 pairs of legs to crawl and pull their case slowly along the lake bottom and amongst the aquatic vegetation. Their cases leave telltale tracks in the marl or mud flats that cover shoal areas of the lake.

Caddis pupa develop within the larval case. The fully developed pupae break out of their larval cases and swim quickly to the surface of the lake. Their ascent is aided by an elongated and feathered hind pair of legs which act as oars to quickly move them through the water. Pupae are often swimming up through 20 feet of water and are thus extremely vulnerable to predation by fish. The adult emerges from the pupal stage as soon as the pupa breaks the surface film. The actual pupal swimming action consists of relatively fast 3 to 6 inch long movements followed by brief pauses.

MAYFLIES

Mayflies (Order Ephemeroptera) are another common inhabitant of productive stillwaters. Most lake dwelling species are good swimmers that are capable of fairly quick but short bursts of speed. Developing nymphs seek food and cover amongst the vegetation growing up from the shallow shoal areas of the lake. When mature, the nymphs will swim to the surface to emerge into the dun or non-reproductive adult stage. The mature nymphs typically swim on a 20 to 30 degree angle to the surface of the lake. They swim in a sinusoidal motion in moderately fast 2 to 4 inch bursts. The nymphs will rest or pause regularly during their ascent. Once at the surface, a split forms along the back of the nymphal shuck to allow the adult form to emerge.

MIDGES OR CHIRONOMIDS

Midges of the Order Diptera are unquestionably the most diverse group of insects inhabiting fresh water. A typical small productive lake in western North America could be host to a hundred different species. Midges or chironomids also undergo a complete metamorphosis or life cycle. Larval and pupal stages are poor swimmers but both require an ability to move in order to complete their respective life stages.

Most stillwater midge larvae live in the benthic or bottom substrate of the water body. In many instances the larvae construct simple "mud tubes" in which they peer out from to feed on detritus. Larvae do leave their tubes in search of more ideal habitat and in other instances can be swept out of their tubes by strong currents formed during a turnover event or a rapid fluctuation of water levels. In either case the movement of the larvae is restricted to an ineffective wiggling motion.

Midge pupa typically develop within the larval tube. Fully developed pupa exit the larval tube and ascend vertically to the surface of the lake. The pupa do not actually swim but wiggle and squirm up through the water. Their ascent is aided by gases trapped beneath the pupal shuck. Often, movement is rather rapid during the last couple of feet before reaching the surface. It is not uncommon for pupa to ascend through 40 feet of water making it a very slow and vulnerable emergence process.

STILLWATER RETRIEVES

Overall, the movement of these aquatic insects is slow or at best they can move at a moderate speed for a very short distance. Some, like midges, only rise or elevate through the water column. It can be hard for many anglers to slow their retrieves down and then maintain the slower pace long enough to effectively retrieve the entire length of a cast. The strip and hand twist are two basic retrieves that all stillwater fly fishers should be comfortable using with the various food sources being imitated. The strip retrieve is as simple as it sounds and it is the basic retrieve for almost all fly fishing situations. The rod hand lightly pinches the fly line between the thumb and index (first) finger while the retrieve hand uses the same two fingers to pull or strip in a set amount of fly line.

It takes a lot of practice to conduct a 4 inch slow strip retrieve cast after cast for several hours but if that is how the damselfly nymphs are swimming then it is best to go with the flow. Without patience and practice, those 4 inch strips soon become 8 then 12 inch pulls! How slow is slow? For instance, to imitate a slow 2 inch strip retrieve count up to 2 seconds while the 2 inches is being retrieved. A fast 4 inch retrieve would count between 1 and 1.5 seconds per strip. Caddis pupa probably swims the fastest of these aquatic insect food sources. A fast 6 inch strip retrieve should take approx. 2 seconds to complete. The hand twist is the other basic retrieve that should be mastered by all stillwater fly fishers.

This is a busy retrieve, meaning that it requires more hand movement while at the same time helps to slow down the retrieve. The other advantage of this technique is that the angler has more control over the retrieve as the retrieve hand is always in contact with the fly line.

The rod and retrieve hand start off in the same positions as the strip retrieve. The fly line is pulled down by the thumb and index finger of the retrieve hand while at the same time rotating the retrieve hand wrist outward thus leaving an open palm. The fly line is then hooked around the bay finger of the retrieve hand and then brought back to the thumb and index finger. This completes a full hand twist. As the manoeuvre is repeated coils of line will collect in the palm of the retrieve hand. It is best to release the fly line after 2 or 3 coils have accumulated.

The hand twist is certainly the preferred retrieve to use when imitating the midge pupa either rising up through the water column or very slowly moving the pupal pattern horizontally through the water. In these instances a complete hand twist could take upwards of 8 seconds to complete. The strip and hand twist retrieves will allow the angler to match the movement of these essential stillwater food sources. It can also be beneficial to impart a bit of irregular movement to your fly as it is being retrieved. In other words, try adding several very short, quick one inch strips in succession during any normal strip or hand twist retrieve. I routinely mix in 3 or 4 of these quick strip tactics into any stillwater retrieve. Sometimes this off tempo movement is enough to commit an otherwise slightly skeptical trout into taking the fly.

Wind drifting from an anchored boat is also an effective technique to present flies being fished with floating fly lines. It is used most commonly with mayfly nymphs, damselfly nymphs, caddis pupa and midge pupa. Although technically not a retrieve, wind drifting takes advantage of gentle wind conditions to naturally move the sunk fly in a subtle undulating motion horizontally through the water column. A cast is made sideways or perpendicular to the direction of the wind. A large arching bow develops in the fly line as the waves carry it downwind.

The length of the leader and weighted versus un-weighted flies determine where in the water column the fly will travel. Make sure the rod is secure when using this tactic! It takes practice in order to get comfortable with these retrieves. While on the water watch for the real insects and note their swimming and movement patterns. That will be the best check as to how realistic your retrieves are being conducted.

NUISANCE AQUATIC INSECTS

Common Name	Scientific Name
Dobsonfly	Corydalus cornutus (L.)
Stonefly	(Plecoptera)
Mayfly	(Ephemeroptera)
Water Boatman	Sigara atropodonta (Hungerford)
Backswimmer	Notonecta undulata (Say)
Giant Water Bug	Lethocerus americanus (Leidy)
Water Strider	Gerris remigis (Say)
Water Scavenger Beetle	Hydrophilus triangularis (Say)
Whirligig Beetle	Dineutus americanus (Say)
Predaceous Diving Beetle	Dytiscus verticalis (Say)

Several kinds of aquatic insects may become a nuisance by their presence in homes, backyards and swimming pools. Many are attracted to lights and may accidentally enter structures or fall into swimming pools, while others may actually live in neglected pools. They do not cause harm to humans, pets, domestic animals, structures, possessions or crops. They will not reproduce or establish themselves indoors. Some such as the dobsonfly, backswimmer and giant water bug can inflict a painful bite if handled carelessly.

DOBSONFLY

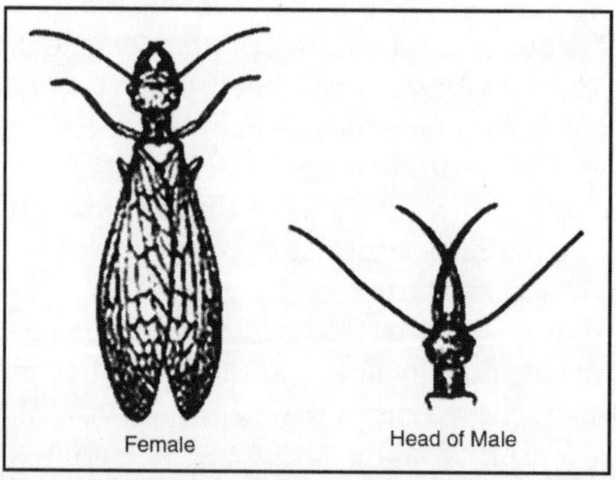

Adults are quite large, measuring two to four inches from the front of the head to the wing tips. They are soft-bodied, brownish body (faintly marked with brown of a different shade), with many cross veins and a white dot in each cell of the wings. The wings are held rooflike over the body.

Males have long, sickle-shaped mandibles (jaws) about half as long as the body, while females have short inconspicuous jaws. They are attracted to lights, have a fluttery flight and are usually found near or some distance from a body of water. Adults are not active during the day, live briefly and are not believed to feed. Larvae, known as "hellgrammites," are aquatic, occurring under stones in fast-flowing shallow water (streams), and make excellent fish bait. Larvae can swim, but usually crawl, live three years before adulthood, grow up to two to three inches, feed on small aquatic organisms and can inflict a painful bite when mishandled.

STONEFLY

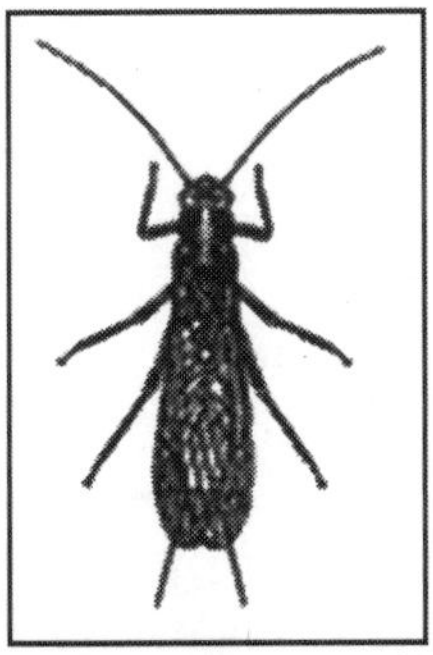

Adults are soft-bodied, between 1/4 to 1-1/2 inches long, greenish, yellow, brown or black, somewhat flattened and wings lay flat on the body when at rest. Wings have many veins and both pairs are about equal in size. The abdomen is soft, bearing a pair of cerci at the end. They have a long tapered antennae. They are poor fliers, found near water and are attracted to lights. Some adults emerge in cold weather from November to March, become active on warmer days (early in the year) when they mate and live only a few weeks. The nymphs (naiads) live in flowing streams, under stones or gravel (one to three years). They feed on vegetable matter and other insects. Stoneflies in the immature stage are fed upon by game fish, being of great economic importance. Anglers refer to stoneflies as "browns" and copy them in wet and dry flies.

MAYFLY

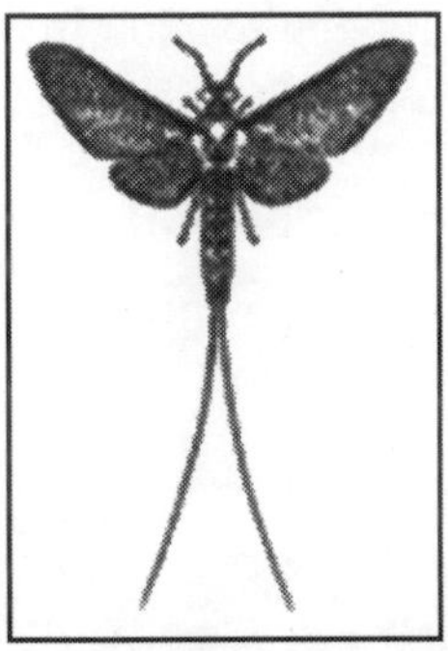

Adults are between 1/4 to 1/2 inches long, soft-bodied, elongate and have two or three long, many-jointed, threadlike "tails" at the end of the abdomen. They are pale yellow with brownish stripes with a reddish-brown thorax. The large triangular, membranous forewings have many longitudinal veins and cross veins with the hindwings quite small or lacking. At rest, the wings are held together, erect and above the body. Antennae are very short (inconspicuous) and legs are short and weak. They are attracted to lights and usually found around streams. Adults do not feed and live only a few hours or at most a day or two. They may emerge in huge numbers from lakes and rivers, piling up along the shore or nearby roads sometimes a few feet deep. The nymphs (naiads) live in clear, fresh water feeding on vegetable matter for a year or more. Mayflies are very important, serving as food for fish and birds. Anglers imitate the adults in dry flies, referring to them as "spinners" or "duns" and pattern wet flies after the nymphs.

WATER BOATMAN

Adults are about 1/4 inch long, somewhat flattened, oval, gray-brown or mottled with the top surface often finely cross-lined and the hind legs elongate and oarlike (flattened and fringed with hairs) for swimming. Eyes are black (space between yellow) and legs and body underparts yellow.

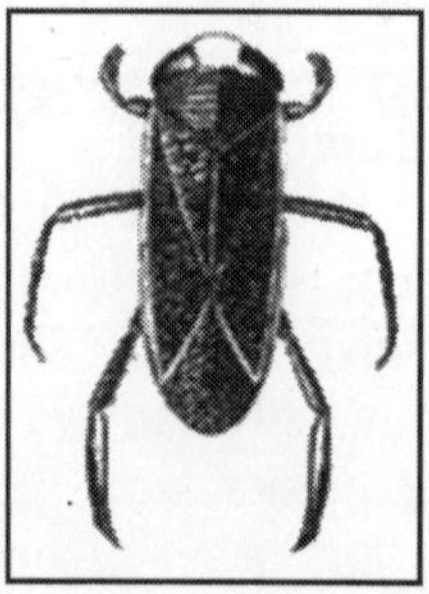

They swim rapidly, usually seen at the bottom in shallow water of ponds and lakes, and feed on algae, protozoa, plankton or dead organic matter on the bottom of pools and roadside ditches.

Some species are attracted to lights and swarm around them. Their life cycle takes about six weeks from egg to adult in the life cycle. They overwinter as adults. Some are sold for bird and fish food. They do not bite.

BACKSWIMMER

Adults are about 1/2 inch long, black and white coloured, swim upside down on their backs, have triangular-like (boat-shaped) bodies, and large eyes. They are kidney shaped, with a four segmented beak, and antennae in three to four segments concealed between the head and thorax. There is an irregular brown and blackish band across the wings at the base of the membranous portion.

The front and middle legs are fitted for grasping while the hind legs are flattened, fringed and fitted for swimming. They are often seen on the water surface with their long hind legs held straight out and pointed forward, poised for a fast start. They live in all kinds of water (fresh to scum-covered stagnant) and overwinter as adults. They fly long distances, often in swarms and are attracted to lights. The life cycle requires about 40 days. They feed on insects, tadpoles and small fish. They can bite, causing pain much like a bee sting when handled carelessly.

GIANT WATER BUG

Adults are large (2 to 2-1/2 inches long), brownish, somewhat flattened, and broad with the middle and hind legs fitted for swimming. The front legs are fitted for grasping (praying mantis-like) with two claws at the tips. The end of the abdomen bears two short, strap-like appendages for breathing. The head is broad and slightly extended beyond the eyes. They are very much attracted to lights and are known as "electric light bugs." Sometimes they leave the water and fly to street lights some distance away.

These bugs feed on other insects, snails, tadpoles and small fish (sometimes several times their size). They inject a poisonous fluid into their prey after capture, sucking them dry. They occur in ponds and quiet pools, feigning death when removed from the water and ejecting a fluid from the anus. Death is quick if not able to return to the water. Eggs are laid in massed rows of 100 or more above the water on cattails and other plants, hatching one to two weeks later. They are vicious biters, inflicting pain when handled carelessly.

WATER STRIDER

Adults are between 5/16 to 5/8 inch long, dull to grayish or reddish brown above and silvery gray on the underside of the body. Many have long slender bodies with long, slender legs and antennae. The front legs are short and modified for grasping while the middle and hind legs are long, like stilts with claws. These insects creep on the surface of running water or pools in a slow deliberate gait where they feed on live and dead insects, crustacea and other organisms. They overwinter as adults. Long, cylindrical eggs are laid during spring and summer in parallel rows glued to objects at the water's edge. Adults are called "skaters" or "Jesus Bugs," apparently for their ability to walk on water of ponds, lakes and streams.

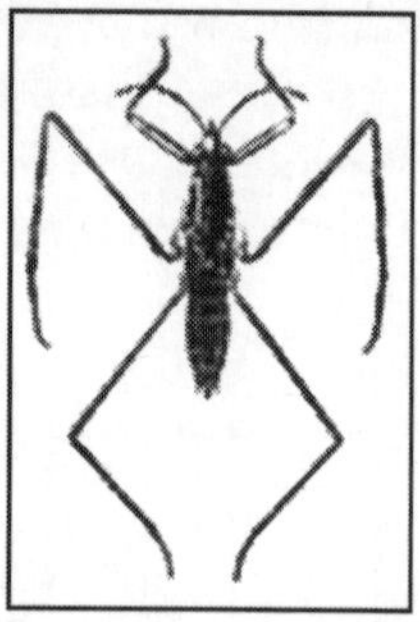

WATER SCAVENGER BEETLE

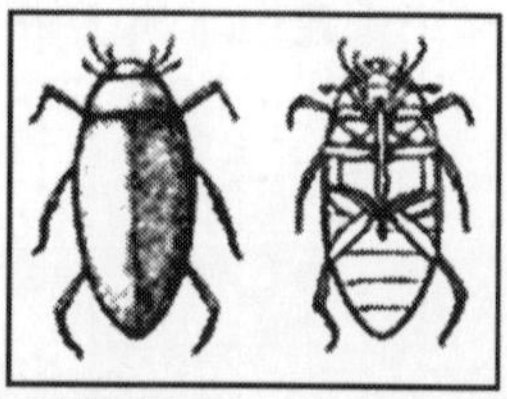

Adults are large about 1-1/2 inches long, hard-bodied and elliptical with triangular yellowish markings on the abdomen sides. The body top is smooth and shiny black with a greenish tinge. A long spinelike keel on the thorax underside between the legs can be jabbed into the fingers when the beetle is handled carelessly.

Antennae are club-shaped and the legs are flattened for swimming. They are good fliers, and are very much attracted to lights in large numbers in the spring. Beetles are commonly seen swimming or crawling among water plants or on the bottom of shallow pools, feeding mostly on dead or decaying vegetation. The hind legs move alternately when swimming. Larvae are predaceous and cannibalistic. There are one to two generations per year.

WHIRLIGIG BEETLE

Adults are 3/8 inch long or slightly longer, oval, flat, hard bodied with a shiny black bronze sheen. The legs are brownish yellow. Eyes are divided into two widely separated parts; the upper one is oval and remains above the water and the lower one is somewhat smaller and remains below the water. These fast-moving beetles are often seen swimming in groups in endless gyrations or circular "skating" movements, in a spot on the water surface. The outer margins of the wing covers are curved inward near the tip. The legs are flattened, fitted for swimming and the antennae are quite short with the last segment enlarged. They are found in lakes, ponds and streams.

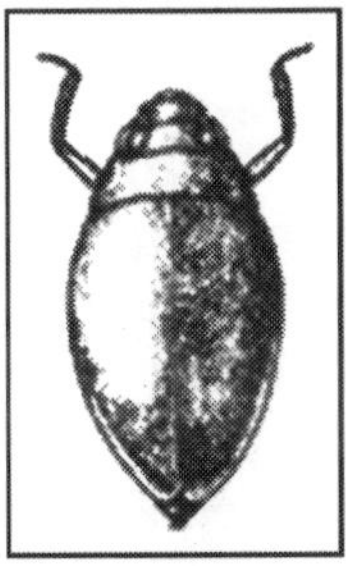

The adult scavengers and the larvae are predaceous and aquatic. Larvae leave the water to pupate on bordering plants and overwinter as adults. There is one generation per year.

PREDACEOUS DIVING BEETLE

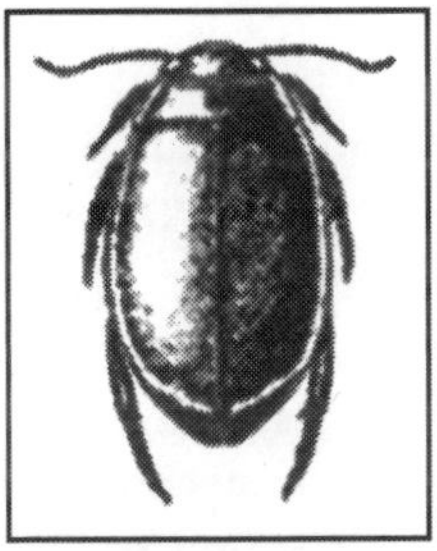

Adults are 1/4 to 1-3/16 inches long, ovate, smooth, and have shining black bodies. There is a yellow margin around the front of the head, sides of the thorax and wing covers. Antennae are threadlike and prominent, with the hind legs longest and flattened to serve as oars in swimming. They are attracted to light, often migrate from one pond to another and are strong fliers.

Unlike other water beetles which move their legs alternately when swimming, these beetles stroke them together like oars as the backswimmer bugs do. They can be collected by sweeping submerged vegetation in streams and ponds with a dip net. Both adults and larvae are predaceous. Sharp hollow mandibles (jaws) are used to inject paralysing and digestive fluids into their prey and to draw out the liquid contents. They attack aquatic insects and fish, and have the ability to regenerate lost parts to some extent. Larvae crawl to shore to pupate in the soil and hibernation occurs in the water as an adult or larvae. There is one generation per year, and adults can live several years.

CONTROL MEASURES OF AQUATIC INSECTS

PREVENTIVE

Since many nuisance aquatic insects are attracted to mercury vapour lights, it is important to use lights sparingly in and around the swimming pool and other areas where insects are not wanted. Use special high-pressure sodium vapour yellow lights which are less attractive to insects and more energy efficient. Set lights 20 to 30 feet from the swimming pool (directing the light to the pool), if practical. Remember that less light equals less insects. Never use 150 watts of light when 25 watts are adequate. All light attracts insects for warmth and ultraviolet energy. Insects are cold-blooded and they see and respond best to ultraviolet energy mercury vapour light. Use long-handled dip nets daily to gather insects, leaves and other debris falling into the water. Maintain good sanitation through proper filtration, chlorine treatment and cleaning. Prevent slime buildup and empty small children's pools when not in use. When on vacation for two or more weeks, either drain the pool or have someone provide maintenance. Some use pool covers. Keep shrubs and grass trimmed closely, especially adjacent to the pool. Increasing the chlorine content to kill the insects could make it harmful for swimmers to use.

INSECTICIDES

Most aquatic insects die rather quickly unless they can return to water. Insecticide control is rarely needed. Never use insecticides in swimming pools due to possible harmful effects on the swimmers and wildlife. Pyrethrin spray used around lights will give temporary knockdown of some flying insects and some repellency. Collect and discard dead, dying and stunned insects with a broom and dust pan, vacuum cleaner, etc. Shrubs and bushes away from the swimming pool can be sprayed lightly with resmethrin plus piperonyl butoxide a few hours before any recreational event to reduce nuisance insects. Avoid contamination of the swimming pool with insecticide spray drift. Before using any pesticide, always read the label. Follow directions and safety precautions.

5

Waste Aquatic Treatment by Microbes

MICROORGANISMS

Microorganisms are of major important in industrial wastewater treatment, agricultural and aquaculture. They reside in the sediment and other substrates, and in the water of aquaculture facilities, as well as in and on the cultured species.

Microorganisms may have positive or negative effects on the outcome of aquaculture operations. Positive microbial activities include elimination of toxic materials such as ammonia, nitrite, and hydrogen sulfide, degradation of uneaten feed, and nutrition of aquatic animals such as shrimp, fish; production of aqua-farmer. These and other functions make microorganisms key players in the health and sustainability of aquaculture. Yet, microorganisms are among the least known and understood elements in aquaculture. Like other areas in aquaculture, microorganisms require management and manipulation.

MAJOR MICROBIAL GROUPS

The world of microorganisms is made of bacteria, fungi, algae, protozoa, and viruses. They are group together only because of their small size, and not by their function. If, for example, the same taxonomical rules were applied to larger animals, some fish, shrimp, green plants, birds and mammals would be grouped together. Some microorganisms such as viruses, bacteria, and protozoa are notoriously small, under one mm. Others, like algae and fungi, have large size relatives (such as the brown algae that is among the largest living organisms). Unlike larger organisms, the morphology of microorganisms is relatively poor and is confined to few shapes and colours. However, their poor morphology is compensated by great physiological versatility.

VIRUSES

Viruses are very small, ranging between 0.01 and 0.03 microns, and only visible by using an electron microscope. They cannot live independently, and only multiply inside the cells of other organisms. However, their demand for a host is fairly specific. For example, it is unlikely that a crustacean virus will attack humans or fish. Viruses are also the simplest of all organisms and are made of nucleic acid (either DNA or RNA), frequently coated with a protein layer.

ALGAE

Algae are photosynthetic organisms (contain chlorophyll) and obtain their energy from the sun and their carbon from carbon dioxide. Their size ranges from one micron to many meters. All organisms that use carbon dioxide for their carbon requirement are called autotrophs. Algae are generally beneficial in aquaculture by supplying oxygen and a natural food base for the cultured animals, such as dinoflagellates that cause the red ties.

FUNGI

Fungi are similar to algae, but they do not contain chlorophyll and require pre-formed organic matter as energy and carbon sources (*e.g.,* sugars, fat, protein, and other carbohydrates). Such organisms are called heterotrophs. Fungi, ranging in size from a few microns to several centimeters, grow either independently by feeding on decaying matter, or in association with plants and animals.

PROTOZOA

Protozoa are heterotrophs, mostly free-living, feeding mainly by devouring smaller microorganisms. Their size ranges between two and 200 micron meters. A large group of

protozoa, the Sporozoa, are parasites. Small numbers of protozoa contain chlorophyll and can switch between autotrophic and heterotrophic modes of feeding, based on light conditions.

BACTERIA

Bacteria range in size from 0.1 to 15 micron, with some giants that may reach half a millimeter. They make up the most metabolically diverse group of living organisms. Although some are parasitic to animals and plants, the majority of bacteria are free-living, having either a neutral or beneficial relationship with humans and other animals and plants. Their metabolic versatility is incredible: while most are heterotrophs, using either light or chemical energy. One of their most remarkable characteristics is their ability to multiply rapidly, with generation times usually ranging between minutes to hours.

MICROBIAL PROCESS

Bacteria and other microorganisms, most notably fungi, are able to metabolise and transform numerous organic and inorganic compounds.

Therefore, man has used them for thousands of years for making yogurt, pickles, bread, cheese, wine, and more recently for waste purification and wastewater purification. Process controlled by microorganisms can occur aerobically (in the presence of oxygen) or anaerobically (with no oxygen present). The starting materials and the end products of such processes vary based on the microorganisms' capabilities (as reflected in their genetic makeup), and the environment in which these processes occur (*e.g.,* availability of oxygen, temperature, salinity, pH, etc.

AEROBIC MICROBIAL PROCESS IN AQUACULTUTRE

Generally, aerobic microbial processes yield compounds which can be beneficial, and are either not toxic or have lox toxicity levels in aquaculture ponds or tanks. Oxidation of organic matter to carbon dioxide, a process which is the main consumer of oxygen in aquaculture ponds or tanks.

Oxidation of ammonia to nitrate via nitrite, which also consumes large quantity of oxygen. xidation of reduced sulfur compounds (such as hydrogen sulfide and elemental sulfur) to sulfate, a process that generally has low oxygen demand in aquaculture. Conversion of carbon dioxide to biomass by autotrophic bacteria (such as the nitrifying bacteria) with a relatively mall amount of biomass produced in aquaculture facilities, when compared to the conversion of carbon dioxide to biomass by algae. Conversion of carbon dioxide to biomass by algae depending on the availability of light. Excluding feeding, the photosynthetic process in aquaculture is the main input of carbon source and natural food for aquatic animals.

ANAEROBIC MICROBIAL PROCESSES IN AQUACULTURE

Microbial anaerobic processes, if not controlled, can produce compounds that are highly toxic to cultured animals.

These processes include:

Consumption of organic matter, without the utilisation of free oxygen, resulting in products which are usually not fully oxidised (such as alcohols, organic acids). Reduction of nitrate and nitrite, which can yield either nitrogen gas or ammonia. In aquaculture, due to the toxicity of ammonia and nitrite, ammonia production is not welcomed, while nitrogen gas production is beneficial. However, in agriculture, the opposite is true - the conversion of nitrate and nitrite to nitrogen das result in a loss of fertilizers. Reduction of sulfur compounds to hydrogen sulfide as a final product, a compound, which is toxic to most animals at even very low concentrations.

FUNCTION OF ENZYMES

Enzymes are "biological catalysts." "Biological" means the substance in question is produced or is derived from some living organism. "Catalyst" denotes a substance that has the ability to increase the rate of a chemical reaction, and is not changed or destroyed by the chemical reaction that it accelerates.

Generally speaking, catalysts are specific in nature as to the type of reaction they can catalyse. Enzymes, as a subclass of catalysts, are very specific in nature. Each enzyme can act to catalyse only very select chemical reactions and only with very select substances. An enzyme has been described as a "key" which can "unlock" complex compounds. An enzyme, as the key, must have a certain structure or multi-dimensional shape that matches a specific section of the "substrate" (a substrate is the compound or substance which undergoes the change). Once these two components come together, certain chemical bonds within the substrate molecule change much as a lock is released, and just like the key in this illustration, the enzyme is free to execute its duty once again.

Many chemical reactions do proceed but at such a slow rate that their progress would seem to be imperceptible at normally encountered environmental temperature. Consider for example, the oxidation of glucose or other sugars to useable energy by animals and plants. For a living organism to derive heat and other energy from sugar, the sugar must be oxidised (combined with oxygen) or metabolically "burned"

However, in a living system, the oxidation of sugar must meet an additional condition; that oxidation of sugar must proceed essentially at normal body temperature. Obviously, sugar surrounded by sufficient oxygen would not oxidise very rapidly at this temperature. In conjunction with a series of enzymes created by the living organism, however, this reaction does proceed quite rapidly at temperatures up to 100°F (38 °C). Therefore, enzymes allow the living organism to make use of the potential energy contained in sugar and other food substances.

Enzymes or biological catalysts allow reactions that are necessary to sustain life proceed relatively quickly at the normal environmental temperatures. Enzymes often increase the rate of a chemical reaction between 10 and 20 million times what the speed of reaction would be when left uncatalysed (at a given temperature). Nutrients locked in certain organics are complex macromolecules, or in hard-to-digest matrices may be released or predigested by a high degree of heat or concentrated acid treatment. In an alternative manner, specific enzymes can promote the pre-digestion of certain complex nutrients and facilitate the release of highly digestible nutrients in organics during processing without the need of excessive heat or rigourous chemical treatment.

NAMING OF ENZYMES

One researcher reports treating grain, sorghum or barley with the enzyme "gumase" while another reports the same with the enzyme "beta-glucanase" When methodologies are examined, it is discovered that both of these preparations are the same product. Unfortunately, this apparent contradiction in terms happens often. Enzymes have been named by several methods and this fact has been known to cause confusion in their classification. For example, common or "trivial" names of enzymes, generally contain a prefix representing the name of the substance or substrate upon which they act or affect, followed by the suffix "ase". The "ase" simply denotes or identifies that the substance is an enzyme. Examples of this system of nomenclature includes the enzyme that catalyses the conversion of proteins into their component amino acids, the name of this enzyme is "protease" or "proteinase".

Another example is the enzyme that accelerates the breakdown of the two components of starch into sugars. The components of starch are known as "amylose" and "amylo-pectin", thus, the enzyme helping to break them down is called "amylase". Confusion may exist, however, when older names of enzymes are used. Included in these older terms are ficin, pepsin, bromelin and trypsin, which are older trivial names of individual types of protease preparations, the enzymes that accelerate digestion of proteins. There are also many sub- classes of enzymes. Amylases are a prime example; subclasses of amylase include: alpha-amylase, beta-amylase, and gluco-amylase, to name a few. All these enzymes do is accelerate the digestion of starch and are broadly classified as amylases, but their actions are all slightly different in nature.

To help sort this out, the International Union of Biochemistry in 1961 proposed a system for enzymes' classification and naming which is finding acceptance mainly in this discussion. One example of this system, however, is the term: "alpha 1, 4-glucan glucanohydrolase" which is a name for alpha-amylase.

All these systems of nomenclature may become confusing to someone who has use for only a few types of enzymes or uses them for industrial or agricultural purposes. Therefore, the use of the more widely known terms such as "amylase" and "protease" are more or less universally in these fields. It should be remembered, however, that there are many types of enzymes that fit into these broad categories that may be more or less suitable for specific agriculturally related application. The final selection for a specific application should be made only after consulting a knowledgeable individual well versed in the technical aspects of the particular enzyme requirements and applicable characteristics.

SOIL-AQUIFER TREATMENT (SAT)

Where soil and groundwater conditions are favourable for artificial recharge of groundwater through infiltration basins, a high degree of upgrading can be achieved by allowing partially-treated sewage effluent to infiltrate into the soil and move down to the groundwater. The unsaturated or "vadose" zone then acts as a natural filter and can remove essentially all suspended solids, biodegradable materials, bacteria, viruses, and other microorganisms. Significant reductions in nitrogen, phosphorus, and heavy metals concentrations can also be achieved.

After the sewage, treated in passage through the vadose zone, has reached the groundwater it is usually allowed to flow some distance through the aquifer before it is collected. This additional movement through the aquifer can produce further purification (removal of microorganisms, precipitation of phosphates, adsorption of synthetic organics, etc.) of the sewage. Since the soil and aquifer are used as natural treatment, systems are called soil-aquifer treatment systems or SAT systems. Soil-aquifer treatment is, essentially, a low-technology, advanced wastewater treatment system. It also has an aesthetic advantage over conventionally treated sewage in that water recovered from an SAT system is not only clear and odour-free but it comes from a well, drain, or via natural drainage to a stream or low area, rather than from a sewer or sewage treatment plant. Thus, the water has lost its connotation of sewage and the public see it water more as coming out of the ground (groundwater) than as sewage effluent. This could be an important factor in the public acceptance of sewage reuse schemes.

SAT SYSTEM LAYOUTS

Various types of SAT system, the simplest being where the sewage effluent is applied to infiltration basins on high ground from where it moves down to the groundwater and eventually drains naturally through an aquifer into a lower area. This lower area can be a natural depression or seepage area, a stream or lake, or a surface drain.

SAT systems as in Figure also serve to reduce the pollution of surface waters. Instead of discharging wastewater directly into streams or lakes, it is applied to infiltration basins at a higher elevation so that it receives soil-aquifer treatment before entering the stream or lake.

Where the groundwater is too deep to collect the renovated sewage water by gravity, pumped wells must be used and there are two basic layouts. In one, the infiltration basins are arranged in two parallel strips and the wells are located on the line midway between the two strips. In the other, the infiltration basins are located close together in a cluster and the wells are on a circle around this cluster. The wells pump essentially all renovated sewage water and no native groundwater from the aquifer outside the SAT system. A mixture of renovated sewage water and native groundwater. Systems 11C and 11D can be used both for seasonal underground storage of sewage water, allowing the groundwater mound to rise during periods of low irrigation water demand (winter), and for pumping the groundwater mound down in periods of high irrigation water demands (summer). The type of SAT system would be suitable for small systems where there are only a few basins around a centrally located well.

WASTEWATER REUSE FOR AQUACULTURE

Wastewater reuse for aquaculture has been practised in many countries for a considerable period of time. It has the potential of wider application in the tropics. There is great diversity of systems involving cultivation of aquatic species, (mainly fish) and plants (mainly aquatic vegetables such as water spinach).

Farmers and local communities have developed most reuse systems; the primary motivating factor has been reuse of nutrients for food production rather than wastewater treatment, and with scant attention to either waste treatment or to public health. In most aquaculture systems, wastewater is not reused directly in aquaculture and the nutrients contained in the wastewater are used as fertilizer to produce natural food such as plankton for fish. These nutrients, mainly nitrogen and phosphorus, are also taken up directly by large aquatic plants such as duckweed which is cultivated for animal feed, and aquatic vegetables such as water spinach and water mimosa cultivated for human food.

As wastewater provides a source of nutrients for aquaculture, it is technically feasible to link it up with most sanitation technologies, providing that land is available at reasonable cost. Farmers have learned by experience how to culture fish, first in static-water nightsoil-fed ponds and more recently in conventional wastewater-fed fishponds. Research has provided a scientific basis for the key parameters in wastewater-fed aquaculture practice developed earlier by farmers and these can be found in the Source Book, published by IWA and IETC.

BIOTA IN AQUACULTURE PONDS

FOOD CHAINS

The objective in fertilizing an aquaculture pond with excreta, nightsoil or wastewater is to produce natural food for fish. Since several species of fish feed directly on faecal solids, use of raw sewage or fresh nightsoil as influent to fish ponds should be prohibited for health reasons.

Edwards has represented the complex food chains in an excreta-fed fish pond, involving ultimate decomposers or bacteria, phytoplankton, zooplankton and invertebrate detritivores. Inorganic nutrients released in the bacterial degredation of organic solids in sewage, nightsoil or excreta are taken up by phytoplankton. Zooplankton graze phytoplankton and small detritus particles coated with bacteria, the latter also serving as food for benthic invertebrate detritivores. Plankton, particularly phytoplankaton, are the major sources of natural food in a fish pond but benthic invertebrates, mainly chironomids, also serve as fish food, although they are quantitatively less important. To optimise fish production in a human waste fed pond, the majority of the fish should be filter feeders, to exploit the plankton growth.

A wide range of fish species has been cultivated in aquaculture ponds receiving human waste, including common carp *(Cyprinùs carpio),* Indian major carps (*Catla catlax, Cirrhina mrigala and Labeo rohita),* Chinese silver carp *(Hypophthalmichthys molitrix),* bighead carp *(Aristichthys nobilis),* grass carp *(Ctenopharyngodon idella),* crucian carp (*Carassius auratus),* Nile carp *(Osteochilus hasseltii),* tilapia *(Oreochromis spp.),* milkfish *(Chanos chanos),* catfish *(Pangasius spp.),* kissing gouramy *(Helostoma temmincki),* giant gourami *(Osphronemus goramy),* silver barb *(Puntius gonionotus)* and freshwater prawn *(Macrobrachium lanchesterii).*

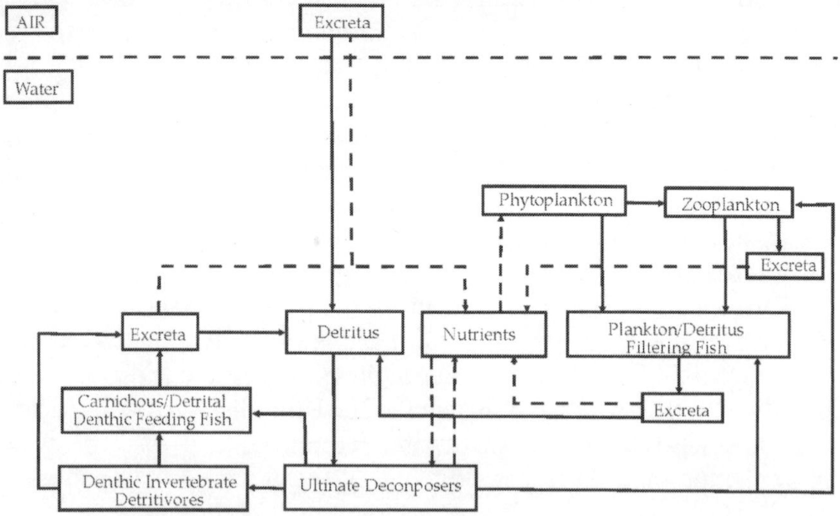

Fig. Food chains in an excreta-fed aquaculture system

FISH SPECIES

The selection reflects local culture rather than fish optimally-suited to such environments. For example, Chinese carps and Indian major carps are the major species in excreta-fed systems in China and India, respectively. In some countries, a polyculture of several fish species is used. Tilapia are generally cultured to a lesser extent than carps in excreta-fed systems although, technically, they are more suitable for this environment because they are better able to tolerate adverse environmental conditions than carp species. Milkfish have been found to have poorer growth and survival statistics compared with Indian major carps and Chinese carps in ponds fed with stabilisation pond effluent in India.

Edwards gives a thorough review of current knowledge on the various fish species which can be cultured in ponds fed with human waste. It would appear that considerable confusion still exists with regard to fish feeding on natural food. Although fish are generally divided into types according to their natural nutritional habits - those that feed on phytoplankton, or zooplankton or benthic animals - several species are known to feed on whatever particles are suspended in the water. There is also uncertainty about the types of phytoplankton fed upon by filter-feeding fish. For example, although blue-green algae are thought to be indigestible to fish, Tilapia have been shown to readily digest these algae and there is evidence that silver carp can do the same.

AQUATIC PLANTS

Aquatic macrophytes grow readily in ponds fed with human waste and their use in wastewater treatment has been discussed. Some creeping aquatic macrophytes are cultivated as vegetables for human consumption in aquaculture ponds and duckweeds are also cultivated, mainly for fish feed. Among the aquatic plants grown for use as vegetables are water spinach *(Ipomoea aquatica),* water mimosa (*Neptunia oleracea),* water cress *(Rorippa nasturtium-aquaticum)* and Chinese water chestnut *(Eleocharis dulcis).* The duckweeds *Lemna, Spirodela* and *Wolffia* are cultivated in some parts of Asia in shallow ponds fertilized with excreta, mainly as feed for Chinese carps but also for chickens, ducks and edible snails.

WASTEWATER TREATMENT

Wastewater is water that has been used and must be treated before it is released into another body of water, so that it does not cause further pollution of water sources. Wastewater comes from a variety of sources. Everything that you flush down your toilet or rinse down the drain is wastewater.

Rainwater and run-off, along with various pollutants, go down street gutters and eventually end up at a wastewater treatment facility. Wastewater can also come from agricultural and industrial sources. Some wastewaters are more difficult to treat than others; for example, industrial wastewater can be difficult to treat, whereas domestic wastewater is relatively easy to treat (though it is increasingly difficult to treat domestic waste, due to increased amounts of pharmaceuticals and personal care products that are found in domestic wastewater.

RESPONSIBILITY FOR MAKING SURE THAT WASTEWATER IS TREATED PROPERLY

Similar to drinking water provisions, the federal government has delegated responsibility for wastewater treatment to the provinces and territories.

There are two federal acts, however, that may apply to wastewater. The Fisheries Act prohibits the release of harmful substances into waters that fish live in. The Canadian Environmental Protection Act governs the release of toxic substances into the environment and allows the federal government to develop regulations for the use of toxic substances.

Most provincial and territorial governments have legislation regarding wastewater treatment standards and requirements. Operators of wastewater treatment facilities must obtain permits or licenses from the provincial or territorial government, and these permits may also require additional treatments or limits on effluent discharges. For example, in British Columbia, all municipalities are required to have a Liquid Waste Management Plan; without an approved plan, discharges are illegal. Provincial and territorial governments generally assist municipal governments with funds to build and maintain infrastructure.

Municipal governments directly oversee the wastewater treatment process, and are able to pass additional by-laws. For example, the Regional Municipality of Ottawa-Carleton has developed a programme to eliminate toxic substances from the wastewater treatment system, requiring all industrial, institutional and commercial facilities to limit the amount of certain pollutants that are allowed into sewers.

Treatment of Wastewater, to Make it Safe for Discharge

There are several levels of wastewater treatment; these are primary, secondary and tertiary levels of treatment. Most municipal wastewater treatment facilities use primary and secondary levels of treatment, and some also use tertiary treatments. The type and order of treatment may vary from one treatment plant to another, but this diagram of the Ottawa-Carleton wastewater treatment plant illustrates the basic components.

The primary level of treatment uses screens and settling tanks to remove the majority of solids. This step is extremely important, because solids make up approximately 35 per cent of the pollutants that must be removed. The screens usually have openings of about 10 millimetres, which is small enough to remove sticks, garbage and other large materials from the wastewater. This material is removed and disposed of at the landfill.

The water is then put into settling tanks (or clarifiers), where it sits for several hours, allowing the sludge to settle and a scum to form on the top. The scum is then skimmed off the top, the sludge is removed from the bottom, and the partially treated wastewater moves on to the secondary treatment level.

The primary treatment generally removes up to 50 per cent of the Biological Oxygen Demand (BOD; these are substances that use up the oxygen in the water), around 90 per cent of suspended solids, and up to 55 per cent of fecal coliforms. While primary treatment removes a significant amount of harmful substances from wastewater, it is not enough to ensure that all harmful pollutants have been removed.

Secondary treatment of wastewater uses bacteria to digest the remaining pollutants. This is accomplished by forcefully mixing the wastewater with bacteria and oxygen. The oxygen helps the bacteria to digest the pollutants faster. The water is then taken to settling tanks where the sludge again settles, leaving the water 90 to 95 per cent free of pollutants. The picture below shows the settling tanks in the Winnipeg Wastewater Treatment Plant. Secondary treatment removes about 85 to 90 per cent of BOD and suspended solid, and about 90 to 99 per cent of coliform bacteria.

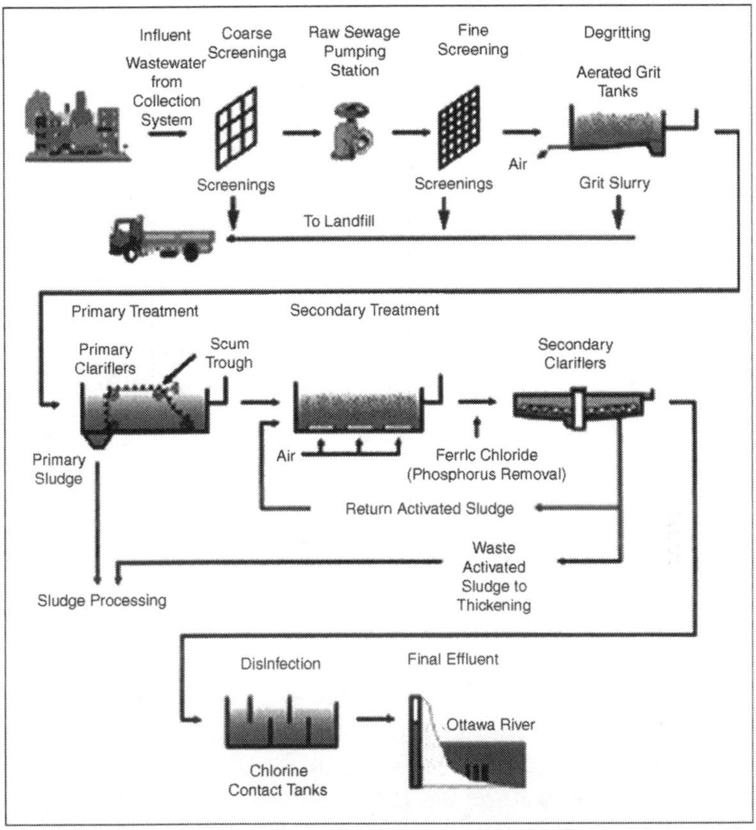

Some treatment plants follow this with a sand filter, to remove additional pollutants. The water is then disinfected with chlorine, ozone, or ultraviolet light, and then discharged.

The sludge that is removed from the settling tanks and the scum that is skimmed off the top during the primary steps are treated separately from the water. Anaerobic bacteria (anaerobic bacteria do not require oxygen) feed off of the sludge for 10 to 20 days at temperatures around 38 degrees Celsius.

This process decreases the odour and organic matter of the sludge, and creates a highly combustible gas of methane and carbon dioxide, which can be used as fuel to heat the treatment plant. Finally, the sludge is sent to a centrifuge, like the one shown in the picture below. A centrifuge is a machine that spins very quickly, forcing the liquid to separate from the solid. The liquid can then be processed with the wastewater and the solid is used as fertilizer on fields.

Tertiary (or advanced) treatment removes dissolved substances, such as colour, metals, organic chemicals and nutrients like phosphorus and nitrogen.

There are a number of physical, chemical and biological treatment processes that are used for tertiary treatment. One of the biological treatment processes is called Biological Nutrient Removal (BNR). This diagram shows the treatment steps that Saskatoon wastewater goes through.

In this treatment plant, wastewater first undergoes primary and secondary treatment. For the tertiary treatment, the BNR process occurs in the bioreactors. The BNR process uses bacteria in different conditions in several tanks, to digest the contaminants in the water. The three tanks have unique environments, with different amounts of oxygen. As the water has passes through the three tanks, the phosphorus is removed and the ammonia is broken down into nitrate and nitrogen gas, which other bacterial processes can not do. The BNR process can remove over 90 per cent of phosphates, while traditional processes remove much less than 90 per cent. The water spends approximately nine hours in the bioreactors, before entering the secondary clarifier, which is a settling tank, where the bacteria-laden sludge settles to the bottom of the tank.

Small Communities Treat Waste Ewater, to Make it Safe for Sischarge

In small communities, wastewater treatment facilities may consist of individual septic systems, simple collection systems that directly discharge effluent to surface waters, or municipal lagoons that are emptied annually. These facilities usually treat and disperse the waste as close as possible to its source, thus minimizing operational costs and maintenance requirements. The longer the waste can sit in a lagoon before being discharged, the less likely it will be to contaminate drinking water sources. Some communities store the waste in lagoons, but others release the waste directly into water sources.

Lagoons are reservoirs in the ground that store waste for a time until it is discharged, either to the soil or a water body. Shallow lagoons, that are less than 1.5 metres deep, are used for primary treatment, which allows the solid waste to settle to the bottom of the lagoon over a period of 6 to 20 days. Shallow lagoons, however, cannot effectively remove the majority of contaminants that pose problems for ground and surface waters. Deep lagoons, which are more than three metres deep, can provide long-term storage and treatment for six months to one year.

Many lagoons in small communities are emptied once per year. Rural communities often make use of surrounding land to dispose of wastewater. When the soil is adequate, and there are no water sources nearby, the bacteria in the soil can remove and break down the contaminants in wastewater. Due to the availability of land in many rural areas, this can be an effective method to treat wastewater. However, there are other communities that dispose of waste in a way that threatens the quality of the lake, river or groundwater source that provides drinking water. The Environmental Protection Agency estimates that between 10 and 20 per cent of small community wastewater treatment facilities in the United States are not operating properly; state water quality agencies have identified malfunctioning wastewater treatment systems as the second greatest threat to water quality (after underground storage tanks). When the inadequate wastewater treatments are combined with ineffective drinking water treatment, the result is a serious contamination issue for a great number of rural communities.

Rural communities typically find it difficult to install and maintain wastewater treatment operations. And while many communities have inadequate methods of treating wastewater, there are some communities that are leading the way with innovative methods of treatment and water conservation measures. In several Arctic communities, including Iqaluit, Nunavut, the high cost of water has led to wastewater treatments that allow water to be reused. Wastewater is passed through a septic tank, filtered, and disinfected with ozone treatment; it is then reused for non-consumptive uses, such as toilets and laundry. These conservation measures allow them to reuse up to 55 per cent of wastewater, while decreasing pressure on wastewater treatment and storage processes. For more information about water conservation, including the ways in which rural and First Nations communities are leading efforts to reduce water use.

WORKING PROCESS OF SEPTIC SYSTEMS

There are many people living in rural areas that are not served by wastewater treatment plants. In fact, according to Environment Canada, as of 2000, only 57 per cent of Canadians were served by wastewater treatment plants, compared with 74 per cent of Americans, 86.5 per cent of Germans, and 99 per cent of Swedes. Many people in rural areas use septic systems to safely store waste. Wastewater travels through pipes, from the house to a buried septic tank. The diagram below illustrates the basic components of a septic system.

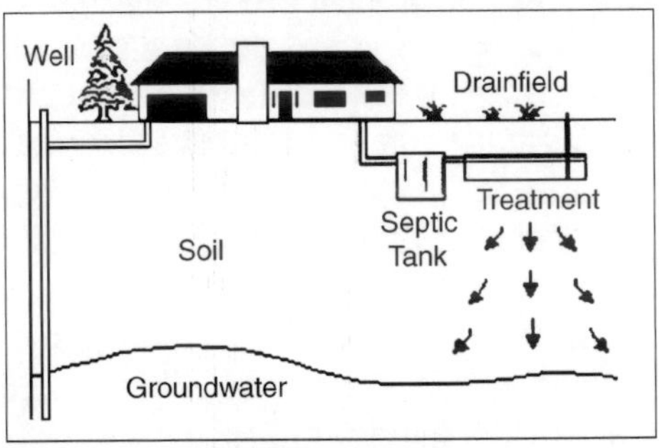

Fig. Septic Tank

The diagram below illustrates the tank in more detail. In the septic tank, the solids settle to the bottom and a scum forms on the top, similar to the process that occurs in settling tanks in municipal wastewater treatment plants. Once separated, the water flows out to the drainfield, and screens and compartments keep the sludge and scum inside the tank, where bacteria begin to partially digest the sludge. When the partially treated wastewater enters the drainfield, it begins to infiltrate the soil and percolate downwards. So long as the soil is appropriate, the microbes in the soil digest the pollutants, removing the bacteria, viruses and excess nutrients by the time the water reaches the source.

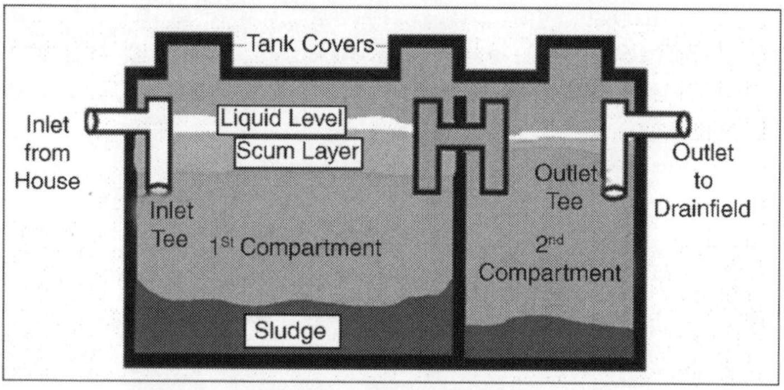

Fig. Inside of Septic System.

When the soil is not suitable, or there are nearby water sources, an alternative system may be used, so that water sources do not become contaminated. The alternative system may use sand, peat, or plastic instead of soil. Constructed wetlands, lagoons, aerators, or disinfection devices are also effective in treating the wastewater.

It is estimated that around half of all rural wells are contaminated, many from septic tanks. The United States Centers for Disease Control and Prevention suggests that wells should be at least 15 metres from septic tanks, but this distance also varies according to the type of soil.

In a recent study of groundwater sources in southeastern Michigan by the United States Geological Survey, 38 wells were tested between 1999 and 2001. Of the wells that were near sewerlines, viruses were detected in only two of the 18 wells. Of the 20 wells that were near septic systems, viruses were found in seven wells. This suggests that septic systems are a major cause of groundwater contamination. Previous studies of contaminated wells that were cited by the USGS study found coliforms in up to 80 per cent of the wells, as well as some with significant numbers of the E. coli bacterium.

If you are using a septic system, it is important to maintain it properly, as failure to do so could result in the leakage of pollutants into the soil or water sources. The United States Environmental Protection Agency (EPA) recommends that you have your septic system inspected at least once every three years, so that it can be checked for leaks and malfunctions, and also be pumped out when it gets full (typically once every three to five years).

Following some water conservation practices can greatly reduce pressure on your septic system.

Here are a few things that you can do to care for your septic system:

- Do not use your drain or toilet as a garbage disposal; avoid putting dental floss, diapers, coffee grounds and paper towel down the drain, as they can clog up your septic system.
- Spread your loads of laundry out over the week. When too much water is added to the septic tank, it does not have time to treat wastes, and you could be flooding your drainfield with wastewater.
- Plant grass on your drainfield, but keep trees and shrubs away from it, because roots can clog the system and cause damage.
- Do not drive on your drainfield, because this can compact the soil and damage the septic system components.

NATURAL WAYS TO TREAT WASTEWATER

If nature itself can cleanse water, then imitating nature's processes may be the most effective and sustainable ways of treating wastewater. A great deal of water renewal occurs naturally in wetlands.

Constructed wetlands consist of a lined cell, which the water flows into. Plants are planted in the cell and the roots filter the contaminants out of the water. Below is a diagram of a constructed wetland. Notice that many of the processes in a wetland are similar to the Biological Nutrient Removal process that was described above.

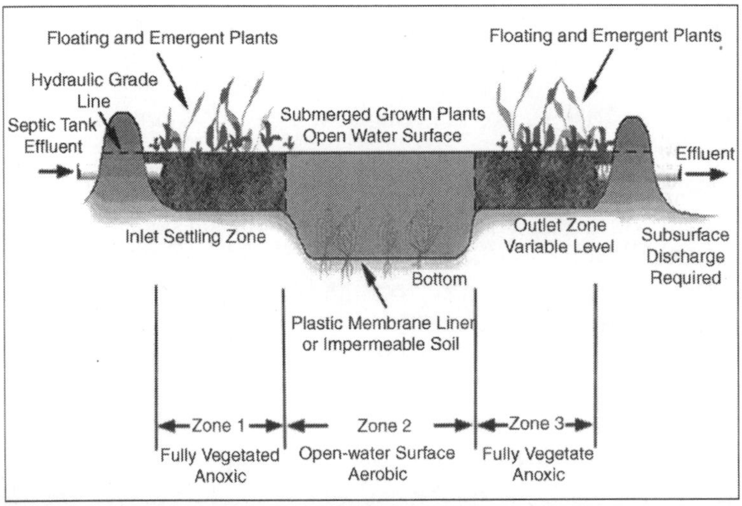

Fig. Constructed Wetland

Another natural method is called rapid infiltration, which is a process where a basin is filled with wastewater, which has already gone through a pre-treatment. The ground acts as a filter and removes the pollutants from the water. This method is similar to what happens in a septic system. A third "natural" process is overland flow, which is used in regions of nearly impermeable ground. The water flows down a sloped surface that is planted with thick grasses. Because the soil is highly impermeable, the water is forced through the vegetation, which effectively removes the pollutants.

Slow rate irrigation is a process that uses a portion of land, and allows the water to flow slowly enough that the land's capacity to infiltrate the water and remove the impurities is not overburdened. Silviculture is similar to slow rate irrigation, in that it uses a large amount of land to treat wastewater, by planting crops or trees that will flourish during the treatment process. Aquiculture uses aquatic plant and animal species to treat wastewater, similar to the constructed wetland process. There are also alternative separation systems that can conserve water. One such system separates blackwater (from toilets) from greywater (from showers and dishwashers), so that greywater can be minimally treated and used for watering the lawn. As well, there are incinerating, chemical, or composting toilets that release the waste when it is safe.

BENEFITS OF WATER-POLLUTION CONTROL

This chapter presents two studies that endeavor to develop methods for assessing the *national* benefits associated with improvements in, or maintenance of, the quality of surface waters. The first focuses exclusively on freshwater fishing and builds up a national total from regional estimates. The second uses a national sample survey technique to elicit individual's valuation of some broad national water-quality goals.

NATIONAL FRESHWATER RECREATION BENEFITS

Among the more important pieces of national environmental legislation created during the 1970s were the comprehensive amendments to the Federal Water Pollution Control Act.

These amendments, signed into law in 1972 and further amended in 1977, in reality constituted a major piece of legislation in their own right, dramatically redirecting the nation's efforts at water-pollution control and setting out ambitious national goals, expressed both in terms of discharge controls and of resulting water quality.

Criticism of the amendments and debate over their goals and requirements began during the legislative process and has continued, with more or less heat, to the present. Some critics argue that the goals are too ambitious, that is, the benefits of meeting the goals (and related requirements) are thought to be too small to justify the costs of compliance. This argument over the balance of benefits and costs can never be resolved entirely by research, but the RFF project described here was undertaken in the conviction that it should be possible to improve methods for estimating at least some of the benefit categories associated with waterpollution control, in this case, the benefits from recreational fishing in freshwater bodies. From the outset the intent was to design a method for estimating benefits for the nation as a whole rather than benefits for particular sites. In this respect, it resembles the study discussed in the last part of chapter. In undertaking this project, a primary question concerned the ways in which water-quality improvement would favourably affect freshwater fishing. Two major ways were identified.

First, it tends to increase the total availability of fishable freshwater bodies by reducing the incidence of conditions such as low dissolved oxygen that results from the bacterial degradation of organic materials and heavy sediment loads that make it difficult for fish to survive.Second, it produces changes in the *types* of fish that can survive in particular water bodies.Simply put, clean water means "game" fish such as trout or bass, and dirty water means rough fish such as carp or buffalo. In general, fishermen prefer game fish. Therefore, pollution control tends to increase the amount of water yielding high-quality fishing relative to that yielding low-quality fishing.Given this view of the benefit-producing mechanisms, one can work towards a methodology for making national benefit estimates based on it. As explained earlier, benefit estimation for environmental improvement requires the understanding of a number of links. For this particular study, the following questions should be considered:

- How will implementation of the law affect pollution discharges by location, quantity, and pollutant type across the entire nation?
- How will the pre- and postpolicy discharge levels affect ambient water quality? Or how does ambient quality change as discharges
- Change not only in terms of such familiar indicators as dissolved oxygen, but also in terms of supportable fish population types?
- How will increases in total amounts of water supporting recreational fishing and shifts in the composition of that water towards more highly valued fish species affect the number of anglers and the amount of time they spend fishing?
- In addition, one needs to be able to value fishing activity of various kinds—that is, for practical purposes, how many days are spent fishing for various species—rough fish versus game fish?

The novelty of this study and its main contribution to methodological development lies in the ingenious way it is able to link models together to structure these linkages and how it is able to take existing and newly developed data sets to estimate them quantitatively. I turn now to a discussion of each step in the procedure.

DISCHARGE REDUCTIONS AND LINKS TO AMBIENT QUALITY AND FISH

Initially, one must have an understanding of the "fishability" of the nation's water prior to the implementation of the Federal Water Pollution Control Act.

A data base is available from the Fish and Wildlife Service that permits estimates of fishable water by state (the state is the basic geographic unit on which this study is operated), but these data do not provide a basis for the breakdown between rough fish and game fish that is basic to the methods used in this study. For this reason, the researchers surveyed state fish-and-game officials asking them for a breakdown by species category within their own states. Using these data, they found that for the contiguous forty-eight states and the District of Columbia, there are about 30.6 million acres of fishable fresh water consisting of about 20.4 per cent cold-water game fisheries, 68.4 per cent warm-water game fisheries, and 11. 2 per cent rough fisheries.

To determine how the implementation of discharge controls would affect the current status requires a knowledge of the amount and location of discharges prior to, and following, the implementation of the 1972 amendments. Then it is necessary to estimate how this change will affect ambient conditions in water courses, and how, in turn, these will affect fishability.

The first three kinds of information have been established by the use of RFF's Water Quality Network (WQN) model. This model, designed specifically to answer those questions, was run for four scenarios representing—albeit roughly in some cases—stages in the implementation of the law. In what follows, I will focus on only one of these stages—the Best Practical Control Technology Currently Available.

This is for simplicity and also because the quantitative benefits still must be regarded as experimental. The Best Practical Control Technology Currently Available (BPT for short) requirement was to be achieved by all point sources (that is, discharges from confined channels, such as pipes) of wastewater discharge. This unmet goal may reflect where we currently are in our control efforts. At best, the WQN model provides a reasonable estimate of the impact of policy changes on one important aspect of ambient conditions: dissolved oxygen. However, it does not translate directly into fishability. Indeed, making that step is an undeveloped discipline, calling for heroic measures.

Fortunately, a fisheries biologist, willing to use his knowledge and skill to survey the literature, developed a set of rules that appear to capture whatever consensus exists on the water-quality conditions appropriate to the survival and reproduction of various fish populations. These rules can be applied to the results of the WQN model to provide estimates of the acreages of different kinds of fishing availability by state, and by aggregation, for the nation as a whole. The reader may be struck by how small the increases in total fishable water are—only about 100,000 acres from a base of more than 30 million. This is because a very large proportion of U.S., fresh waters already was fishable before implementation of the water-pollution law. However, at the same time, it is projected that the waters regarded as unfishable or capable of supporting rough fish only will decline dramatically.

This does not mean a proportionate decline in rough fish populations, but rather a large increase in the water that rough fish will share with warmand cold-water game fish. The next step is to devise ways of converting the water-quality results into changes in fishermen's participation in various kinds of fishing. Before proceeding, however, it is pertinent to note that what has been discussed so far is not types of research and modeling that are in the usual purview of economics. But the situation here, is reflective of the fact that existing models of natural systems rarely fit the needs of the economist who would estimate the benefits of environmental improvement. Accordingly, he is often forced into disciplinary imperialism.

Behavioral Economic Aspects of the Study

I now turn to steps in the analysis that are more clearly economic in character. In order to estimate total activity in various types of fishing, the individual fisherman's chain of decision about recreational fishing must be broken down into several logical stages.

The first choice is whether to do any fishing at all. The researchers' hypothesis is that the decision of whether to fish is sensitive, among other things, to the opportunity to fish, represented by the quantity of fishable water. The object of this first stage of the research is then to quantitatively estimate how the decision to fish is influenced, in the population at large, by the availability of fishable water. Regression analysis is the method used to determine the separate influences of availability of fishing opportunity and those other factors that might affect the decision.

The indicators of existing availability of fishable water are the statelevel estimates divided by the state population to get a per capita measure. This is rather crude, but a more refined indicator was not available at the time. The other data needed for this stage of the research were obtained from a very large survey conducted by the U.S. Department of the Interior, Fish and Wildlife Service. The first or screening stage of this survey was conducted by telephone interview of more than 100,000 households (300,000 individuals). Its primary intent was to determine whether individuals participated in hunting, fishing, and other recreational activities associated with wildlife.

The survey also contained information on other pertinent variables such as age, sex, income, and other factors so that it was possible to include them in the regression analysis and control for their possible effects on participation. The dependent variable was the decision to fish or not to fish. Since the availability of fishable water was included among the independent variables, once the coefficients of the equation have been estimated, the size of the availability variable can be changed and the corresponding change in participation calculated. We have seen regression analysis results used in a similar way in other chapters, for example, in projecting the effect of air-quality improvement.

So far, all the analysis permits us to do is to project fishing in general as a function of water quality. But since, as I have indicated, different types of fishing (warm-water game fishing, cold-water game fishing, and rough fishing) probably differ in value, we must also be able to project how likely a representative individual is to pick each of these types if he or she does decide to fish.

For this purpose, data obtained by the Fish and Wildlife Service in the second stage of the 1975 survey was used. A questionnaire was mailed to more than 50,000 persons who had declared themselves to be hunters or fishermen in the screening stage. For this subgroup, detailed information was gathered on their participation patterns, socioeconomic characteristics, and preferences. Data for the fishermen only was used in analysing the second stage in the decision chain—namely, once a person has decided to fish how likely is he or she to participate in each of the three types of fishing given the availability of water suitable for each type?

Because doing some trout fishing, for example, does not rule out doing some bass or rough fishing as well during the course of the year, the regression equations for the type of fishing decision might best be characterised in "some-of" terms. Either a person did some cold-water game fishing or he or she did not. But the individual also might have done some bass fishing. In any case, whether some of a particular kind of fishing was done was hypothesised to be a function of the availability of water suitable for that kind of fishing as well as other characteristics of the participant.

The final stage in the decision chain is the decision on how much time (how many days) will be spent in that activity. The same set of survey data was used in the analysis of this question, and regression analysis was the tool for connecting reported decisions on days of fishing to the independent variables, including the availability of water suitable for the fishing in question. The drift of the analysis is now clear. The steps are as follows: the amount of increase in total fishable water and fishable-type water associated with water-pollution control is given for the nation as a whole from the models of the previous sections. Given this, the results of Stage 1 are used to calculate how much fishing participation will increase in general.

Then, the results of Stages 2 and 3 are used to calculate how this increase in participation will be distributed across the fishing types and how many days of increased fishing of each type will occur nationally as a result of the pollution-control policy. The final problem confronted by this research on the benefits from improved freshwater fishing opportunities is how to assign dollar-value benefits (that is, willingness to pay) to the increase in each category of fishing activity. The approach adopted estimated a demand curve for fishing days for each category and used those to calculate the average consumer's surplus per day. The travel-cost method, described generally in earlier chapter, was the technique selected. I now turn to a brief discussion of how it was applied in this study. Recall that the basic assumption of the travel-cost method is that higher costs of access, as reflected in distance from a recreational site, will have the same effect on visitation as an equivalent admission fee assuming zero distance from the site. I presented a very simple example of how this relationship can be used to develop a demand curve by assuming successively higher admissions fees and using information on access costs to estimate their effects on visitation.

This establishes points on a demand curve, that is, the relationship of price to the number of visitor-days. The area under the demand curve, by principles discussed in chapter, is the total willingness to pay of participants for the total number of visitor-days to the site, say, a trout fishery. If one then divides the number of visitor-days into this number, one obtains the average willingness to pay for a day of fishing for trout. The researchers who conducted the study collected data from a large number of fishing sites around the country which permitted them, by statistical means, to make exactly such a calculation yielding average willingness to pay per visitor-day for each type of fishery.

ESTIMATING NATIONAL WATER-QUALITY BENEFITS

We are now at a point where a national benefits estimate can be made. All the earlier machinations were designed to estimate how many days of increased recreational fishing of each type would correspond to the water-quality changes resulting from a reduction of wastewater discharges corresponding to the implementation of a pollution-control policy. Having these numbers in hand, it is a simple matter to multiply them by average willingness to pay for a day by fish type and get a total benefit number for freshwater fishing in the United States. When this is done, the following results are obtained for Best Practical Control Technology Currently Available.

Valuation base	*Total Annual Benefits over base (millions)*
Low	307
High	683

A few words of explanation are needed about the difference between the low and the high estimates. For the low estimate, travel cost is based only on out-of-pocket expenses—gasoline, restaurant food, motels, and others. This is the conventional method. The higher estimate takes account of the fact that the fisherman may also attach a cost to the time it takes to get to the site. For the higher figure, an estimate of this cost is made by attaching average wage rates to the travel time needed to reach the site.

Needless to say, large uncertainties attend these numbers and, because of this, they must be regarded as largely experimental. Nevertheless, in view of the heavy costs of the national programme for water-quality improvement they may strike the reader as being quite low. There are several things to be said in this connection. First, the reader should recall that in terms of the availability of fish species the vast majority of the nation's fresh water was already fishable prior to the 1972 Amendments. Second, these estimates are partial in the sense that they consider only the fresh waters and even then they do not include values that may accrue to fishermen from the possible effects of pollution control on the aesthetic aspects of the fishing experience. At present, research is under way to extend the methodology developed in this study to effects of pollution control on marine (saltwater) recreational fisheries, and on both marine and freshwater swimming and recreational boating.

METHOD FOR ESTIMATING NATIONAL WATER-QUALITY BENEFITS

The research reported in the previous section was designed to yield national recreational fishing benefits of water-quality improvement. Basically, it used subregions as units of analysis and aggregated them by adding up the results. Thus it can be described as a large-scale simulation falling somewhere between a particular site (or micro) study and a national survey that asks respondents directly about their willingness to pay for national programmes of pollution control. This last procedure has been called the "macro" approach. Among other potential advantages of such an approach, two are especially important.

First, a randomised national sample of persons can be interviewed which permits the use of well-established statistical procedures to extrapolate the results to the entire population. Second, one can enquire about "intrinsic" or existence benefits as well as user benefits. The second reason invites more explanation. Because the U.S., population politically supports very expensive programmes of water-pollution conrol—much more costly than the benefits estimated for recreational users in the previous section—the researchers were led to believe that there must be some form or forms of benefits accruing to persons who do not actually use particular water bodies. We have termed such benefits variously as intrinsic or existence benefits. These benefits may accrue because persons value the options for possible use that are opened to them when water bodies are cleaned up.

This type of value, discussed widely in the economics literature, has come to be called the option value. Other intrinsic values may accrue from a sense of national pride or rectitude associated with having clean waters. One of the main conclusions of the research reported in this chapter and in the following one is that intrinsic benefits definitely exist with respect to environmental improvements or maintenance. Moreover, and with the usual caution about accuracy of results, not only do they exist, but they are large, perhaps even larger than user benefits in some instances. Some aspects of water quality make it more appropriate than air quality for an experimental application of the macro approach. Chiefly, goals of our national policy are set out in a manner that would let most of the population understand what they mean in terms of ordinary experience.

The objectives are to make all the nation's water fishable and swimmable in successive stages. Furthermore, much of the cost of these programmes is to be paid from taxes levied at the national levels (taxes financing subsidies to local governments) so that respondents can be realistically asked how much in added tax burden they are willing to pay for improved water quality across the whole nation. Neither one of these situations holds with respect to air quality, so it would be much harder to pose understandable and realistic alternatives in a national clean air survey.

A macro study, then, is potentially useful for doing a benefit-cost analysis for national water programmes. It should be noted, however, that it is *not* a substitute for site-specific studies in other applications. For example, determining whether the benefits outweigh the costs of a programme for water-quality improvement in the Potomac estuary would require a site-specific study.

RESEARCH PROCEDURES

One problem with national surveys is that they are quite expensive. What made it possible to conduct an experiment with the macro approach, given available resources, was that the researchers were able to add some water-quality questions onto a survey being funded by another source. After the interview for the other survey was completed, the interviewers administered a sequence of benefits questions that had been carefully pretested by researchers on the benefits project. From the respondents' perspective, the two interviews appeared as one long interview. In all, 1,576 personal interviews of a national probability sample of persons eighteen years of age and older were completed. The sample was designed and the interviews were conducted by Roper and Cantril, a national polling firm.

A penalty of this add-on approach proved to be that an unfortunately large number of persons failed to complete all of the questions. In part this was because they came at the end of an already fairly lengthy survey and in part because it was not possible to undertake special training of the interviewers to administer the benefits section. Because of the likelihood of item-response bias (caused by respondents failing to answer individual items), the researchers regard their estimates as only suggestive and warn against accepting them as definitive. The main intent of the experiment was not to develop definitive estimates at this stage but to test whether a macro approach is applicable to an investigation of waterquality benefits. The low response rate presumably can be cured by an improved questionnaire and by training of the interviewers. A study is currently being planned in which both of these elements will exist.

THE WATER-POLLUTION LADDER AND VALUE LEVELS

The levels of water quality for which the research team sought willingness-to-pay estimates are "boatable, " fishable," and "swimmable." These levels are described in words and depicted graphically by means of a "water-quality ladder". Use of these categories, two of which are embodied in the law mandating the national programme for water-pollution control, permitted avoidance of the communications problems associated with describing water quality in terms of the numerous abstract technical measures of pollution (oxygen depletion, for example).

Although the boatable-fishable-swimmable categories are widely understood by the public, they did require further specification to ensure that different people perceived them in a similar fashion. *Boatable water* was defined as an intermediate level between water which "has oil, raw sewage and other things in it, has no plant or animal life and smells bad" on the one hand, and water which is of fishable quality on the other. As discussed earlier, *fishable water* covers a fairly large range of water quality.

Game fish such as bass and trout cannot tolerate water in which certain rough fish such as carp and catfish flourish. In pretests, experiments were made with two levels of fishable water-one for rough fish like carp and catfish, and the other for game fish like bass—but a single definition of fishable was adopted as water "clean enough so that game fish like bass can live in it," under the assumption that the words "game fish" and "bass" had wide recognition and denoted water of the quality that Congress had in mind. *Swimmable water* appeared to present less difficulty for popular understanding since the enforcement of water-quality standards for swimming by health authorities has led to widespread awareness that swimming in polluted water can cause illness.

Because willingness-to-pay questions have to describe in some detail the conditions of the "market" for the good, they are inevitably longer than the usual survey research questions.

Respondents quickly become bored and restless if material is read to them without giving them frequent opportunities to express judgements or to look at visual aids. The questionnaire for this experiment was designed to be as interactive as possible by interpreting the text with questions which required the respondents to use the newly described water-quality categories. They were also handed a card depicting the water-quality ladder which was referred to constantly during the sequence of benefits questions.

WILLINGNESS-TO-PAY QUESTIONS AND ANSWERS

Questions about willingness to pay should seem realistic to respondents. Accordingly, they were couched in terms of annual household payments in higher prices and taxes because this is the way people do pay for waterpollution control. A portion of each household's annual federal tax payment goes towards the expense of regulating water pollution and providing construction grants for sewage-treatment plants. Local sewage taxes pay for the maintenance of these plants. Those private users, such as manufacturing plants, who incur pollution-control expenses ultimately pass much or all of the cost along to consumers in higher prices. Thus, this payment method has a ring of truth to the respondents. As explained earlier "starting-point bias" can be an important problem in bidding games and surveys. That is, a high starting bid from an interviewer may elicit a higher bid from a respondent than a low starting bid.

A major methodological innovation of the research reported in this chapter is the development of a device for eliminating such a bias, the "payment card." In this technique, the respondent is given a card which contains a menu of alternative amounts of payment beginning at $0 and increasing by a fixed interval until an arbitrarily determined large amount is reached. When the time comes to elicit the amount one is willing to pay, the respondent is asked to pick a number from the card (or any number in between) which "is the *most* you would be willing to pay in taxes and higher price *each year*" (italics in the questionnaire) for a given level of water quality. Thus, the interviewer suggests no bid at all. It turns out, however, that this presents some problems of its own. In initial pretests, it was found that the respondents had considerable difficulty in determining their willingness to pay when a card was used which only presented various dollar amounts.

A number of them expressed embarrassment, confusion, or resentment at the task, and some who gave amounts indicated they were very uncertain about them. The problem lay with the lack of benchmarks for their estimates. People are not normally aware of the total amounts they pay for public goods even when that amount comes out of their taxes, nor do they know how much such goods cost. Without a way of psychologically anchoring their estimate in some manner, they were not able to arrive at meaningful estimates. They needed benchmarks of some kind which would convey sufficient information without biasing their responses. Their most appropriate benchmarks for willingness to pay for water-pollution control would appear to be the amounts they are already paying in higher prices and taxes for other non-environmental, publicly provided goods and services.

Amounts were identified on the card for several such goods, and further pretests were conducted, indicating that the benchmarks made the task meaningful for most people. But the use of payment cards with benchmarks raises the possibility of introducing its own kind of bias. Are the respondents who gave amounts for water-pollution control using the benchmarks for general orientation or are they basing their amounts directly on the benchmarks themselves in some manner? In the former case, respondents would be giving unique values for water quality; in the latter case, they would be giving values for water quality relative to what they think they are paying for a particular set of other public goods. If the latter case holds and their water-quality values are sensitive to changes in the benchmark amounts, or to changes in the set of public goods identified on the payment card, their validity as estimates of consumer surplus for water quality are suspect. A test for this kind of bias was conducted in the pretest by using different versions of the payment card with the amounts paid for other publicly provided goods changed by modest amounts. No bias was found, and so the "anchored" payment card was deemed to be a suitable device for the full-scale experiment. Tests were also conducted to attempt to discover if any of the other sorts of bias were inherent in the questionnaire. Again, none was found.

A final point should be made regarding the payment card. What people actually pay for publicly provided goods varies with their income. To correct for this, four different payment cards were developed corresponding to four income classes. At the appropriate point in the interview, the interviewer gave the respondent the payment card for his or her income category, which had been established by a prior question.

As already discussed, the respondents valued three levels of water quality which were described in words and depicted on the water quality ladder. They were first asked how much they were willing to pay to maintain national water quality in the boatable level. Subsequent questions asked them their willingness to pay for overall water quality to fishable quality and swimmable quality. The average willingness-to-pay amounts given by the respondent for the two higher levels consists of the amounts they offered for the lower levels plus any additional amount they offered for the higher level.

The average annual amounts per household for those respondents who answered the willingness-to-pay questions turned out to be:

Water quality	Total $	Marginal $
Boatable	152	152
Fishable	194	42
Swimmable	225	31

The most substantial benefit is for boatable water. The respondents are willing to give about 20 per cent more for fishable water than boatable water, but only an additional 15 per cent to make the water swimmable.

The data also permitted one to make a rough distinction between the type of recreation and the intrinsic values discussed earlier. Since the willingness-to-pay questions measure the overall value that respondents have for water quality, the amount given by each respondent represents the combination of recreational and intrinsic values held by that person. But it was possible to tell from the questions whether a person actually engaged in water-based recreation. It was reasoned that the values expressed by the respondents who do not engage in in-stream recreation should be almost purely intrinsic in nature. In calculating the average willingness-to-pay amount for the non-recreationists alone, therefore, we get an approximation of the intrinsic value of water quality. By subtracting this amount from the total the recreationists are willing to pay, one can estimate, in a rough way, the portions of the recreationists' benefits which are attributable to recreation and intrinsic values.

When this is done, it is found that intrinsic value constitutes about 45 per cent of the total value for recreationists, 100 per cent for the non-recreationists (by assumption), and about 55 per cent for the sample as a whole. If this is a correct reflection of reality, it is a major finding and may have large implications for the future study of benefits from environmental improvement. It was noted earlier that, while the sample of persons interviewed was initially chosen at random, quite a few respondents failed to give usable answers. Any aggregate national benefit estimate based on these data therefore could not be put forward as accurate. Thus, I make such an estimate simply to illustrate that the results of this experiment imply very large values.

There are about 80 million households in the United States. Assume that the sample results imply that to have high-quality recreational waters throughout the country there is an annual willingness to pay of $200 per household. This would imply a total willingness to pay of $16 billion. According to the earlier discussion, this would divide about equally between user and non-user values. At first this might seem out of line with the value of well under the billion dollars that was calculated for recreational fishing. But this is not necessarily the case. Recall that that estimate is for a relatively small *increase* in the nation's fishable waters over the actual conditions of the early 1970s, and that the estimate from the national survey is the value people attach to making and *maintaining* the whole of the nation's fresh waters of high recreational quality where the alternative is almost total degradation of most of the nation's watercourses. In other words, both the baselines and the routes of benefit accrual considered are different in the two studies.

A somewhat closer comparison, though still not a perfect one, is between the survey's reported willingness to pay for an improvement from boatable to fishable water and the largest value found in the fishing study for essentially complete cleanup (in fishing terms) of the nation's *fresh* water—roughly $1 billion. The objective of this experiment was not to produce an accurate estimate of national benefits, rather it was to test the feasibility of using a macro approach to the estimation of water-quality benefits. In that, it succeeded.

TECHNICAL ASPECTS OF FISH CULTURE

ENVIRONMENTAL FACTORS

In a successful aquaculture system there must be both an organismic balance, to produce an optimal supply of natural food at all levels, and a chemical balance, to ensure sufficient oxygen supply for the growth of fish and their natural food organisms and to minimize the build-up of toxic metabolic products.

Chemical balance is usually achieved through organismic balance in waste-fed ponds because the most important chemical transformations are biologically mediated. It is now recognised that depletion of dissolved oxygen in fertilized fish ponds is due primarily to the high rates of respiration at night of dense concentrations of phytoplankton. Romaire *et al.* introduced Equation to cover the factors influencing waste-fed fish pond dissolved oxygen (DO) at dawn:

DO_{dn}	=	$DO_{dk} \pm DO_{df} - DO_m - DO_f - DO_p$
where:		
DO_{dn}	=	DO concentration at dawn
DO_{dk}	=	DO concentration at dusk
DO_{df}	=	DO gain or loss due to diffusion
DO_m	=	DO consumed by mud
DO_f	=	DO consumed by fish
DO_p	=	DO consumed by plankton

Bacterial respiration is not specifically mentioned in this equation but is included in the mud consumption of DO and in the planktonic DO consumption. In a well-managed waste-fed fish pond the DO in the morning should be only a few mg/l whereas in late afternoon the pond should be supersaturated with DO.

Mud respiration probably lowers DO by less than 1 mg/l overnight and a fish population weighing 3000 kg/ha would also lower DO by only about 1 mg/l overnight. Phytoplankton photosynthesis is the major source of oxygen during daylight hours and, during the night, the major cause of oxygen depletion is respiration. It has been estimated that respiration of plankton (bacterioplankton, phytoplankton and zooplankton) can lower pond DO by 8-10 mg/l overnight. By far the greatest proportion of the DO depletion overnight is caused by the respiration of the phytoplankton that develops as a result of the nutrients contained in the waste.

Phytoplanktons provide feed for the largest percentage of fish farmed in Asia. They also exhibit a positive net primary productivity on a 24-hour basis and are net oxygen contributors to a fish pond. The objective in a waste-fed fish pond should be to maintain an algal standing crop at an optimum level for net primary productivity by balancing the production of phytoplankton biomass, in response to waste fertilization, with the grasing of phytoplankton biomass by filter-feeding fish.

Fish mortality in a waste-fed pond can result from at least three possible causes. First, the depletion of oxygen due to bacterial oxygen demand caused by an increase in organic load. Second, the depletion of oxygen overnight due to the respiratory demand of too large a concentration of phytoplankton, having grown in response to an increase in inorganic nutrients, caused by an organismic imbalance. The third possible cause is high ammonia concentration in the waste feed. All three causes of fish mortality have been reported in respect of sewage-fertilized fish ponds. The sensitivity of fish to low levels of DO varies with species, life stage (eggs, larvae, adults) and life process (feeding, growth, and reproduction). A minimum constant DO concentration of 5 mg/l is considered satisfactory, although an absolute minimum consistent with the presence of fish is probably less than 1 mg/l.

Fish cultured in waste-fed ponds appear to be able to tolerate very low DO concentrations, for at least short periods of time, with air-breathing fish (such as walking catfish *(Clarias batrachus)* being the most tolerant, followed in decreasing order of tolerance by tilapia, carps, channel catfish and trout. Reducing phytoplankton biomass to maintain a reasonable DO in the early morning hours might well depress fish growth more than exposure to a few hours of low DO. A wastewater fertilized aquaculture system might occasionally require a stand-by mechanical oxygenation system for use during periods when DO would otherwise be very low. However, if the system is well managed to avoid overloading, this expense can be avoided.

Unionised ammonia (NH_3) is toxic to fish in the concentration range 0.2 - 2.0 mg/l. However, the tolerance of different species of fish varies, with tilapa species being least affected by high ammonia levels. Bartone *et al.* found that satisfactory growth and survival of tilapia was possible in fish ponds fed with tertiary effluent in Lima, Peru when the average total ammonia concentration was less than 2 mg N/l and the average unionised ammonia concentration was less than 0.5 mg N/l, with the latter only exceeding 2 mg N/l for short periods. In ponds receiving large quantities of organic matter, sediments tend to accumulate and release anaerobic breakdown products, such as methane and sulphides, which can inhibit fish growth. Bottom feeding fish, such as the common carp (*Cyprinus carpio),* are most affected by such conditions, especially if the macrozoobenthos disappear.

FISH YIELDS AND POPULATION MANAGEMENT

A wide range of yields has been reported from waste-fed aquaculture systems, for example: 2-6 tons/ha yr in Indonesia, 2.7 - 9.3 tons/ha yr in China and 3.5 - 7.8 tons/ha yr in Taiwan. Although the majority of waste-fed fish ponds stocks carps, research in Peru and Thailand has demonstrated the potential of tilapia for such systems. Management of fish ponds can have a significant effect on fish yields but the maximum attainable yield in practice is of the order of 10 - 12 tons/ha yr.

Increase in weight of small fingerlings stocked in a pond follows a sigmoidal curve. The first phase of growth is slow, so a high stocking density can be adopted to better utilise the spatial and nutritional resources of the pond. Alternatively, this can be achieved by stocking with larger fish having a higher initial weight, following growth in nursery ponds.

Fish yield is positively correlated with the size of the stocked fish at a given stocking density. In South China, tilapia are stocked once a year at rates of either 30g fish and $0.15/m^2$ or 1.3g fish at 2.3 - $3.0/m^2$ stocking density. An increase in weight of fish in a pond leads initially to an increase in yield or production but there is subsequently a reduction in the growth rate of individual fish because of the limitation of natural food production in the system.

The third phase of slow growth in Figure is because the total weight of fish in the pond is approaching the carrying capacity. Intermediate harvesting when the rapid growth ceases, at the end of phase 2, should lead to significant increases in total yield. The high yields of tilapia reported in South China sewage-fed ponds are due to high stocking density and frequent harvesting.

Clearly, the key to achieving high yields in a waste-fed pond is to determine the carrying capacity of the pond, the maximum standing stock of fish.

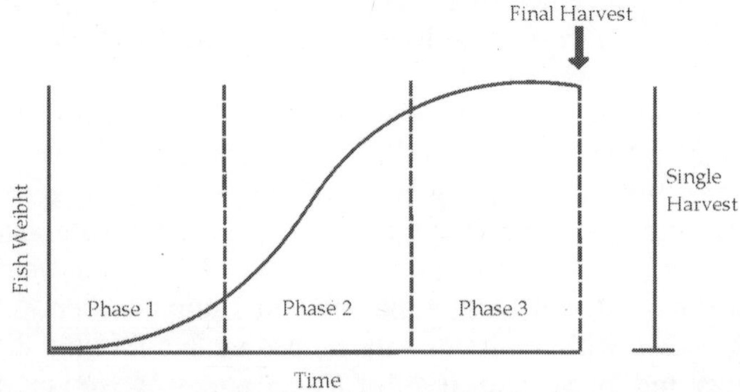

Fig. Fish growth cycle

This can be assessed by varying the waste load and determining the maximum production of natural food consistent with satisfactory water quality, sustainable through a fish culture cycle. Fish stocking density is related to carrying capacity according to the desired weight of individual fish at harvest, as follows:

Fish stocking density,

$$\text{(number/ha)} = \frac{\text{Carrying capacity (kg/ha)}}{\text{Harvesable weight of individual fish (kg)}}$$

Experience has shown that there is a limit to the fish yield attainable from a waste-fed fish pond. Higher yields can be achieved by addition of energy-rich supplementary feed, such as cereals, cereal brans or pelleted-feed. The highest yields are only achieved with a sufficiently high fish stocking density to benefit from the improvement in pond nutrition. There appears to be increased efficiency of utilisation of supplementary feed by fish in ponds receiving sewage effluent. Marketable weights of fish vary with species and local market preferences but, in general, desirable sizes of the following fish range from 0.25-0.6 kg for tilapia, 0.5-1.5 kg for Indian major carps (mrigal 0.5, rohu 1.0, catla 1.5 kg) and perhaps 1-2 kg for Chinese carps. Thus, for a particular carrying capacity, Chinese carps should be stocked at an intermediate density. The length of culture cycle, or frequency of harvesting, depends on the time it takes stocked fish to reach marketable size. It should be recognised that the size of individual fish is only significant if the product is to be consumed by humans. When fish are raised as high-protein feed for carnivorous fish or livestock, size is relatively unimportant. Nevertheless, it is now appreciated that sustainable yields of even high densities of small-size fish with a high specific growth rate are not significantly different from the yield of table fish for human consumption (6.2-7.8 tons/ha yr).

HEALTH RELATED ASPECTS OF FISH CULTURE

Although it is good practice to limit the discharge of toxic materials to sewerage systems, inevitably some of these materials gain access and heavy metals and pesticides are frequently present in municipal sewage. This gives rise to concern about bioaccumulation when sewage effluent is used in aquaculture. Algae are known to accumulate various heavy metals but, with the possible exception of mercury, fish raised in sewage-fed ponds have not been observed to accumulate high concentrations of these toxic substances.

It would appear that the concentrations of heavy metals in the pond water may be accumulated at slower rates than new tissues develop in rapidly growing fish, such as tilapia. In the case of mercury, the position of fish in the food chain seems to be important in determining their mercury uptake, with carnivorous fish accumulating more than herbivores.

Fish, apparently, have the ability to regulate the heavy metal content of their tissues, except for mercury, and tend to accumulate metals in parts other than muscle tissue. There is little information on the uptake of toxics other than heavy metals but a high phenol content in the sewage fed to fish ponds in Wuhon, China caused the fish flesh to become unpalatable due to the odour of phenol. Weis *et al.* have reported on the effects of treated municipal wastewater on the early life stages of three species of fish and indicated that moderately toxic effluent (organic fractions) caused cardiovascular and skeletal defects, depression of heart rate and poor hatching, larval and juvenile growth rates. The health effects of aquacultural use of human wastes in respect of pathogenic organisms have been discussed. Depuration was mentioned as a means to decontaminate fish grown in waste-fed aquaculture. It is generally believed that holding fish in clean-water ponds for several weeks at the end of the growing cycle will remove residual objectionable odours and pathogens and provide fish acceptable for market.

MAJORITY OF FRESHWATER FISH PONDS

RENOVATION OF EXISTING PONDS

The majority of freshwater fish ponds in the Indian subcontinent are the dugout ponds of an undrainable nature which at times lack proper embankments.

During the course of culture operations, such ponds receive huge amounts of feed, fertilizers and manures as critical inputs and sediment particles carried down by rain water from the catchment area. A portion of the organic production in the pond also undergoes death and decay and gradually adds to the pond bottom sediment.

Thus, with the advancement of time, a thick sediment layer is formed reducing the depth of the pond. They are quite rich in organic and inorganic nutrients, but due to slow bacterial action under prevailing anaerobic conditions the nutrients are almost locked up in the sediment and are not available for primary production. Further, the anaerobic decomposition of the organic matter accumulated in the sediment releases harmful gases and depletes dissolved oxygen level in the water. Thus, it becomes necessary to renovate the existing ponds periodically every 4–6 years by removing sediment from the pond bottom, redressing and repairing the dykes, etc., in order to make the ponds more suitable and to regain their fertility. For this, the following practical measures are recommended.

DEWEEDING

It has been observed that most of the rural ponds are not properly managed and become weed-infested in course of time. Before dewatering the pond, large floating weeds such as water hyacinth should be eradicated by pulling them out manually or mechanically. Otherwise, collection and removal of such weeds will require more labour and time. Other rooted emergent or submerged weeds can be taken care of only after draining the pond.

WHEN TO TAKE UP THE RENOVATION WORK

As soon as the water table of the area surrounding the pond goes down the renovation work can be initiated. Summer is the most suitable period for this Purpose as complete drying of the water body is possible. In this period pond renovation can be carried out efficiently and economically. Removal of slushy silt from the partially dried pond bottom is difficult, laborious and expensive.

DEWATERING AND DRYING

Dewatering of the existing pond is possible either by draining the water after cutting a portion of the embankment or by pumping out. If the water table in the surrounding area is high, there is considerable inflow of water from the pond bottom. This phenomenon of sub-surface secretion is called percolation. In case the rate of percolation is high, several furrows or ditches may be made towards the lowest contour point where a pit may be dug out to drain all the percolated wall. Periodical pumping of water from the pit facilitates keeping the bed dry.

DESILTING

After complete dewatering, the pond bed is allowed to dry and develop cracks in the silt mass. The texture of silt is different from that of the bottom hard soil and cracks quickly. At this stage dried silt is cut and removed manually or mechanically and heaped at a suitable place for its utilisation in agricultural fields. Where complete drying is not possible due to high rate of percolation, walking platforms made up of bamboos or wooden planks may be put on the slushy bed to facilitate desilting work.

In some larger ponds it becomes difficult to dry the central portion of the pond bottom as it is nearer to underground water table. In such cases the slushy and loose silt should be scrapped and spread to the sides with the help of wooden planks tied with ropes for pulling. This helps in drying the silt and easy removal thereafter.

CONTOURING

Where the bed is found to be uneven, contouring is necessary to estimate the amount of silt to be removed. Itis done by taking the level measurements at certain spots on the pond bed. It will also help in redesigning the pond taking into consideration the highest flood level and maximum rain water level.

MAINTENANCE OF DYKES

In general, rural ponds lack proper embankments. During high rainfall or peak irrigation periods in canal-irrigated areas such ponds get inundated with water from the neighbouring agricultural fields causing stocked fish to escape, predators and unwanted species to enter and at times results in mass fish kills due to pesticide pollution. Hence, provision of proper dykes is a must. Existing pond dykes should be repaired every year after the monsoon.

Rats and crabs cause great harm to pond dykes by making holes. Such holes allow serious leakage and if not checked immediately, may endanger the stability of the dykes. Periodically and especially at the time of renovation, such spots should be properly repaired by stuffing binding clay, claylime mixture or any other locally cheap cementing material.

Due to poor consolidation, erosion from the top of the dyke during heavy rains usually results in grooving out of small channels. These areas should be covered with earth, levelled, thoroughly rammed and grass turfed. In relatively larger ponds, wave action due to wind also causes large-scale dyke erosion.

By putting large floating aquatic plants such as water hyacinth along the sides of the dykes exposed to wave action during the windy season such erosion can be checked. Frequent erosion in steep dykes during heavy rain or wind can be avoided by strengthening the inner sides of the dykes with poles or bamboos or corrugated cement planks. Most of the traditional pond dykes are below the required height; as a result, overflow of water occurs during heavy rains or flood. These dykes should be properly raised and the height may be kept at a minimum of one meter above the maximum water level recorded in that area. While raising the dykes, the top width may be kept at a minimum of 1.5 m with 2:1 slope (horizontal: vertical). Cutting the dyke to allow water into the pond from the surrounding area without any secured screening is a normal practice, which however creates many management problems. The silt mass is very rich in organic and inorganic nutrients making it most suitable for application in agriculture and horticulture. Being non-cohesive and unstable, it is unsuitable for making dykes as it may be washed back in the pond.

RECLAMATION OF DERELICT WATER BODIES

Derelict waters in millions of hectares, lying unutilised, are common sights in most of the South Asian countries. Such untapped water bodies with potential for aquacultural production may be reclaimed and made suitable for fish culture by adopting more or less similar procedures.

In case of larger water areas, it would be better if they are connected temporarily to nearby natural or man-made drainage systems having relatively lower bed level for complete dewatering by gravity and making the entire area completely dry. However, if such topographic facilities are not available, heavy duty water pumps may be put into use for quicker dewatering.

In extensively large areas dewatering by draining or by pumping is not feasible. Moreover, the dry period of the year also may not last long enough to permit the work to be completed. It has been experienced that such areas can also be successfully reclaimed and renovated by partitioning into smaller units by raising cross bundhs, farm roads, etc. Each newly formed unit then can be dewatered, dried and desilted.

FRESHWATER FISH

Freshwater fish are fish that spend some or all of their lives in fresh water, such as rivers and lakes, with a salinity of less than 0.05 per cent. These environments differ from marine conditions in many ways, the most obvious being the difference in levels of salinity. To survive fresh water, the fish need a range of physiological adaptations. 41.24 per cent of all known species of fish are found in fresh water. This is primarily due to the rapid speciation that the scattered habitats make possible. When dealing with ponds and lakes, one might use the same basic models of speciation as when studying island biogeography.

PHYSIOLOGY OF FRESHWATER FISH

Freshwater fish differ physiologically from salt water fish in several respects. Their gills must be able to diffuse dissolved gasses while keeping the salts in the body fluids inside. Their scales reduce water diffusion through the skin: freshwater fish that have lost too many scales will die. They also have well developed kidneys to reclaim salts from body fluids before excretion.

MIGRATING FISH

Many species of fish do reproduce in freshwater, but spend most of their adult lives in the sea. These are known as anadromous fish, and include, for instance, salmon, trout and three-spined stickleback. Some other kinds of fish are, on the contrary, born in salt water, but live most of or parts of their adult lives in fresh water; for instance the eels. These are known as catadromous fish. Species migrating between marine and fresh waters need adaptations for both environments; when in salt water they need to keep the bodily salt concentration on a level lower than the surroundings, and vice versa. Many species solve this problem by associating different habitats with different stages of life. Both eels, anadromous salmoniform fish and the sea lamprey have different tolerances in salinity in different stages of their lives.

STATUS IN NORTH AMERICA

About four in ten North American freshwater fish are endangered, according to a pan-North American study, the main cause being human pollution. The number of fish species and subspecies to become endangered has risen from 40 to 61, since 1989.

ISOLATED FORM OF ENZYMES

Enzymes have been isolated from every type of living organism. Many of these biological catalysts are significant only from an academic or medical standpoint, but some of the available enzyme from this vast repertoire have been utilised for agricultural and industrial purposes for years. The table below lists several of the industrially on sequential enzymes and their sources in nature. It is significant to note that animals plants and microorganisms all yield industrially important enzymes. Some enzymes of animal or plant origin have been used in agricultural applications; however, those enzymes most broadly used are of microbial origin.

Source of Enzyme

Plant

- Malted grains or tubers Amylase
- Pineapple Bromelin (Protease)
- Fig Tree Ficin (Protease)
- Papaya Papain (Protease)

Animal

- Liver Catalase (Peroxide Breakdown)
- Calf Stomach Rennet/Chymosin (Milk Clotting)
- Hog Stomach Pepsin (Protease)

- Hog Pancreas Pancreatic Enzymes (Several)
- Digestive Tract Trypain (Protease)

Microbial

- Fungi (Molds and Yeast) amylase, beta glucanase, hemicellulase, protease, cellulase, pectinase, lipase, (many types of each), lactase

Bacteria

- Amylase, protease, isomerase, lactase (many types of each), rennet, oxidase, catalase, beta-glucanase, hemicellulase.

One encounters many digestive or hydrolysing enzymes in the digestive tract of human and other animals. These biological catalysts are necessary for the full utilisation of foods ingested. Microorganisms, many being as small as 1/10,000 th of an inch in length, are much too minute to have complicated digestive systems as animals do. Therefore, these microbes must predigest their potential foods outside of their cell boundaries so that they can absorb the very small nutrient compounds of predigested foods. In order to predigest the potential food sources outside their cell boundaries, many microbes excrete enzymes out through their enveloping membrane with its supportive cell wall and into the surrounding environment. Hence these "extra-cellular enzymes" must function in the environment outside the protection of the cell's wall and membrane, they must be reasonably stable and have relatively high resistance to chemicals and must function over a relatively broad temperature range.

To realise the effects of the enzymes they produce, microorganisms also must produce relatively large quantities of these catalysts. All of these factors contribute to the industrial significance and durability of extra-cellular microbial enzymes. It should be noted that most of the agriculturally and industrial important enzymes, are those that catalyse the digestion or "hydrolysis" of certain large organic molecules like starch, cellulose, and protein. The enzymes actually attack these complex molecules, accelerating their digestion and yielding simpler substances. Hence this process of digestion is referred to as hydrolysis, the enzymes that catalyse the process are considered to be "hydrolysing enzymes" or "hydrolases".

The hydrolysing enzymes include:

- Amylases, which catalyse the digestion of starch into small segments of multiple sugars and into individual soluble sugars.
- Proteases, (or proteinase), which split up proteins into their component amino acid building blocks.
- Lipase, which split up animal and vegetable fats and oils into their component part: glycerol and fatty acids.
- Cellulase (of various types) which breaks down the complex molecule of cellulose into more digestible components of single and multiple sugars.

- Beta-glucanase, (or gumase) which digest one type of vegetable gum into sugars and/or dextrins.
- Pectinase which digests pectin and similar carbohydrates of plant origin.

ACTIVITY AND STABILITY OF ENZYMES

Wastes processing levels usually dictate some variation in physical conditions under which the enzyme products must function. In order to utilise enzymes to their optimal potential in catalytic ability, we must be familiar with the basic principles that can affect the activity and stability of these enzymes. Enzymes, being biological compounds and being comprised of a high percentage of protein, are subject to many environmental effects. Although the following principles hold true for most biological enzymes produced for commercial agricultural use: The pH of the environment has a profound affect on enzyme activity and stability.

Activity optimal for pH's of various enzymes vary; however, the optimal pH's for the biological catalysts produced by most commercial strains of microorganisms lies between pH 4.0 and 7.5. This range is from moderately acidic to mildly alkaline in nature. These are the pH levels normally encountered.

Another major affect of enzyme activity and stability is temperature. Hence enzymes are biochemical catalysts, made up at least partially of protein, they are sensitive in varying degrees to heat. Raising temperatures of the environment generally multiplies the degree of activity by the enzyme.

Once an optimum temperature has been reached, however, even higher temperatures cause rapid degradation of the enzyme with concurrent and irreversible loss in activity. Optimal temperatures generally range from 98 °F to 140 °F (37 °C to 60 °C) for most hydrolytic enzymes. High temperatures (over 150 °F, 66 °C) generally have detrimental effects on the enzymes.

However, there is broad variation in resistance and sensitivity to heat among the enzymes' types. Bacterial enzymes such as those from Bacillus subtilis are less sensitive to heat than are the fungal enzymes of *A oryzae.* Some amylase preparations prepared by the fermentation of Bacillus species can withstand even boiling for short periods and have optimal activities in the 158 °F - 176 °F (70 °C - 80 °C) range.

Our laboratory has determined that approximately 85 per cent of the activity from Bacillus subtilis/licheniformis alpha-amylase survives high heat. A. oryzae amylases, however, showed a greater than 90 per cent loss activity in high heat. When the enzyme-bearing, dried fermentation products of these two microorganisms are kept dry, they are much more resistant to environmental temperature stress than if they are moistened. In fact, very few stability problems are encountered with most enzymes in typical situations.

Working System of Bacteria and Enzymes to Digest Organic Wastes

The following discussion outlines the biological process. This process is responsible for the digestion of organic waste, no matter where it occurs.

With minor variations, this same process digest waste in:

- Aquaculture,
- Agriculture,
- Composting,
- Livestock,
- Municipal,
- Industrial Wastewater Treatment

BACTERIAL/ENZYME DIGESTION

Bacterial digestion is the process of bacteria, consuming organic matter. Enzymes act to break the organic matter into water soluble nutrients, which the bacteria digest. Using complex chemical reactions, the organic waste is metabolised down to water and carbon dioxide (the final metabolic waste products), providing the bacteria with energy for growth and reproduction.

It may be simply shown by the following equation:

Aerobic Digestion:

$$\text{Organic waste + water} \xrightarrow{\text{Enzyme}} \text{water soluble nutrients + oxygen} \xrightarrow{\text{Bacteria}} \text{water + carbon dioxide}$$

Anaerobic Digestion:

$$\text{Organic waste + Water} \xrightarrow{\text{Enzyme}} \text{Water Soluble Nutrients} \xrightarrow{\text{Bacteria}} \text{Water + Carbon Dioxide}$$

Organic waste is consumed by the bacteria, used as nutrients by the bacteria, and is no longer present to produce odours, sludge, pollution, or unsightly mess.

SELECTED BACTERIA CULTURES

Thousand of different types of bacteria exist everywhere in our world, and most of them carry on bacterial organic digestion in some way. However, some of them are found only in a particular environment, require specialised types of food, and/or have very unique biological roles (niches). A bacteria is a single cell life of form - each individual cell is a separate, unique organism. Bacteria often grow into colonies that appear as jelly-like masses, but each cell remains as independent. Bacteria reproduce by a process called cell division. A mature bacteria reproduces by dividing into two cells, each identical to each other and the parent bacteria. Under ideal conditions, bacteria can reproduce very rapidly, producing a new generation every 20 to 30 minutes. Following this reproduction process, we see that the number of individual bacteria doubles with each generation. The population explodes as the number of organisms increases logarithmically. This population boom begins soon after the bacteria is introduced into a favourable environment, after a short lag time when the bacteria becomes acclimated to the new conditions.

Obviously, this population cannot increase forever. At some point, the food source will be depleted, waste products will accumulate, or some other change in the environment. Will cause the population to level off or decrease (such as a change in pH, temperature, or oxygen content of the environment). Also, introduction of any poisons into the environment may have negative effects on the population, as well as competition from other types of bacteria.

Bacteria can be classified into different types:

- Aerobic types (which require oxygen to live)
- Anaerobic (which can live without oxygen)
- Facultative types can thrive under both aerobic and anaerobic conditions.

For waste digestion, we can identify several beneficial characteristics that we want our chosen bacteria to have.

The "good" bacteria that we will choose must:

- Consume (digest) a wide variety of organic material that are present in wastes.
- Digest waste quickly and completely, without producing significant odours of noxious gas.
- Not cause any disease in man or animals - they must be *non-pathogenic.*
- Grow and reproduce quickly and readily in the environmental conditions found in waste disposal systems.

Certain bacteria belonging to the Bacillus species have these desirable characteristics. They consume organic waste thousands of times faster than the bacteria that are naturally present in the waste. They grow and reproduce easily, are non-pathogenic, and do not produce foul odours or gas as they digest waste. These "good" bacteria are cultured (grown by artificial means) on liquid or dry nutrient medium. These cultured bacteria are then freeze dried to put them in a state of suspension. They remain alive, ready to swim, eat, and reproduce as soon as they are activated (rehydrated) and put into the proper environment.

The proper environment needed for rapid growth and reproduction of these good bacteria must have these characteristics:

- A water medium containing food (organic waste).
- Dissolved oxygen (for the aerobic types that require it) in sufficient quantities.
- Proper pH—not too acid nor too alkaline—between 6 and 9 on the pH scale.
- Moderate temperature, between 50 °F and 110 °F.

ENZYMES AID IN DIGESTION

An enzyme is a chemical catalyst that breaks up long, complex waste molecules (Hydrolytic Reaction) into smaller pieces, which can then be digested directly by the bacteria. Enzymes are simply chemicals - they are not living things, and they cannot grow or reproduce themselves. Enzymes are manufactured by bacteria, and used by the bacteria in order to digest waste. The enzymes that are mixed into the products are actually produced by special bacteria, extracted from them in dry form, and blended into the mixture.

Enzyme are added to help them go to work faster. When added to the organic waste, the enzymes immediately go to work breaking down the waste into water - soluble nutrients for the bacteria to digest. The enzymes break the large, complex molecules of starches, proteins, carbohydrates, and cellulose into smaller, simpler pieces. These enzymes act like chemical "knives", chopping the large molecules of waste into smaller pieces of water - soluble nutrients for the bacteria. The growing bacteria will then start to produce more enzymes on their own, creating a continuing cycle of enzyme production.

THE ROLE OF BACTERIA IN THE SEWAGE TREATMENT PROCESS

One area of sewage treatment that is not well understood is the bacterial decomposition process. Bacteria may be aerobic, anaerobic or facultative. Aerobic bacteria require oxygen for life support whereas anaerobes can sustain life without oxygen. Facultative bacteria have the capability of living either in the presence or in the absent of oxygen. In the typical sewage treatment plant, oxygen is added to improve the functioning of aerobic bacteria and to assist them in maintaining superiority over the anaerobes. Agitation, settling, pH and other controllable are carefully considered and employed as a means of maximizing the potential of bacterial reduction of organic in the wastewater.

Single-celled organisms grow and when they have attained a certain size, divide, becoming two. Assuming an adequate food supply, they then grow and divide again like the original cell. Every time a cell splits, approximately every 20 to 30 minutes, a new generation occurs. This is known as the exponential or logarithmic growth phase. At the exponential growth rate, the largest number of cells are produced in the shortest period of time. In nature and in the laboratory, this growth cannot be maintained indefinitely, simply because the optimum environment of growth cannot be maintained. The amount of growth is the function of two variables: - environment and food. The pattern which actually results is known as the bacterial growth rate curve. Initially dehydrated products (dry) must first re-hydrate and acclimate in a linear growth phase before the exponential rate is reached.

Microorganisms and their enzyme systems are responsible for many different chemical reactions produced in the degradation of organic matter. As the bacteria metabolise, grow and divide they produce enzymes. These enzymes are high molecular weight proteins.

It is important to recognise the fact that colonies of bacteria are literally factories for the production of enzymes. The enzymes which are manufactured by the bacteria will be appropriate to the substrate in which the enzyme will be working and so you have automatic production of the right enzyme for the biological reduction of any waste material, provided you have the right bacteria to start with. Enzymes do not reproduce whereas as bacteria do.

Enzymes in biochemical reactions act as organic catalysts. The enzymes actually become a part of the action, but after having caused it, split off from it and are themselves unchanged. After the biochemical reactions are complete and products formed, the enzyme is released to catalyse another reaction. The rate of reaction may be increase by increasing the quantity of the substrate or temperature up to a certain point, but beyond this, the rate of reaction ceases to increase because the enzyme concentration limits it.

All treatment plants should be designed to take advantage of the decomposition of organic materials by bacterial activity. This is something you can equate to lower costs, increased capacity, and an improved quality of effluent; even freedom from bad odours which may typically result when anaerobe bacteria become dominant and in their decomposition process, produce hydrogen sulfide gas and similar by-products. Consider the fact that the total organic load of wastewater or sewage is composed of constantly changing constituent, it would be quite difficult to degrade all of these organics by the addition of one enzyme, or even several enzymes. Enzymes are specific catalysts and do not reproduce. What is needed is the addition of an enzyme manufacturing system right in the sewage that can be pre - determined as to its activity and performance and which has the initial or continuing capacity to reduce waste.

At the present time, the addition of specifically cultured bacteria seems to be the least expensive and most generally reliable way to accomplish desirable results. When you add the right bacteria in proper proportions to the environment, you have established entirely new parameters of potential for the treatment situation.

6

Respiration and Importance in Aquatic Insects

One of the major issues in the physiological ecology of aquatic aerobic animals is the availability of oxygen. Compared to atmospheric chemists and other people along those lines, physiologists take a rather simple view of air. We generally consider air is composed of something like this:

Element	Proportion
Oxygen	21 %
Nitrogen	79 %
Carbon dioxide	0.03 %

We can quickly gather a couple of points from this information. First, there are several trace elements in air that are lumped into nitrogen for purposes of discussion. This is acceptable because the quantitatively minor gasses, such as argon, are also inert, and physiologically not different from nitrogen. The second point is that we can note from the percentage composition that a litre of dry air at sea level contains about 210 ml of oxygen. Just from watching animals moving around on land, 210 ml of oxygen per litre of air seems to be plenty of oxygen.

Well, Ok, let us ask how much oxygen is in water? The answer is not quite so simple, even for physiologists. To correctly look at availability of oxygen to animals living in water, we need to consider several points. First, gases are not equally soluble in water. Hence, the solubility of individual gases will partly determine availability of oxygen. Here are a few data on solubility of pure gases in distilled water at 15°C and 760 mm Hg (1 atmosphere of pressure):

Element	Solubility
Oxygen	34.1 ml/litre
Nitrogen	16.9 ml/litre
Carbon dioxide	1019.0 ml/litre

We can use these solubility data to calculate volumes of gas available in water when the water is in equilibrium with air at one atmosphere:

Volume of gas = SOLUBILITY × per cent COMPOSITION

Element	Solubility × % Composition	Volume
Oxygen	34.1 × 21%	7.16 ml/litre
Nitrogen	16.9 × 79%	13.35 ml/litre
Carbon dioxide	1019 × 0.03%	0.31 ml/litre

Now with just a little bit of high-powered mathematics, such as long division, we can calculate that the 7.16 ml/l of oxygen in water is about 30 times less than the 210 ml/l of oxygen in air. We can also calculate the percentage composition of gases in water at 15°C and one atmosphere pressure by adding the three volume values and dividing each by the total:

Element	Volume/Total	Percentage Composition in Water
Oxygen	7.16/20.82	34%
Nitrogen	13.35/20.82	64%
Carbon dioxide	0.31/20.82	0.01%

These values hold for water at 15°C, equilibrated with air at one atmosphere, typically regarded as sea level. The amont of gas in water, however, varies with water temperature, among other physical parameters. The following table will make the point:

This table shows an inverse relationship between water temperature and oxygen contents, that is to say, there is a constraint on oxygen availablity in water due to the influence of temperature. This temperature constraint presents a problem because most aquatic animals, particularly small animals such as insects, are poikilothermic. The body temperatures of poikilothermic animals vary with ambient temperature: as water temperatures increase, so do body temperatures. The problem stems from the influence of temperature on metabolic rates. Metabolic rates are directly related to body temperatures. As metabolic rates increase, oxygen consumption rates increase. The environmental influence of water temperature is seen in body temperature, which influences metabolic rates. These factors intersect at a crucial, yet paradoxical point: increasing oxygen requirements are attended by decreasing availability of oxygen in water. Respiration in aquatic animals, then, is a look at physiological solutions to the problem of obtaining sufficient oxygen to support metabolic needs from an environment where oxygen is in short supply. From the perspective of environmental scientists, we note oxygen is one of the physical parameters of surface fresh water that change markedly in reaction to the various of the polluting chemicals regularly dumped into lakes, rivers and streams. Many aquatic insects and certain of our favourite fresh water fishes can not survive in waters with reduced oxygen levels. The presence or absence of these animals that require higher levels of disolved oxygen can be used to assess the relative environmental health of a water system. Such species are referred to as indicator organisms.

More germane to our topic, however, is understanding that insects and other aquatic animals have potential problems of low oxygen availability. One solution to the problem of obtaining oxygen while living in aquatic environments is to directly breathe atmospheric air. Many insects have respiratory siphons. In insects with this sort of anatomical arrangement only the posterior spiracles are functional, and they are located at the end of a siphon such that only the posterior portion of the insect penetrates the surface film of water, while the rest of the body is submerged, often suspended from the surface film. The respiratory siphon on larvae of drone flies, *Eristalis*, is telescopic, not meant to suggest one can see through it. Mosquito larvae also have respiratory siphons, but they are not so long as in drone flies.

One constraint on the function of respiratory siphons is that the tips of siphon tubes must be arranged such that air can move into the siphon while water is excluded. Commonly, siphons have hydrofuge hairs associated with their spiracles. When their spiracle is submerged the hairs close over the spiracle and water is kept out of the siphon. At the surface, the hairs are separated by surface tension forces, allowing air to pass to and fro within the siphons. There is typically a funnel shape at the surface.

Respiratory siphons have certain shortcomings. The siphons tend to limit insects to the upper levels, or strata, of water, so that they are close to the surface. Insects require other adaptations to achieve some independence from the surface strata. Some insects manage to exploit deeper reaches of water by carrying air stores along with them during their dives. This strategy is well developed in the predacious diving beetle *Dytiscus*. By keeping air stores in contact with spiracles during a dive, these beetles have their trachael systems plus air stores to draw upon to meet oxygen requirements.

Spiracles are typically located along the sides of many terrestrial insects, but this arrangement does not help much with carrying supplies of air under water. Many beetles have their abdominal spiracles on the dorsal surface, under the elytra. This needs be only slightly adjusted for carrying air stores during diving. Many other insects, especially aquatic true bugs also carry air stores under water.

The efficacy of the air store is vastly improved because the bubble itself can act as a physical gill. That is, it can extract oxygen directly from the surrounding water. The following work on the backswimer *Notonecta* illustrates the point. When this bug was held under water that had been equilibrated with pure nitrogen, it survived for only 5 minutes. When the water was equilibrated with air the insect could survive under water for 6 hours, 72 times as long. These two observations are consistent, and they can be understood with respect to extraction of oxygen from the water. The importance of nitrogen in the air bubble can be shown by the following experiment. If *Notonecta* is given a gas bubble of pure oxygen, instead of an air bubble, the bug can survive for about 35 minutes under water. The role of nitrogen in performance of a physical gill can be explained as follows:

Soon after a bug dives, a disequilibrium occurs. Removing oxygen from the bubble by metabolism decreases the partial pressure of oxygen compared to water and increases the partial pressure of nitrogen. Oxygen tends to move down its concentration gradient into the bubble while nitrogen moves down its gradient from the bubble into water. The solubilities of gases become important. Nitrogen has low solubility in water, and it diffuses out of the bubble more slowly than oxygen diffuses from the water into the bubble. The nitrogen is essential to the bubble as physical gill because without it there would be no change in partial pressure of oxygen; the differential partial pressures drive oxygen from the water into the bubble.

This system can provide enough oxygen to small, inactive insects to allow them to remain submerged for months. More active insects require more oxygen. The diving beetles have to surface every few minutes to replenish their oxygen stores. In addition, certain physical parameters assert themselves. For one, as we saw earlier, water temperature effects the amount of oxygen that can be dissolved in water. At higher temperatures there is less oxygen in water, and at the same time an insect requires, and consumes, more oxygen.

A physical gill may be of small value under such circumstances. Another factor impacting on air bubble life spans is depth in water. For every meter of depth in water there is about a tenth of an atmosphere increase in pressure. Increased pressure speeds diffusion out of air bubbles, and shortens their useful life span. Here is a summary of several factors that impact on the useful life spans of air bubbles, and thereby determine how long an insect may survive under water:

- Temperature of water
- Amount of O_2 in water
- Metabolic rate
- Size of bubble
- Depth of dive

Now, when an insect dives with an air bubble, nitrogen slowly diffuses into the water, and the bubble becomes progressively smaller until it collapses. The collapse is delayed because nitrogen moves into water only slowly, but sooner or later, air bubbles will collapse. Insect using air bubbles as physical gills must return to the surface. If an air bubble were non-collapsible, many aquatic insects would not have to surface. Some insects, such as the water bug *Apohelocheirus* are equipped with a plastron, which holds a non-collapsible thin film of air along the outer surface of the body. A plastron is continuous with the trachael system. A plastron is a dense (up to 2.5 million hairs/ square mm of surface area) layer of stiff hydro-phobic hairs. Plastrons typically run along the body surface, continuous with the spiracles. There are cuticular openings which lead into air channels, which in turn lead to spiracles and trachael systems. The hydrophobic quality of hairs along the air channels serve to keep water out of trachael systems. The most efficient arrangement of plastron hairs would be to have them parallel to a body surface. Insects approximate this by having the tips of plastron hairs bent, or else they slope as some angle, so that air films resist collapse when subjected to higher pressures. Plastrons function in the same patterns as air bubbles when they are extracting oxygen from water, that is to say, they act as physical gills. Plastrons are found on surfaces of many insect eggs and on many species with aquatic pupal stages.

We have seen two approaches to organismal respiration in aquatic environments. One is to breathe air directly by way of siphons and another is to extract oxygen from water using a sort of physical gill. Several groups of insects express a third approach, which is to develop a true physiological gill to extract oxygen from water. The term gill, with no prefix such as "true" or "physiological" is sufficient because other sorts of gills are linked to an appropriate modifier, such as anal gill or physical gill. Gills are appendages for gas exchange, and tracheae branch into these appendages. Oxygen diffuses from water into tracheae of gills, and from there moves under the usual motivation by way of trachael systems to tissues.

The locations of gills varies in larvae of several insect species. Stone fly nymphs, *Pteronarcys,* have gills on the thorax and first two abdominal segments. Gill plates are located at the end of abdomens of damselfly larvae. Finally, mayfly nymph have gills on thoraces and abdomens. Just in passing, the gills of mayfly nymphs are equipped with nervous and muscular connections. The gills can be swept back and fro to mix water, and thereby alleviate the problem of local oxygen depletion. This arrangement is the source of the tracheal gill theory in arguments over the evolution of insect flight.

Gills extract oxygen from surrounding water. Due to its low diffusion rate in water, oxygen can be depleted from water immediately local to respiratory surfaces. This local oxygen-poor water must be replaced by fresh, oxygen-rich water. Moving water for this purpose is called ventilation of the gills. Some insect ventilate their gills by selecting a microhabitat where water is flowing. Still others ventilate by muscular action.

This can require considerable energy expenditure because water has greater density and viscosity than does air. The oxygen costs of ventilation must be more than offset by oxygen gains. Work on caddis fly larvae illustrates the point. In the experiments, metabolic rates of caddis fly larvae increased when water current velocities decreased. These data suggest the larvae can somehow register changes in oxygen availability, then respond by ventilation. You can try it yourself. Hold your head underwater for some time.

Insects use a variety of methods to ventilate their gills. Mayfly nymphs beat their abdominal gills in the absence of water movement. Stonefly nymphs ventilate by doing sort of "push-ups" at a rate related to amounts of oxygen in water. Larval dragonflies combine ventilation of gill surfaces with locomotion.

Dragonfly larvae aspirate water into their rectums, which feature rectal gills. These larvae can locomote by jetting themselves forward. They do this by rapidly squirting aspirated water back out of their rectum. Damselfly larvae use their abdominal gills as paddles to propel themselves while simultaneiously ventillating their gills. In vertebrate respiratory systems the respiratory pigment hemoglobin serves three functions. First, it withdraws oxygen from the air at respiratory surfaces in lungs. Second, it transport oxygen throughout bodies. Third, hemoglobin releases oxygen to tissues at controlled rates. Some insects also contain hemoglobin molecules, but the role of insect hemoglobin is storage, rather than transportation, of oxygen. Insect hemoglobin is about half the molecular weight of vertebrate hemoglobin, and it has a much higher affinity for oxygen. That is, it can be saturated at much lower oxygen tensions. Finally, insect hemoglobin is disolved in hemolymph; it is not carried in blood cells.

The larvae of certain midges living in anoxic mud have hemoglobin in their hemolymph. To emphasize the point, these larvae tend to look sort of pink when they are well oxygenated. They live in burrows, in which they periodically wiggle, setting up currents of water. The fresh water contains oxygen, which is taken up by the insect hemoglobin. The hemoglobin can store enough oxygen to support about 9 minutes of metabolism. Two aquatic hemipterans, *Buenoa* and *Anisops*, also store use hemoglobin to store oxygen. Part of their oxygen stores for dives is in a small bubble, the rest is in hemoglobin. Their bubbles are small enough the insect can have neutral buoyancy and remain suspended in water while awaiting prey. It has been suggested that with larger bubbles they would not be able to suspend themselves in water, and without hemoglobin they would have to surface much more often. In this case, the hemoglobin seems to be an adaptation for predation in water.

Finally, another adaptation for obtaining oxygen while remaining submerged is to extract oxygen from aquatic plants. Some insects do this by thrusting their spiracles into the aerenchyma of plants; their spiracles are located at the tip of a sharp-pointed post-abdominal siphon. Larvae of two dipterans, *Chrysogaster* and *Notiphila* do this, as well as larvae of the beetle *Donacia*. These insects live in oxygen-poor mud.

RECTUM OF INSECTS

The Malpighian tubules and rectum of insects form a physiological complex that serves as the functional kidney at the organismal level. A primary urine is formed in the Malpighian tubules, followed by movement to the rectum where selective resorption of ions and water can take place. These secretion and resorption processes are responsible for maintaining osmotic balance between the intracellular and extracellular compartments.

In addition to maintaining osmotic balances, most organisms must also metabolize and eliminate the ammonia nitrogen that is formed by routine amino acid metabolism within cells. The normal dietary intake of protein by most animals is considerably different from ordinary protein needs. Typical diets provide amino acids in excess of the amounts required for synthesis of new protein, and often in quite different compositions. Cells are nearly always in the processes of protein turnover.

This may not be an intuitively clear situation. We most often read about the nutritional importance of amino acids and proteins with respect to satisfying needs for essential amino acids. The point to be taken, however, is that dietary supplies of amino acids are rarely in optimal balance for protein synthesis within individual cells. Considerable levels of interconversions and other metabolic alterations of dietary amino acids follow from these imbalances between dietary supplies and instantaneous amino acid needs. As in regulation of salt balance, it might be thought that insects regulate amino acid balance at the level of the alimentary tract, then just excrete unnecessary amino acids. There are cases where this sort of regulation does occur, but for the most part amino acids are transported across the midgut epithelium into the hemolymph, and thence to various body cells where diverse metabolic fates ensue.

The protein amino acids have a backbone of two carbon atoms and at least one nitrogen atom. Amino acids are distinguished by an R group, where R stands for some side-chain that gives the individual amino acid its identity and particular chemical properties. Once the alpha-amino group is removed from the amino acid, by a metabolic process known as deamination, a carbon skeleton remains. The carbon skeleton can undergo a number of metabolic fates.

Possibilities include reamination, thereby becoming an amino acid again, and, alternatively, entering pathways of oxidative metabolism, mainly at various points in the glycolytic and Kreb's pathways. By entering these metabolic pathways, a potentially important energy source is not lost to the animal. Depending on the energy state of a cell, carbon skeletons from amino acids can be taken into a reverse of glycolysis called gluconeogenesis in which glucose can be made form non-carbohydrate precursors. The ultimate fates of the individual carbon skeletons from amino acids is often the domain of a nutrition course given over to "nutritional metabolism". Most of you will probably be taking such a course next spring.

We want to focus our attention on the alpha-amino group that is removed from amino acids prior to carbon metabolism. All amino acids have an alpha-amino group; some amino acids feature an amino group in the side chain as well as the alpha-amino group in the main skeleton.

Most of the protein amino acids undergo transamination reactions. Here the amino acid alanine reacts with the Kreb's cycle intermediate alpha-keto-gluterate to yield the transamination product glutamate plus pyruvate. The central point is that the amino acid donates its amino group to something else. It does not just drop the nitrogen from the skeleton. The donation makes the remaining carbon skeleton available for other metabolic fates. All but two of the major amino acids are subject to transamination reactions. The next metabolic step is the removal of the amino group, known as a deamination reaction. This step regenerates the alpha-keto-gluterate and releases free ammonia. Just recently, in 1937, it was first suggested that these two events are elegantly combined within cells. The most general expression, which has been given the term transdeamination. First, in the cytosol the amino group is transferred from the amino acid to alpha-keto-gluterate, yielding a carbon skeleton and glutemate. As shown, this reaction most often occurs in cytosol, although there are transaminases within mitochondria. The glutamate then crosses the mitochrondrial membrane where the release of the ammonia group attends regeneration of alpha-keto-gluterate and the reduction of NAD^+ to yield $NADH^+$ H^+.

Please note that we have another example of the biological significance of the vitamin niacin, again playing its essential role in electron transfer. This biochemistry has been most well studied, as usual, in various laboratory mammals. But it has also been worked out in a few of our own sorts of lab animals, namely, in the cockroach *Periplaneta americana* and the fruitfly *Drosophila melanogaster*. It has also been shown in the flight muscles of several other insects. The main point here is to appreciate the metabolic source of ammonia in insects.

A fundamental precept in nitrogen metabolism is that nitrogen is toxic to animal cells. For example, in an atmosphere of 500 ppm of ammonia humans die within 60 minutes. Your basic, off-the-shelf laboratory rat dies within 16 hours in an atmosphere containing 1000 ppm of ammonia. Ammonia tolerance is less well known in insects, but in the blowfly *Lucilia cuprina* the amount of ammonia in the body fluids is only slightly higher than the range of body fluid ammonia found in many mammals. Why is ammonia toxic, at one level of exposure or another, to so many organisms, including fungi, plants and animals? We consider the two leading hypotheses.

One mechanism of toxicity appears to be general pH effects. Some enzyme systems release ammonia in the unprotonated form, NH_3. At physiological pH within most cells, about 7.0 to 7.4, most ammonia takes on a proton to become $NH4^+$. This has the effect of removing hydrogen ions from the cytosol, and creates a local increase in pH, which can in turn negatively effect enzyme activity and membrane function. Please note this line of reasoning will probably not hold as a general mechanism when we realise that in the most common reaction, the glutamate dehydrogenase system we just looked at, the proton comes from the reactants and there is no local pH change. Hence, general pH effects probably account for some but not all of ammonia toxicity.

The main effects of ammonia are probably their direct action on membrane biology. Ammonia salts can inhibit the active transport of sodium and of chloride, and can inhibit water resorption in several tissues. Oddly enough, ammonia does not seem to effect ion transport in nervous tissue. Drawing on our treatment of salt and water balance, it is clear that inhibition of transport in excretory tissues could have very profound effects on physiological function.

In addition to general membrane effects, there is one specific effect. NH4+ directly inhibits the formation of ATP by the electron transport chain within mitochondrial membranes. We can understand this in terms of the driving force behind oxidative phosphorylation in mitochondria and chloroplastal membranes. The electron transport chains are found within the membranes of these organelles. A hydrogen ion gradient is maintained across these membranes; the hydrogen ion gradient is thought to drive formation of ATP. It is thought that ammonia moves across these membranes, which abolishes the hydrogen ion gradient necessary for ATP formation. This has yet to be definitively proven, but it will most likely turn out to be the most direct mechanism of ammonia toxicity in cells. The specific mechanisms differ among toxic compounds, but oxidative phosphorylation is interrupted by a number of poisons, including carbon monoxide and cyanide as well as ammonia.

We have, then, some idea of how ammonia arises is general metabolism within cells and why it is poisonous to animals. Our next concern will be how ammonia is detoxified so that it is not poisonous while it is in an insect body. There are two major pathways, and several minor ones that we will not worry about. The first pathway is the formation of urea, a very important pathway in mammals. This is a relatively minor pathway in insects. The second pathway is formation of uric acid, which is very common in birds, reptiles and insects. In point of history, the uric acid pathway was first worked out using avian liver slices.Ammonia is linked to carbon dioxide and a phosphate at the expense of energy in the form of ATP. This new nitrogen compound, carbamoyl phosphate, joins with the non-protein amino acid ornithine to form citruline. The citruline combines with another amino acid, aspartate to form the larger molecule arginosuccinate. This large compound breaks down into the Kreb's cycle intermediate fumerate and the amino acid arginine. Splitting arginine releases urea and regenerates ornithine.

Here are the key points to remember:

- Urea is formed by a cyclic pathway called the urea cycle and
- The urea cycle is linked to the Kreb's cycle by shared intermediates

This cycle occurs in mitochondria and part of it occurs in cytosol. As with all the pathways presented in this course, you are urged to resist the temptation to memorize the steps. Instead, please note that ammonia enters the cycle in the mitochondria and that urea exits the cycle in the cytoplasm. This simply reiterates a common theme of recognising the important pathways, knowing where they are located within cells, and knowing what chemicals are taken into the pathway and what chemicals are released.

The second major pathway for detoxification of ammonia is formation of uric acid. The glutamate is transdeaminated to yield a free ammonia molecule. In this case the ammonia is joined to a second molecule of glutamate, thereby producing the common protein amino acid glutamine. The glutamine effluxes from mitochondria to cytosol, wherein a complex pathway leads to xanthine, which is oxidized to uric acid by xanthine oxidase, the terminal enzyme in biosynthesis of uric acid from glutamine. There are four nitrogen atoms in the ring structures of uric acid. These all derive from various amino acids: glycine contributes N-7; aspartate donates N-1 and the last two nitrogens, N-3 and N-9, derive from the amide nitrogen of two gluatamine molecules. The amount of energy that is invested in biosynthesis of uric acid is five ATPs per uric acid molecule. There are two parallels with the urea cycle.

- The free ammonia originates inside mitochondria and
- The detoxified end-products are released in cytosol

Pulling together, we have looked at the major sources of ammonia, at why it is toxic to cellular functions and how it is metabolized into non-toxic forms. Our next point is to consider excretion of nitrogen. The most general statement is that urea and uric acid diffuse out of cells into circulating hemolymph. Uric acid and urea are secreted into lumens of Malpighian tubules, possibly by active transport processes, and follow the now familiar anatomical route to excretion. Let us consider environmental correlates with respect to nitrogen metabolism.

Fishes and many aquatic insect larvae live in a situation where water tends to enter a body and needs be secondarily removed to maintain optimal osmotic balances. These animals generally excrete ammonia *per se* without further metabolism. Such animals are called ammonotelic. The vast majority of nitrogen in the excreta of several aquatic larvae is in the form of ammonia. Following release of ammonia within mitochondria, ammonia diffuses out of cells, and into hemolymph. The ammonia probably enters Malpighian tubules with the water that is constantly removed from hemolymph.

Since it takes considerable amounts of energy to metabolize ammonia to uric acid or to urea, an advantage in terms of energy economy attends direct elimination of ammonia. There are, however, some general requirements for the ammonotelic strategy. First, direct elimination of ammonia requires simultaneous elimination of substantial amounts of water, so ammonotelic animals are mostly aquatic or amphibious. Second, these insects have relatively low metabolic rates, and hence, low rates of ammonia production. Finally, ammonotelic insects probably have a higher tolerance of tissue ammonia than do many terrestrial insects. Higher ammonia tolerance is certainly true of fishes.

Most terrestrial insects are what we call uricotelic. The vast majority of their excreted nitrogen is in the form of uric acid. Uric acid can diffuse from cells into hemolymph, whence it is taken up into Malpighian tubules, most likely by active transport, although some textbooks say it is by passive transport.

The uric acid moves towards the alimentary canal along with the flow of water. Depending on the histological structure of Malpighian tubules, the proximal portion of the tubules may be involved in the resorption of some water and salts.

When the water is resorbed, either in proximal portion of Malpighian tubules or in rectum, the uric acid precipitates out of solution and the uric acid crystals can be seen in hindgut or rectum. Since uric acid crystals fall out of solution in water, the water can be resorbed and the ammonia metabolite excreted with minimal loss of water. Here, then, is the main environmental correlate with uricotelism: in those insects (and other animals) where water conservation is important, the toxic ammonia can be ultimately excreted in a form that does not require loss of body water.

A strategy called storage excretion occurs in some insects. Since uric acid is a harmless repository of ammonia, and it can be stored in the body. *Periplaneta* and *Apis* have specialised fat body cells called urate cells. Uric acid accumulates in these cells, often to the point that it can be seen under dissecting microscope. Uric acid also accumulates in non-specialised fat body cells of other insects, for example the mosquito *Culex*. Finally, uric acid may be involved in the coloration of some insects. The white wings of *Pieris* butterflies is due to uric acid storage in the wing scales. Urea makes up about 10 per cent of the excreted nitrogen in larval mosquitoes, and is also found in the excreta of other insects. It is much more solubile than uric acid and more water is required to eliminate urea in feces. We use the term ureotelic for animals in which urea makes up a large proportion of the excreted nitrogen. In general, insects can not be called ureotelic because urea makes up a relatively small proportion of their excreted nitrogen. Non-etheless, the pathway for urea synthesis is present in many insects.

In general terms, most insects can be easily classified as ammonotelic or uricotelic. Chapman tells us that most insects do excrete minor proportions of their nitrogen as other than the main excretory forms, but for the most part, they will fit into the general scheme. On the other hand, a good number of insects have adapted strategies that are outside the main line. For example, the pyrrhocorid *Dysdercus* excrete most of their nitrogen as allantoin. Allantoin is a single metabolic step beyond uric acid.

Still other insects excrete a good bit of amino acids without further modification, after first absorbing them. In tsetse, arginine and histidine from the blood of hosts are absorbed and excreted without metabolism. This may be quite adaptive because these compounds are relatively high in nitrogen. Arginine, for example, has a total of four nitrogens per molecule. These nitrogens would require considerable amounts of energy to metabolize along the normal pathways. The required energy is greater than the amount of energy that could be derived from the carbon skeletons.

ECOLOGICAL IMPORTANCE OF AQUATIC MACROINVERTEBRATES

When one looks out on the St. Lawrence waterway, they normally see large container ships passing by. However, the traffic below the surface is far greater, as it is teeming with life. Although they constitute a large part of this underwater community, aquatic macroinvertebrates are often overlooked. Despite this, they represent a key link between primary producers and other consumers, breaking down organic matter and making it available for other organisms. Aquatic macroinvertebrates are also greatly affected by watershed conditions, making them ideal bioindicators for pollution and other changes in water conditions.

Aquatic macroinvertebrates form a rather large group of organisms, which includes representatives from multiple phyla, such as arthropods, annelids, and molluscs. The aquatic arthropods, when considering freshwater species, are comprised of multiple orders of insects, spiders, and crustaceans. Insects constitute the most numerous arthropods, and orders such as Odonata (dragonflies and damselflies), Trichoptera (caddisflies), Plecoptera (stoneflies), Coleoptera (beetles), Diptera (flies), Ephemeroptera (mayflies) and Hemiptera (water bugs) are common inhabitants of freshwater bodies. Annelids (worms and leeches), though less represented, are also present in freshwater ecosystems. Molluscs are represented in freshwater ecosystems mainly by the gastropod class. Most aquatic gastropods are snails and are found near the substrate or on supporting structures like rocks. Aquatic freshwater invertebrates can live in nearly any freshwater body, with the exception of very highly polluted or deep ground waters bodies. Aquatic macroinvertebrates can be divided into two different groups: benthic and nektonic organisms. Benthic organisms are those organisms that live near the substrate. Leaf litter material and other organic matter that falls to the substrate significantly affect these organisms. Nektonic organisms reside in the water column above the substrate, swimming freely. These organisms are generally more affected by factors such as water current.

Aquatic macroinvertebrates play a crucial role in nutrient cycling, decomposition and translocation of materials. Watson-Ferguson *et al.* Described five major functional groups of aquatic macroinvertebrates: shredders, collectors, scrapers, piercers, and engulfers. Each occupy a particular niche within an ecosystem. Shedders, like some caddisfly larvae, feed on coarse organic material such as wood and leaf litter and create fine particles, which are readily available for collectors to feed on. Collectors, such as mayfly larvae, are not limited to eating these particles, as they also feed on bacteria and waste from other organisms. Snails and water pennies are scrapers, which are adapted to feed on algae and plant matter attached to substrate. Piercers, such as water scorpions, obtain food by injecting their prey with a digestive enzyme, which turns the internal organs of prey into a consumable liquid. Engulfers, which include dragonfly larvae, feed on other invertebrates, and can even consume small fish. Many insect larvae undergo metamorphosis and leave the water as adults, translocating nutrients to terrestrial ecosystems.

Understanding the habitat preferences of aquatic macroinvertebrates can aid in the preservation and monitoring of aquatic habitats. Their large taxonomic diversity has allowed aquatic macroinvertebrates to adapt to a wide variety of environmental conditions and factors. The type of river habitat being studied can be used to predict the traits of macroinvertebrates within them, as well as the taxonomic composition present. Factors such as pH, elevation, water depth, and substrate type affect how different organisms adapt to their aquatic habitats. For example, in sand and silt substrate, deposit feeders are responsible for detritus, whereas it is done by filter feeders within water bodies containing high amounts of submerged vegetation. Dragonfly larvae are an example of aquatic macroinvertebrates that can be used as biological indicators for changes in aquatic habitats. Knowledge concerning these adaptations can provide important information about habitat conditions. For example, dragonfly abundance increases with a higher amount of aquatic plant density. Dragonfly larvae use plants for a number of purposes, such as for the protection of larvae from predators. Additionally, the pH of an aquatic habitat has been shown to affect dragonfly larvae abundance, with more acidic habitats decreasing the amount of dragonflies present. With knowledge about how aquatic factors affect dragonfly abundance, habitats can be monitored for changes in conditions, such as pH, by observing changes in dragonfly populations.

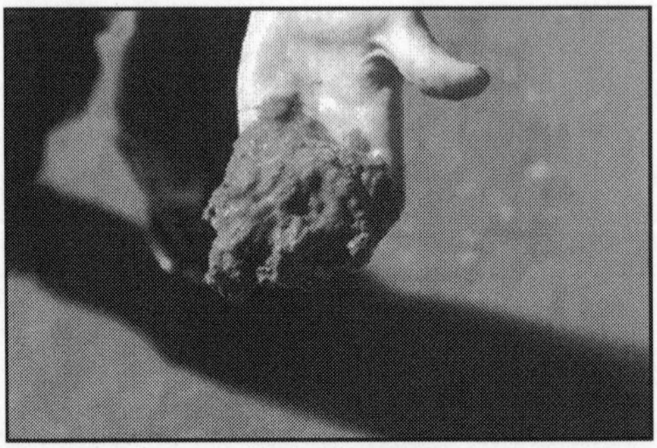

Fig. Different Types of Substrates Lead to different Macroinvertebrate Diversities.

Currently, researchers are investigating the effects of aquatic macroinvertebrates on other species, as well as on their environment. Many are investigating the potential use of predatory invertebrates as biological controls for mosquito populations. This includes studies by Banerjee *et al.* and Saha *et al.*. Downing and Leibold found that increased macroinvertebrate diversity accelerated the recovery rate of ecosystems following environmental shocks. Additionally, the potential impacts of new nanotechnologies on aquatic ecosystems are being studied. Pokhrel and Dubey researched the effect of silver nanoparticles on the predator-prey relationship of *Daphnia magna* and larval dragonflies and found that these nanoparticles reduced the ability of the Daphnia to detect predators.

Though aquatic macroinvertebrates are physically small, they are a major link in the energy exchange from producer to consumer. The biodiversity of aquatic invertebrates within watersheds is an important indicator of the ecosystem's function. Knowing the types of aquatic invertebrates present in a watershed allows for proper environmental assessments to be made.

Benthic organisms are especially important for this, as they reside on the substrate, which is where debris and pollutants often collect. Having a detailed natural history of these species is important for seeing the changes happening in the populations. With these kinds of records, changes in diversity caused by pollutants or litter collection can be correlated more easily. Aquatic macroinvertebrates play important roles within their ecosystems, while also providing valuable data for humans to use.

OSMOREGULATION AQUATIC ENVIRONMENTS

One perspective on the physiology of organisms is to regard the question of how organisms are physiologically adapted to live and thrive in the particular place they happen to occupy. Elimination of ammonia in the form of uric acid, for example, is regarded as one of several physiological or biochemical mechanisms that help insects, as well as reptileS and birds, make their living in very arid environments. We will address nitrogen metabolism later in the series. The ability to maintain thoracic temperatures well above ambient temperatures is another physiological adaptation that enables many insects to operate in thermal regimes that would otherwise restrict activity.

Physiological ecology directs attention to those areas of physiology that can be perceived as a sort of template, variations upon which can convey special adaptations to environment. These areas can be seen in contradistinction to other areas of physiology, ones relatively fixed across a wide range of phylogeny and environment.

The physiology of protein targeting is a good example of a process seen in most eukaryotic cells. The study of protein targeting would not fall under the mantle of physiological ecology. Aquatic insects share with their terrestrial counterparts the problem of maintaining an optimal osmotic balance with respect to their intracellular and extracellular body fluid compartments. We review the potential effects of osmotic imbalance. Cells placed in an environment of far higher salt concentration than themselves often lose water, becoming shrunken and deformed. To the other side, cells placed in an environment of far lower salt concentration than themselves often gain water, even to the point of bursting. Again, maintaining an osmotic balance is important to survival of organisms.

The situation for aquatic insects is easy to apprehend. Insects living in fresh water are generally hyperosmotic to their environment. That is, extracellular compartments of aquatic insects have higher osmolyte (ions and amino acids) concentrations than the surrounding water. Under this condition water tends to diffuse into aquatic insects. If left unattended this process would effectively place cells in a very dilute environment. Insects living in salt water, such as brine flies, are generally hypoosmotic to their environment. These insects tend to lose water to the environment. Again, if left unattended, this would put the cells within the insect in a state of water loss and shrinkage.

Equipped with a feeling for the nature of the problem, let us begin with a consideration of fresh water insects. The hemolymph, or extracellular compartment, of these insects is hyperosmotic to the water comprising their habitats. Water tends to pass through the cuticle and into insects. Some adult forms, such as the predatory beetle *Dytiscus*, have cuticles relatively impermeable to water. However, the majority of aquatic insects in their larval stages are quite permeable to water. Often respiratory gill structures are even more permeable to water, and additional water is taken in with food. We gather that fresh water insects have a problem of taking in too much water. The relatively large amounts of water invading fresh water insects are removed from their bodies by formation of copious amounts of dilute urine. What we want to know, then, is how the urine is produced.

The timing of this lecture is designed to provide a review of the general information seen just lask week, while adding a bit more information on osmoregulation generally. We begin by revisiting the structural features of the insect renal system. Malpighian tubules arising from the alimentary canal. Again, these are thin, close-ended tubes floating free in the hemocoel where they are continuously bathed in hemolymph. A single layer of cells is attached to a basement membrane; often muscles spiral about the outside of this membrane, and the tubules can wiggle within the hemocoel. Small tracheae run along the length of the tubes, which supply oxygen to mitochondria located in muscles and in Malpighian tubules. Fluid within the lumen Malpighian tubules is isomotic to the hemolymph, but of different composition. Hemolymph is generally high in sodium, chloride and amino acids, while the intratubular fluid is high in potassium. The general expression is that potassium is actively pumped from hemolymph to tubule lumens and water passively follows the ion movement. Recall there are homogeneous and heterogeneous Malpighian tubules. They differ in resorptive properties, but the general patterns apply to both types of Malpighian tubules. The flow of water into Malpighian tubules increases local hydrostatic pressure within the tubules, resulting in a bulk flow of fluid. This is the primary urine flow. The primary urine is moved into hindgut, then to rectum. After selective resorptive processes in the rectum, the fluid is voided into the fresh water environment. Then we can imagine water moving from outside insects into their bodies by crossing their cuticles, and in some inseccts, the gills. The water is transported to rectums by way of Malpighian tubules at considerable expense of metabolic energy. The energy cost is related to actively pumping potassium ions. The active pumping processes also help us understand the many tracheal connections to Malpighian tubules.

Let us allow ourselves a slight diversion for just a minute. We want to be sure our language is clear. When we consider any two fluid compartments, three and only three osmotic conditions can exist. One compartment can be more dilute than the other, or hypo-osmotic. A compartment can be exactly the same as the other compartment, or isosmotic. Finally, one compartment can be more concentrated than the other compartment, that is, hyperosmotic. You will recall from your mathematics courses that any number can only be less than, equal to, or greater than another number. There are no other possibilities. Then what we need to keep in mind is that an osmotic situation of one compartment is always relative to another. Hence, fresh water is likely to be hypoosmotic to an organism and the organism would be hyperosmotic to the fresh water.

We are considering insects living in fresh water. In this situation ions inside insects tend to move down their concentration gradients to their relatively dilute, or hypoosmotic environment. There is a general requirement that this loss of ions be offset. More immediate to our story, however, is the observation that if the potassium ions in the urine were not somehow recovered, an insect would soon be unable to continue excreting excess water. Why would this be so? Because the movement of excess body water follows passively from the active movement of potassium ions. The physiological solution, as you already well know, is that the complete excretory system of insects includes the rectum. Secretion is by way of Malpighian tubules and reabsorption by way of rectum.

Here are some data from larvae of the trichopteran *Limnophilus* reared in a salinity equivalent to 0.01 per cent NaCl. The osmolarity of the extracellular compartment and of the fluid within the Malpighian tubules is about 0.031 per cent NaCl equivalents. The osmolarity of the rectal fluids is about 0.009 per cent NaCl equivalents. Similarly, in the fresh water mosquito larvae of *Aedes aegypti* we find something like 0.63 per cent NaCl equivalents in the extracellular compartment and 0.57 per cent NaCl equivalents in the fluid within the Malpighian tubules. Again, the rectal fluid is considerably different at 0.07 per cent NaCl equivalents. In both of these exemplars the rectal fluids are very dilute relative to the Malpighian tubules and hemolymph.

What can we make of this information? The general idea is potassium is actively pumped into the lumen of Malpighian tubules. Water accompanies this movement, along with some amounts of other salts, and there is a concomitant bulk flow into hindgut and thence to rectum. Before the water is voided, salts are again actively pumped back into the extracellular compartment by specialised regions of rectum. This second pumping of ions produces a very dilute urine which can be voided while ions are kept within bodies. Recall that salts are lost across cuticles and, to lesser extent, in urine. The physiological consequence of losing body salts is similar to gaining excess water: extracellular compartments become too hypo-osmotic with respect to intracellular compartments, thereby presenting a deleterious osmotic imbalance between the two. Insects living in fresh water, then, actually have two problems. First, they must produce copious amounts of dilute urine so that extracellular fluid compartments are not diluted by incoming water. Second, they must somehow replace salts that are inevitably lost, also to keep extracellular compartments from becoming too dilute. Some salts will be gained from food, but in addition to this, some aquatic larvae are able to take up salts from very dilute solutions in the water they live in. Larvae of the dipteran genera *Aedes*, *Culex* and *Chironomus* take up salts though specialised structures called anal papillae, also known as anal gills. This uptake requires energy expenditure. The enery is exerted in pumping sodium, potassium, chloride and phosphate ions against gradients from dilute fresh water to hyperosmotic extracellular compartments. The anal papillae hypertrophy as media become more dilute, presumably giving greater surface area for transport of salts. Some insects are able to extract ions from very dilute solutions. For example, larvae of the yellow fever mosquito *Aedes aegypti* are able to maintain steady osmotic states in media containing only 1 X 10–6 moles per litre of sodium. Other aquatic larvae, such as the neuropteran *Sialis*, have no specialised surface for salt uptake. These larvae produce a urine very low in salts, but higher in other osmolytes, mostly ammonium carbonate.

As shown here:

Table. Ionic Composition

Fluid	Na	K	Cl	NH_4	HCO_3
Hemolymph	10	5	31	- -	- -
Urine	12	4	0	106	91

In addition to their ability to resorb salts in rectums, these insects are able to compensate for osmotic imbalance by maintaining higher levels of hemolymph amino acids. These amino acids serve as osmolytes and thereby help maintain the crucial balance between the major fluid compartments. This emphasizes the point that osmotic compositions of the two major fluid compartments can be quite different while osmotic concentrations can be quite similar. In contrast to fresh water insects, those living in salt water face problems more like terrestrial insects. Their hemolymph is often hypoosmotic to their media and they tend to lose body water across their cuticles. We know that this can lead to extracellular fluid compartments that are hyperosmotic with respect to intracellular compartments. Water lost to hyperosmotic environments can be replaced by controlled drinking of saline media followed by physiological uptake of water across midgut epithelia and from rectums, yielding a very concentrated urine. Salt water ingestion can disrupt midgut cells in fresh water insects, but salt water insects can drink saline water without injury.

Here are some data to clarify the point:

Table. Equivalent % NaCl

	Urine	Hemolymph	M.t.
A. aegypti (dH_2O)	0.07	0.63	0.57
A. detritus (sea H_2O)	3.50	0.93	1.20

These are data from larval mosquitoes. Although both species are in the same genus, *Aedes*, one species lives in fresh water and the other lives in brackish water. What we might see here is that the hemolymph in both species is fairly similar in osmotic concentration, while the urine of the salt water species is much more concentrated than the urine of the fresh water species.

It is appropriate to summaries the basic physiological adaptations to aquatic life:

Table. Aquatic Habitat Type

Parameter	Freshwater Habitat	Saltwater Habitat
Ionic concentration	Hypoosmotic	Hyperosmotic
Extracellular compartment environment	Hyperosmotic to environment	Hypoosmotic to
Water movement	From media to body	From body to media
Ion movement	Lose body salts	Gain body salts
Malpighian tubule fluid	Isosmotic to hemolymph	Isomotic to hemolymph
Rectal fluid hemolymph	Hypoosmotic to hemolymph	Hyperosmotic to
Rectal uptake	Salts	Water

Let us look at the nature of aquatic environments just a little more closely. Fresh water habitats, depending upon the local geology, may have total ion concentrations ranging from less than 0.1 mM to more than 10 mM in NaCl equivalents. Moreover, salt concentrations can vary widely due to evaporation. In addition to total salinity, the relative amounts of particular ions may vary over a considerable range. This is markedly true in waters with unusual substrates such as limestone.

Likewise, salt water habitats may be extremely variable in ionic composition. Brackish water occurs in coastal regions where sea water and fresh water are mixed. Tidal flows can vary salinities of such areas over a wide range of concentrations. Because environments are variable, we achieve far greater insight into physiological adaptations of aquatic insects if we know the range of salinity that insects are adapted to tolerate. Organisms that can tolerate wide variations in salt concentrations in the water they live in are called euryhaline from the Greek words for wide and salt.

An extreme example is the brine fly *Ephydra* which lives in the Great Salt Lake. These flies can tolerate periods in distilled water and periods in water concentrated to 20 per cent NaCl equivalents. In low environmental salt concentrations the larvae are able to hyperosmoregulate, that is, able to maintain extracellular fluid compartments at higher osmotic concentrations than the medium.

When their environment becomes highly concentrated these larvae can hypoosmoregulate, or maintain extracellular fluid compartments at an osmotic concentration lower than their environment. These insects are adapted to tolerate a very wide range of salt concentrations in their environments. At the other end of the spectrum, some aquatic insects are stenohaline, indicating a narrow range of tolerance to variations in environmental salinity. Larvae of the mosquito *Aedes aegypti* represent a good example. These larvae maintain hyperosmotic hemolymph concentrations over a narrow range of salt concentrations. When environmental salt concentrations exceed osmotic concentrations in hemolymph, however, control breaks down, and the larvae become osmoconformer. Their hemolymph salt concentrations go far beyond appropriate levels, and the larvae fail to survive.

Many insects are adapted somewhere between these extremes of highly euryhaline and very stenohaline. These insects are able to hyperosmoregulate in fresh water and to a limited extend they can hypoosmoregulate in slightly brackish water. Beyond some limit of concentration, control is lost and the larvae become osmoconformers. You can appreciate the adaptive significance of nonconformity.

WATER BALANCE IN TERRESTRIAL INSECTS

We have spent a little bit of time considering aspects of movement. We've gotten an idea of the contractile process in muscle, a process driven by hydrolysis of ATP. We saw that ATP is generated as an end product of energy metabolism and considered gas exchange as a requisite for energy metabolism. Flight is very important to many aspects of insect biology, and I purposefully emphasize flight. Let us also emphasize the broader biological issue in flight. Flight can be taken under the mantle of motility, or movement. In this view, flight is one of many kinds of movement, all fundamental processes in biology.

Let us now turn attention to still another are of physiology confronting nearly every life form: osmotic balance. Osmotic regulation is the business of maintaining optimal salt and water concentrations in an organism. We begin by considering the biological significance of this. About 50 to 90 per cent of the body weight of an insect, or nearly any other animal, is water. Some of this fluid is inside the various cells that make up the body while the remainder is in the spaces outside of the cells. As a unit, these spaces are regarded as a collective known as the extracellular space.

The extracellular space is filled with fluid, for obvious reasons called the extracellular fluid. Among the dissolved constituents of the extracellular fluids are the ions and the nutrients needed by the cells for maintenance of life. The extracellular fluid is in constant motion throughout the body, and it is rapidly mixed to near homogeneity by hemolymph circulation and by diffusion. Because of these mixing processes, we can say that all cells in the body live in essentially the same environment. The great French physiologist, Bernard, called the extracellular fluid the melieu int'erieur or what we would call the internal environment.

The point is to recognise two main fluid compartments in an insect body. One is the intracellular compartment and the other is the extracellular compartment. To appropriate a teaching device from Daniel Matzia, the famed cell biologist retired from UC Berkeley, we can regard the extracellular fluids as the *out environment*, that is, the environment inside the body that cells experience. The intracellular environment, then, could be the *in* environment. However, we know eucaryotic cells have organells and compartments within them. Within cells, the compartment inside any particular organelle, such as a nucleus or mitochondrion, could be the *in-in environment*. At the organismal level, the world we live in is the *out-out environment*. This language allows us to appreciate the world in terms of environments that things experience: the in-in, the in, the out and the out-out environments. Cells experience the out environment, the same thing that Bernard called the *melieu int'erieur*.

Some percentage of the total body fluids of any organism are intracellular fluids. The fluid within each cell is made up of its own individual mixture of various constituents, such as ions, amino acids, proteins, sugars and so forth. On the other hand, within an organism, the concentrations of these things appear to be reasonably similar from cell to cell. For our discussion, the intracellular fluids of all of the different cells of an organism can be considered to be one large fluid compartment: the in environment or the intracellular compartment. Regardless of the particular language or expression that is easiest to keep in mind, we want to come to the point of view that an insect body has two major fluid compartments: the intracellular and extracellular compartments. The extracellular compartment is easy to visualize as the circulating hemolymph.

While the extracellular environment is substantially different in composition from the intracellular environment, the two compartments are in osmotic balance. That is to say that there is rarely a net flow of water from one compartment to another. There are various mechanisms to handle the maintenance of the correct concentrations of the intracellular constituents, including active transport and facilitated diffusion. These mechanisms are a related, but different matter. The thrust for our purposes is that water always tends to diffuse by a process called osmosis across cellular membranes from the side with fewer total dissolved particles to the side with a greater number of dissolved particles. Let us consider some possibilities. First, if a cell in placed in a solution with far fewer osmotic particles than the cell, water will diffuse into the cell. This can result in considerable swelling, even to the point of bursting the cell. Alternatively, if a cell is placed into a solution with a far greater number of particles, water tends to diffuse, once again, down its osmotic gradient, in this case, out of the cell. A badly shrunken cell that may or may not survive may result from this flow.

We gather the significance of a physical phenomenon, osmosis. Osmosis explains the importance of maintaining a proper balance of water and of salts in the extracellular compartment or the melieu int'erieur. During times when, for whatever reason, the extracellular compartment is not correctly regulated, the physiological situation is exactly like placing cells into a solution that is too dilute or too concentrated. This is a common problem faced by insects and most other organisms living in freshwater, salt water or in terrestrial environments.

Again, terrestrial insects are faced with the problem of regulating their levels of body fluids and ionic concentrations such that the extracellular and intracellular fluid compartments are maintained in osmotic balance. Failure in this respect would lead to this situation: the body's cells are effectively placed in a fluid that would cause them to swell or shrink to a point of dysfunction or death. Physiological solutions to the problem of water balance have tremendous evolutionary and ecological significance. First, it can be argued that the spectacular evolutionary radiation of insects occurred after a host of adaptations to terrestrial life. We saw some of these adaptations during our consideration of movement.

Physiological regulation of water balance is another adaptation that supports terrestrial life. The biogeographical distribution of insects reflects the broad range of terrestrial habitats in which insects live and thrive. Many of these habitats are extremely arid and very hot. Physiological adaptations to these sorts of environments have permitted exploitation of such habitats that would otherwise be unavailable to insects. The focus on terrestrial life is not meant to overlook aquatic insects, which we will discuss in future. Furthermore, aquatic insects arose from their terrestrial counterparts. Hence, adaptations to aquatic life can be thought of as a component of successful terrestrialisation.

Two main factors may work to upset the essential osmotic equilibrium within an insect's body:

1. Loss of water by transpiration across the cuticle, or a very large gain of water by drinking or eating.
2. The intake of food whose concentration of inorganic salts is quite different from the ionic composition of the hemolymph.

Today we are going to begin with the first point, the regulation of water balance, which may be insulted by substantial loss or gain of water. Insects are generally small creatures, which tends to put them in an unfavourable setting with respect to the ratio of body surface area to body volume. The idea is to recognise a relatively large surface across which water can be lost by transpiration. We will see that water loss across the body surface can not be entirely stopped; however, if it were not restricted to a level that can be balanced by water uptake, insects would not survive in the more arid environments. The first line of defence was the development of a cuticle relatively impermeable to water.

The outer layer of an insect is the integument, which is made up of the epidermis and the cuticle. The epidermis is the outer cell layer. The epidermis lies on a basement membrante, and secretes the cuticle, which so far as I know is illustrated in virtually every book written on insect biology.

There are three major sections of cuticle, endocuticle, exocuticle and epicuticle. In standard drawings, we note the pore canal, cells of the epidermis and the basement membrane. The endocuticle provides the extensibility of the integument and combines toughness with flexibility. The rigidity of the hard parts of an insect, the appendages and so on, comes from the properties of the exocuticle. The relative impermeability of the cuticle to water is due to the presence of the epicuticle.

The epicuticle is a complex structure that differs among various insects. There is a cuticulin layer made of two components: a thin, dense outer layer and a thicker, still denser inner layer. Cuticulin is generally made of lipoprotein. A waxy layer lies atop the cuticulin. This is made of long-chain hydrocarbons, and esters of fatty acids and alcohols. There is an inner wax layer composed of strongly oriented long-chain molecules that generally form a monolayer only one molecule thick. The strong orientation provides the waterproof properties of the epicuticle. The outer wax layer is composed of wax molecules that are randomly orientated; on top of this is the cement layer. People think that this is a shellac-like substance that may serve to protect the wax. Terrestrial insects lose water by evaporation from the general body surface - a process called transpiration. The rate of water loss from the surface on any given insect is a function of both wind speed and also of the humidity of air. Transpiration is relatively independent of body temperature, up to some point. Water is lost from the body surface at a constant, but rather low, rate through a broad range of temperatures.

At some temperature the cuticle reaches a temperature point called the transition point. This point varies from insect to insect, depending upon the chemical composition of the hydrocarbons and waxes in the epicuticle. Insects living in hot, arid habitats tend to have wax layers with higher transition temperatures than do insects living in cooler or more humid habitats. Some insects have two transitions points. Water loss from the body surface is held to a low rate by the oriented wax layer on the outside of the cuticulin. When the temperature of the cuticle rises to such a point that the van der Waals forces holding the wax molecules together are broken, the molecules move slightly apart and water is allowed to escape. Below the transition temperatures the wax molecules are close together, and at a slight angle so that water loss is restricted. At temperatures above the the transition point, the molecules part, which usually leads to a quick and probably pleasing and comfortable death due to dehydration.

Then we see that osmotic balance in the melieu int'erieur is in part maintained by reduction of water loss from the body surface. Another important avenue of water loss in terrestrial insects is from the respiratory surfaces. In the orthopteran *Gastrimargus*, for example, 70 per cent of the normal water loss is from the trachael system. We have already seen that water loss from the respiratory surfaces is restricted by the invagination, or internalisation, of the trachael system such that it is inside the body. Water loss is also reduced by control of the spiracular valves. Internalisation of the respiratory surfaces in a major adaptation to terrestrial life in vertebrates as well as in invertebrates.

We can model an experiment designed to demonstrate the influence of spiracles on water loss. Consider the rate of water loss from *Tenebrio* larvae as a function of time. *Tenebrio* is well adapted to arid conditions, and its water loss rates are typically quite low. If we impose the condition of open spiracles, by holding the animals in an atmosphere of 5 per cent carbon dioxide, water loss would be far higher than what could be easily balanced by an insect living in an arid habitat. Physiological regulation of the spiracles plays an important part in reducing potential water loss.

Some insects live through long periods of quiescence during which they do not feed or drink water. Control of the spiracles in these species is aimed at keeping them closed as much as possible to minimize water loss while facilitating the gas exchange necessary to support metabolic needs. One strategy is called cyclic release of carbon dioxide. This has been well studied in pupae of *Hyalopohora cecropia*, a large insect suitable for study in this regard. Carbon dioxide is allowed to build up in the trachael system for a long as seven hours, then is released in bursts during which the spiracles are wide open. Between the bursts of carbon dioxide release the spiracles are not closed as tight as possible, but rather flutter so as to allow a continuous uptake of oxygen. While the cuticle and control of the spiracles serve to restrict the loss of water, a certain amount of water is inevitably lost.

Then we can identify three avenues of water loss:

1. Transpiration across the cuticular surfaces
2. Transpiration across respiratory surfaces
3. Loss of water in the excreta

Water that is lost must be balanced by an appropriate level of water uptake. Most insects take in sufficient water with the food they eat. In his book on insect structures and function, Chapman thought some insects may select food with a higher water content under certain circumstances. Still other insect species may consume more food than they actually need in order to realise the potential water in the food. Moreover, many insects drink water. The dipteran *Phormia*, for example, has specific water receptors in the tarsal and labellar sensilla. I've often observed wasps drinking water in desert pools in Mexico.

Some insects have evolved slightly more bizarre adaptations to gain water. Larvae of the dipteran, *Epistrophe,* can extrude an anal pipilla into a water droplet and absorb water. The cockroach *Periplaneta* is among many insects that can absorb water from water droplets on the cuticle. The cuticle is asymmetrical with respect to water movement: water passes inwards more readily than outwards. Also, a few insects can absorb water from water vapor in the air when they are desiccated. *Tenebrio*, for example, can absorb water from air with relative humidities greater than 90 per cent. Prepupa of the flea *Xinopsylla* can absorb water from air at relative humidity above 50 per cent.

A final source of water is called metabolic water. As we saw during our look at metabolism, the final electron acceptor at the end of the electron transport chain is oxygen. Electrons and their hydrogens combine with oxygen at this point, yielding water. The amount of water that can be gained by metabolism depends on the metabolic substrate. A hundred grams of the fatty acid palmitate (16:0) yields 112 grams of water while 100 grams of glycogen yields 56 grams of water. Grain insects, such as *Tribolium*, normally obtain a very large proportion if not all of their water from metabolism of their food. We can guess that metabolism of lipid may be adaptive with respect to water balance during long periods of hibernation in several insect species.

MAINTAINING INTERNAL STABILITY

Mechanisms responsible for maintaining internal stability within organisms are referred to as *homeostatic mechanisms*. Homeostasis is the maintenance of static conditions in the internal environment. Homeostatic mechanisms are active processes because conditions are nearly always subject for forces that will bring about changes. Like virtually all other organisms, terrestrial and aquatic insects confront the problem of maintaining an optimal balance of water and of salts between the intracellular and extracellular compartments. With respect to insects, the most general physiological expression is that ions and water are secreted by Malpighian tubules and they are selectively resorbed in hindguts. Osmoregulation is a complex phenomenon involving mechanisms to reduce and offset water losses, and mechanisms to recover optimal osmotic balances when they are insulted.

Removing excess water from the extracellular compartment is known as diuresis. Diuresis is regulated at the organismal level by the actions of diuretic hormone (DH) and antidiuretic hormone (ADH). Most of our information on the regulation of diuresis by DH and ADH comes from research on adult mosquitoes, on the blood-sucking bug *Rhodnius* and, of course, on the locust.

Hematophagous insects experience a challenging insult to the homeostasis of water and salt balance each time they take a blood meal. Females of the yellow fever mosquito, for example, often consume twice their body weight in a single blood meal.

The vast majority of the weight is water. Imagine, for a moment, what it might feel like to eat about 50 kg of watermelon in less than 5 minutes. Female mosquitoes often begin diuresis within about 2 minutes of starting a blood meal, even before they complete the meal. The rapid diuretic response occurs in response to DH. DH is released from the brain, by way of median neurosecretory cells. The hormone stimulates increased fluid secretion rates in the Malpighian tubules. The diuretic response results in a return to an appropriate balance of salts and water between the intracellular and extracellular compartments.

It might be thought that DHs occur in those insects that are subject to extreme challenges to osmotic homeostasis, but not in other species. This is not the case. Endocrine regulation of Malpighian tubule function was first described in locusts over 20 years ago. The DH activity was found in extracts of the corpora cardiaca. When the extracts were separated by high performance liquid chromatography, two separate peaks were detected. The peptides associated with both peaks were concentrated, and tested for diuretic activity by bioassay. Both peptides stimulated fluid secretion in Malpighian tubules. Similar extracts of other tissues, including brain, sub-esophageal ganglion, and thoracic ganglion all stimulated increased fluid secretion. Extracts from the brain had two active peaks, but all the other ganglia had a single peak of activity, then known as DIURETIC PEPTIDE I.

Mammals also produce DHs, one of which is known as arginine vasopressin, or AVP. A group in Bordeaux, headed by Jacque Proux, used antibody antibodies from a commercial radioimmunoassay kit for AVP to detect a factor in extracts of the sub-esophageal ganglion. The factor was called AVP-like DH due to its antigenic similarities to mammalian AVP. This group recently isolated the AVP-like DH from thoracic and sub-esophageal ganglia. Their work makes an interesting story. The group began by removing the glands from insects, then storing them in acetone. Over a period of some years, they collected a total of 51,000 glands.

The glands were homogenized, and the DH was prepurified by centrifuging the homogenate at 100,000 X g for 1 hour, then filtering the supernatant on silica gel cartridges. The next step was semi-purification by HPLC on a Vydac column. As it turned out, most of the immunoreactivity was associated with the first of the two peaks, but all of the biological activity on actual Malpighian tubules was in the second fraction.

Amino acid studies showed that both factors had the same amino acid composition. The factors were then separated on size exclusion chromatography. This indicated that the first factor had a molecular weight of about 700 daltons and the second factor was about 1400 daltons. This finding suggested that the biologically active factor was a dimer of the first factor. It is not at all unusual to have two peptide chains hooked up to form a single functional entity. The mammalian hormone, insulin, is a classic example.

Very often the chains are connected at cysteine residues by disulfide bridges. The next step was to obtain the amino acid sequence of the first factor. The issue then became a matter of learning how the peptides are connected. They could be connected in a parallel or anti-parallel form. An extremely talented peptide chemist named David Schooley synthesized both forms. Only the anti-parallel dimer had biological activity on Malpighian tubules. The AVP-like DH of *Locusta* is an anti-parallel dimer. This is the first insect DH to be isolated, sequenced and synthesized.

DHs have been detected in aquatic insects. A factor extracted from heads of the water boatman *Cenocorixa blaisdelli* caused an 8-fold increase in fluid secretion rates. Similarly, extracts of thoracic ganglion caused an increase in fluid secretion rates in larvae of the dragonfly *Libellula quadrimaculata*. For many years the received wisdom was ADHs appear to act on hindgut, but not on Malpighian tubules. However, more recently we have learned that ADH factors also act on the Malpighian tubules. ADH appears to increase water resorption rates in isolated recta, and decrease water secretion in Malpighian tubules. Actions at both tissues constitute ADH action.

It is believed that biological epithelia can not directly pump water from one side of the membrane to another. The general expression is that ions can be selectively moved, then water follows a local gradient. If this is so, then how can water be selective moved out of the rectum, leaving salts behind? We consider a model. The model is a schematic of the principal cells of rectal pads in cockroaches and locusts. These cells are characterised by scalariform complexes and intercellular spaces, which are quite rigid. Active transport processes are thought to create very high osmotic concentrations within the rigid scalariform complexes. Water in the lumen of the rectum would then move into these hyperosmotic spaces.

These spaces are rigid, so the osmotic movement increases hydrostatic pressure. The increased hydrostatic pressure forces water to move through the spaces, and finally into the hemocoel. The final openings to the hemocoel are thought to be valve-like to prevent flow in the reverse direction. According to this model, water is moved from one side of a membrane to the other. Here, we emphasize the water movement is powered by the usual physiological actions: ions are pumped, and water follows a local osmotic gradient. The key point is the rigid structures within the cells create a directional water movement. The model has been supported by electron-probe X-ray micro-analysis. This kind of analysis is performed on a scanning electron microscope.

A beam of electrons is focussed on a section of the cell. Ions within the cell release X-rays which are characteristic of the particular ion. The characteristic X-rays are detected by photomultiplier tubes. The movement of water is regulated by moving ions within principal cells of the recta. The ADH of insects may be identical to a hormone known as chloride transport simulating hormone (CTSH). This is also a factor that has been isolated from corpora cardiaca of locusts. CTSH has been shown to cause electrogenic chloride movement within recta. Physiological models are useful tools. Non-etheless, we want to underscore the point that these are models. Models are continuously changed to accommodate new information. Let us revisit our general model of DH and ADH actions. Another anti-diuretic factor has been isolated from the corpora cardiaca, in this case during work on the house cricket. This factor acts on Malpighian tubules and causes a reduction in fluid secretion. Hence, in the final analysis, we might come to the view that DHs and ADHs act on Malpighian tubules and recta in coordinated ways.

As we've just seen in insects, mammalian kidney function is regulated by major hormones. However, other factors also exert regulatory actions. One is a class of molecules known as eicosanoids. These molecules are synthesized from certain polyunsaturated fatty acids. David Petzel and I developed the hypothesis that eicosanoids might also be important elements in regulation of Malpighian tubule physiology. We tested this idea in direct studies of mosquito Malpighian tubule function.

We treated Malpighian tubules with eicosanoid biosynthesis inhibitors, then observing the effects of the inhibition on fluid secretion rates in adult females of the mosquito *Aedes aegypti*. We found that increasing doses of indomethacin (a prostaglandin biosynthesis inhibitor) resulted in decreases in fluid secretion rates. Our current idea is that the prostaglandins modulate the effects of major hormones at the tissue level. It follows that there are at least two tiers, or levels, of regulation of fluid secretion. Future studies will probably make the picture even more complicated.

Malpighian tubule and hindgut functions involve moving ions from place to place. We use the expression epithelial transport to describe the movement of ions from hemolymph into Malpighian tubule lumen or from rectal lumen into hemolymph. Epithelial transport entails moving ions across one cell membrane, through the cell, and finally across another cell membrane. Epithelial transport occurs through a number of processes. One process is movement through ion channels. An ion channel might look like a couple of large proteins assembled such that they form a tube running through a biomembrane. We also can imagine a group of proteins forming a tube or channel allowing passive movement of ions according to the forces of electrochemical gradients. The channels are of different sizes, and it is the size differences that determine movement of specific ions.

Ions can be moved across epithelia against electrochemical gradients by active processes. Primary active transport processes directly the potential energy in ATP to move ions. Examples of active ion pumps include the sodium-potassium ATPase, and the sodium-potassium-2chloride co-transporters. The exact shape and operation of these pumps are unknown, but it is thought that the transporter proteins undergo a change in conformation with the expenditure of energy, and that the altered conformations physically move the ions from one side of a membrane to another.

People have tried to model the sodium-potassium ATPase actions. In the first step, sodium binds to the protein. The protein is then phosphorylated by hydrolysis of ATP. Sodium binding and phosphorylation, both on the cytoplasmic side of the protein, cause a conformational change resulting in transporting the sodium to the other side of the membrane, where it can be released into the medium.

After releasing the sodium, the protein is available to bind potassium. Potassium binding leads to dephosphorylation of the ATPase, which returns the protein to its dephosphorylated conformation. The return transports the potassium ion across the membrane. The potassium ion is released from the ATPase, and the transport cycle can begin again. This model suggests that one sodium and one potassium ion are transported in each phosphorylation cycle. However, the biological sodium-potassium ATPase has three sodium binding sites and two potassium binding sites.

Let's look at another biological significance of secretion in insect systems. Increased Malpighian tubule and rectal fluid secretion rates is thought to serve a function that goes beyond homeostasis of salt and water balance in insects. It has been proposed that DH and ADH could work in concert in increase fluid cycling with the body. The increased fluid cycling is thought to facilitate filtration of metabolic wastes and xenobiotic chemicals from the hemolymph.

AVAILABILITY AND IMPORTANCE OF AQUATIC RESOURCES

Aquatic resources are fundamental to local livelihoods in Attapeu and are a key component in the well being of local people. Their livelihood strategies involve the use of multiple resources.

AQUATIC ENVIRONMENTS

The study areas in Attapeu are characterised by a wide range of aquatic environments. These include permanent and seasonal, flowing and standing, natural and man-made water bodies. Each type of waterbody has its own annual hydrological cycle, is characterised by its own community of aquatic organisms and is subjected to a particular pattern of resource use by local people. The main categories of aquatic environment identified during this study include rivers and perennial streams, perennial ponds, marshes, oxbows and rice fields, seasonal ponds and seasonal streams.

RIVERS AND PERENNIAL STREAMS

Rivers and perennial streams are key features in the lowlands of Attapeu and important sources of fish and other aquatic produce. They sustain a range of aquatic organisms throughout the year. They are also subject to large annual fluctuations in volume and flow between the rainy and dry seasons. Because these are permanent water bodies, they serve as dry season refuges to a broad range of fishes and aquatic animals and are critical habitats for a number of strictly riverine species.

Fishing in rivers requires a certain level of specialisation. The changing nature of the environment caused by the fluctuations in water flow and the seasonal activities of fish, some of which are migratory, requires specific knowledge and equipment. Boats, specialised nets and other gear are needed to fish successfully in rivers and large streams. Much of the simpler, cheaper and more common household gear has limited application in the river. This means that successfully accessing riverine aquatic resources requires financial investments in gear and a certain amount of skilled labour (typically strong males). Households that lack the labour to use the specialised gear or the means to purchase boats and equipment are limited to resources along the edges of the rivers using smaller gear during periods of low water and assisting others during peak fishing periods. Those who can fish in the river are subject to seasonal scarcity, but are able to catch some fish throughout the year.

PERENNIAL PONDS, MARSHES AND OXBOWS

Perennial ponds, marshes and oxbows are fairly common in the lowland-floodplains of Attapeu. They serve an important function by receiving excess water during the rainy season and holding it throughout the dry season. These water bodies are usually shallow and vary greatly in size over the course of the year, expanding during the rainy season and receding during the dry season. In many cases, they function like an aquatic lung, receiving water directly from a rising river or stream during the rainy season, and then draining it back into the river or stream as the water level drops in the river or stream.

These water bodies serve as refuges for fish and other aquatic organisms during the dry season and are the source of many flood plain fishes during the rainy season. The key species are mainly floodplain fishes but some riverine fishes are trapped when the water recedes following periods of flooding. Because the water is relatively fertile and shallow in areas, many types of aquatic plants, molluscs, crustaceans, amphibians and reptiles are abundant. Fishing in perennial water bodies such as ponds, marshes and oxbows is less specialised than river fishing and requires less investment. Because these water bodies are typically shallow (or have shallow areas), aquatic plants and animals are easy to access with simple, small gear or by hand-collecting. These areas are often of special importance to poor people. When the water recedes and seasonal water bodies have dried or been harvested, people fish in permanent water bodies with other types of gear, including small-scale, household gear.

RICE FIELDS, SEASONAL PONDS AND SEASONAL STREAMS

Rice fields are an important and often overlooked source of aquatic resources. Rice fields and constructed seasonal ponds extend the productive phase of the aquatic environment. Seasonal rains inundate wide areas of lowland, typically from June through to October.

Fish from perennial water bodies migrate out to seek food and reproduce in these newly created water bodies. Migration out to rice fields (and return migration at the end of the rainy season) takes place through seasonal streams that drain into the rivers. Rice fields and seasonal ponds play a similar role in local hydrology in that they hold water higher in the watershed for longer periods than would otherwise be possible.

Several species of fish and aquatic animals have evolved to take advantage of these temporary aquatic environments. At the begining of the rainy season they quickly disperse and reproduce (or in some cases reproduce and disperse) to fill the empty ecological niches that quickly appear in newly inundated areas. Almost as soon as the rains begin, rice fields are populated with organisms that have been dormant or relatively inactive in permanent water bodies during the dry part of the year.

Much of the household fishing activity is focused on rice fields from the beginning of the rainy season (June) until they dry up (in early to mid-November to January). Harvesting is done with simple, inexpensive gear and requires few specialised skills, which makes it easily accessible to poor people. Fishing effort focuses on migration pathways to and from water bodies, and can be especially productive at the end of the aquatic cycle when fish are moving out of the flood plain back to permanent water bodies.

ACCESSING AQUATIC RESOURCES

Aquatic resource use patterns are influenced by the type of aquatic environment and a range of socio-economic factors such as value, rules of access, ownership and marketability. This varies from village to village and a wide range of situations are represented in the lowland areas of Attapeu.

Some of the factors that determine aquatic resource use patterns include:

- The types of water bodies accessible
- Natural productivity of water bodies
- Distance traveled to water bodies

- Traditional, cultural and historic patterns of aquatic resource use
- Opportunity cost of aquatic resource acquisition relative to other livelihood activities
- Level of specialisation or investment required
- Proximity to markets
- General status of food security

SPECIES DIVERSITY

Lists of fish and aquatic organisms were compiled from interviews with villagers rather than sampling. Non-etheless, they provide a good indication of the aquatic biodiversity in the lowlands of Attapeu Province. As seen in the lists of aquatic species reported from each village, a wide range of organisms representing different aquatic environments are being caught and consumed or sold. In the fish lists reported from each village in Table, a broad range of species representing different aquatic environments is being caught. The list grows longer when we add other aquatic animals such as crustaceans, molluscs, amphibians, reptiles, insects and aquatic plants.

Fish - An impressively high number of fish species were reported during each village activity, ranging from 66 species reported by Tamoyot to 102 species reported by Saisi. These species fall into two major groups: riverine species and floodplain species. Aquatic animals (non-fish) - Other aquatic animals reported include several species of crabs, shrimp, frogs, shell fish, turtles, and insects. In some cases, these animals (especially frogs, shrimp and crabs) are as important to household consumption as fish.

Aquatic plants - Aquatic or semi-aquatic plants are an important part of rural diets in Attapeu. Aquatic plants include those that float or are submerged and rooted in water. These are most abundant during the rainy season when the area is inundated by flood water. Semi-aquatic plants grow in moist areas along the edges of water bodies and are most abundant during the dry season.

HISTORICAL TRENDS IN AQUATIC RESOURCES ACCESS AND USE

Local people say there is a general trend towards intensification of resource use which includes agricultural development as well as increased efforts to harvest aquatic resources. An important driver of this trend is the ongoing shift from a subsistence-based economy, where people provide for most of their own needs, to a cash-based economy.

Local people participating in this assessment felt that most living aquatic resources, especially the fisheries, are in decline and provided the following explanations:

- Over havesting due to population growth
- The use of modern, more efficient gear such as monofilament nets

- A trend from subsistence to commercial fishing and a growing demand for aquatic products in local markets
- Environmental degradation

Households that routinely suffer from food insecurity in the form of insufficient rice often depend on wild aquatic resources to compensate for this deficiency. Indicators used to determine whether aquatic resources are in decline are based on observations of declines in certain key species as well as a perceived decrease in catch per unit effort (CPUE).

These indicators include:

- More time spent fishing to catch less fish (*i.e.* spending three times as much time to catch the same amount of fish)
- Fishing more frequently to meet demand (catches cover one meal, whereas previously one catch provided enough for several meals)
- More investment required in household fishing gear (*i.e.* must own three gillnets, where-as previously one cast net would suffice)
- Less fish being preserved for future use (*i.e.* one jar of fermented fish prepared in a household, rather than five as in previous times)
- Other aquatic animals increasing in the diet while fish decline (*i.e.* eating more small frogs than fish, rather than the opposite as in previous times)
- Decline in the use of traditional gear and techniques because of diminishing return on effort

Decreases in total fish production, in absolute terms, are difficult to determine. In this case, what is more significant is the availability of fish to individual households. By all the measures listed above, the availability of fish and other aquatic resources at the household level is declining. Poor households are likely to be the most adversely effected because of their limited availability of labour or capital to purchase more effective gear.

Table. Numbers of Aquatic Species Reported

	Tamoyot Village	Saisi Village	Gayeu Village
Fin fish	61	102	95
Crustaceans	3	6	6
Molluscs	4	7	6
Amphibians	8	14	6
Reptiles	8	10	5
Insects	-	7	7
Aquatic plants	19	16	31

NATURAL HISTORY OF THE AQUATIC HEMIPTERA

The Hemiptera belong to the Division Exopterygota. They usually possess wings and go through incomplete metamorphosis. The most notable characteristic of this order is the "beak," a piercing mouthpart which has a suctorial function. One group of Hemiptera, suborder Homoptera, uses this beak to feed on the fluids in plant tissue. This suborder will not be considered here as its only aquatic members are a few semiaquatic aphids. The other suborder, the Heteroptera, contains members which are predatory, parasites or detritivores and aquatic Hemiptera belongs to this group. Heteroptera is further divided into families. Six of these are fully aquatic. These families are Notonectidae (backswimmers), Pleidae (pygmy backswimmers), Nepidae (water scorpions), Naucoridae (creeping water bugs), Belastomatidae (giant water bugs), and Corixidae (water boatmen).

COMMON ADAPTATIONS

The families share some adaptations, which have allowed for successful inhabitance of the aquatic environment. These adaptations include legs, wings, breathing apparatus and communication. The legs of most Hemiptera are modified in two ways.

First of all, the forelegs are modified to be raptorial. Raptorial legs aid the insect in capturing prey. For example, the giant water bugs (*Belastoma* spp.), may use its large hooked and sclerotized forelegs to snatch a minnow or tadpole out of the water. Raptorial legs are important for these insects which are almost exclusively predators.

The second and third pairs of legs (particularly the third) are often modified to be natatorial. Natatorial legs in Heteroptera are lined with stiff cilia which form a horizontal plane. These cilia are analogous to the paddles used by humans to row a canoe. By moving its legs, the insect can propel itself through the water. Besides locomotion, these natatorial legs can be used to increase water flow around the insect or its egg mass which improves diffusion and respiration.

Wings are also an important adaptation shared by aquatic Heteroptera. The wings of aquatic insects probably allow them to have a major advantage over many other water-dwelling animals.

Many bodies of water are ephemeral, meaning they disappear for part of the year due to a dry spell or some other factor. If an ephemeral pond with a population of adult aquatic bugs dries up, the bugs may simply be able to fly to another pool. Additionally, the ability to fly may allow these insects to rapidly colonize newly formed bodies of water, such as puddles. This contrasts with the abilities of animals like fish which must remain in water in order to colonize a new environment.

How do Heteroptera survive underwater? Like all higher organisms, insects require a source of oxygen. Water does not contain as high a volume of oxygen as air, so alternate breathing methods were developed. Aquatic breathing methods may be divided into two types, aeropneustic and hydropneustic. Aeropneustic breathing methods allow the insect to continue to utilise surface air for respiration. The simplest aeropneustic breathing method is breathing with modified cerci, joined together to form a tube, the siphon. The insect sticks this tube out of the water in order to reach air. A good example of this type of respiration is found in the water scorpions belonging to family Nepidae.

Periodic contact breathing is another form of aeropneustic breathing that is slightly more sophisticated. Periodic contact breathers move to the surface to replenish an air supply which they carry with them. Often this is done in a chamber under the wing.

Tiny hairs provide plenty of surface area for bubbles to attach; this forms a thin film of air, called the plastron. The advantage of periodic contact breathing is that it allows for mobility; the insect doesn't have to perpetually sit still at the water's surface, it may dive and swim around. The more advanced form of aquatic breathing is hydropneustic.

Aeropneustic breathing and hydropneustic breathing techniques are not mutually exclusive. Many Heteroptera use the plastron as a physical gill, and can prolong their stay underwater (particularly in fast-moving streams or other areas with high oxygen). Osmosis allows the plastron to remove carbon dioxide and replace it with oxygen from the surrounding water. Another, rarer, form of hydropneustic breathing is cutaneous respiration. Cutaneous respiration occurs when the cuticle is thin enough for air to diffuse directly in and out for the insect.

High oxygen environments, usually lotic (running water as opposed to lentic, still water) are also necessary for cutaneous respiration. Cutaneous respiration allows an insect to remain underwater indefinitely. Adaptations for communication are also extremely important and probably present in many aquatic Hemipterans. Ripple communication is the primary means of communication for several groups of Hemipterans and has been most extensively studied in the Belastomatids and the family Gerridae, the water striders. These insects oscillate their bodies, sometimes just the legs, other times the entire body, to produce ripples.

Other insects have amazing abilities to detect these ripples, with some species able to detect amplitudes of only .002 micrometers. It has been demonstrated that ripple communication has many uses for an insect ranging from attraction of mates, territoriality and induction of oviposition. For underwater insects, ripple communication is probably a most efficient and effective communication device. In addition to ripple communication, stridulation, rubbing two parts of the exoskeleton together to produce noise, can occur in these species.

FAMILY CHARACTERISTICS

The families of these insects have many notable and interesting characteristics. This section is an overview of the unique adaptations and ecology of each family. Nepidae is the family which includes the insects commonly known as "water-scorpions." These insects have this colloquial name because of the long siphon at the end of their abdomen. This siphon allows for aeropneustic breathing. Water scorpions are carnivorous, and because they must stay motionless at the water's surface for long periods of time, they capture prey using the ambush method. They are poor swimmers, and have hindlegs modified for grasping onto plant matter as opposed to natatorial legs. The physiological limitations of having a siphon mean that water-scorpions usually inhabit lentic water where there is plenty of plant material for them to hold onto as they lie in wait.

The Pleidae are small insects that are usually less than 3mm. These short-legged bugs lack the natatorial and raptorial legs of their brethren, and can commonly be found climbing on the stems of aquatic vegetation in lentic waters. The Notonectidae or backswimmers are an interesting group of aquatic Hemipterans. These insects swim with their dorsal side on the bottom, with their wings pointing downwards. They are capable of fast swimming through the use of paddle like legs.

The rowing motions of these legs, which quickly move water over the physical gill, increase the efficiency of their plastron improving their aeropneustic quailities. These species are notable for possessing a toxic bite (like many Hemipterans they not only use their beak for sucking but also for injection of fluids into the prey animal).

The Notonectidae are one of the few insects to posess hemoglobin in the hemolymph. Apparently, this helps with buoyancy. Additionally, these species use stridulation (in particular rubbing the forelegs against the mouthparts) to attract mates.

The Corixidae are superficially similar to the Notonectidae, except that the Corixidae swim facing right-side up. Like the Notonectidae, these insects increase the effectiveness of their plastron by rowing motions. Unlike the Notonectidae and most other insects in the Hemiptera group, the Corixidae are not exclusively carnivorous. Some species may be detritivores feeding upon algae and dissolved organic material. This group is the largest group of water bugs with over 100 species occurring in North America alone. This success is probably in part due to the group's amazing ability to withstand pollution. The Naucoridae, also known as creeping water bugs, have an appearance like thin disks; their bodies are vertically depressed. Like most aquatic Hemiptera, their forelegs are modified to be raptorial. These insects resemble the Belastomatidae to which they are closely related, except for the fact that they are smaller; reaching maximum sizes of 15 mm.

The Belastomatidae, or giant water bugs, are one of the most well known groups of aquatic insect. Their large size (up to 65 mm) makes them highly noticeable. Additional factors which make them noticeable to humans include a very painful bite and strongly positive phototaxis. Their unpleasant bite has led to the name of "toe-biters" to describe these insects which may defensively attack the foot of a wader or hurt someone who picks them up. They are additionally known as "electric light bugs" for their habit of attraction to electric lights. The large size and voracious predatory appetite of these insects makes them one of the few important insect predators on vertebrate animals. They can often be observed feeding upon tadpoles, fish and even animals as large as ducklings. Often these insects hang from the surface of the water breathing in the aeropneustic style from modified cerci, but are also highly capable swimmers.

A unique characteristic of this group is the parental care exhibited for the eggs. Many female Belastomatid species lay eggs on the back of the male who then spends time aerating the eggs and taking care of them until they hatch. This subsocial behaviour is a large sacrifice on the part of the male, reducing foraging ability and increasing susceptibility to predators. In addition, female water bugs may often try to scrape existing eggs off the back of a male in order to lay her. These subsocial behavioral techniques are a good example of the complicated social interactions present in insects.

ECOLOGY

Aquatic Hemiptera hold an important place in the ecology of freshwater systems. They are important food for many organisms, including fish, amphibians, waterfowl and other animals. These insects generally hold an intermediate place in food chains, and apart from being eaten, are often important predators.

For example, in a study of a fishless spring in Arizona, a species of *Belastoma* ate more prey (mostly amphipods) than other top predators, including dragonfly nymphs. In the desert spring it accounted for 25 per cent of the energy in secondary consumer production as well as 72 per cent of production by animals at the top of the food chain.

Another important function of aquatic Hemipterans in ecosystems is to inhabit conditions that would be extremely stressful for other organisms. Many of these insects have the ability to live in extremely adverse environments. Corixidae, in particular, are some of the first colonizers of extreme environments. These insects have been found in German mining lakes with a pH below 3. They have also been present in Finland waters heavily polluted by sewage. In such environments, aquatic Hemipterans may be the first step in redeveloping a healthy ecosystem.

ECONOMIC IMPORTANCE

As could be expected from their important role in ecosystems, this group of insects also has much economic importance. They control mosquito populations, provide food for fish, are used as bioindicators, and even provide food for people in some countries. The economic importance of these insects has probably been underestimated.

These insects are voracious predators and can fly to different bodies of water. This makes them a more important mosquito regulator than even the widely-used mosquitofish which can't move out of one body of water. The World Health Organisation as well as other researchers are looking into use of these predators for mosquito control, and there is even some concern that insecticide run-off could be damaging to populations of these important predators. These insects are also highly important in relation to fish populations, which makes them important to humans. They are a primary food of many wild fish, which makes them valuable for sports fisheries. In fact, many fishing lures have been modeled off of aquatic Hemipterans. As a side note, these predators are also occasional pests in manmade fish hatcheries where they feed off of the young fish. Environmental reclamation of aquatic habitats is aided by aquatic Hemipterans which often can function as bioindicators. A bioindicator is a plant or animal that can be used to measure levels of pollution through its behaviour and population dynamic. These bugs, since they can survive in heavily polluted areas, are often used to gauge the toxins in an environment. Finally, in many cultures these insects are eaten and enjoyed. For example, mangdana, a species of giant water bug, that is enjoyed by Thailand people in stirfiries and salads. Because of the large size of these insects, there's presumably a lot of "meat" from one insect making them a good food source. Perhaps these water bugs could be a good way to ease Americans into entomophagy; if we're willing to eat other freshwater arthropods (crayfish), then eating freshwater insects can't be that great of a step.

RESULTS AND DISCUSSION

The aquatic Hemipterans are a highly important group of insects. Factors of this group such as the high number of species and the unique adaptations of the insects have allowed them to become one of the most significant insect clads in many senses. Their adaptations give examples of the extremes to which insects can evolve. The ecological balance of freshwater systems depends upon the aquatic Hemipterans. Humans benefit from these insects for sport fisheries and mosquito control among other things. The importance of this group has been acknowledged by many scientists and there is a large amount of research that has been completed on aquatic Hemipterans. However, there is always more to learn about these fascinating insects.

IMPORTANCE OF AQUATIC INSECTS

Aquatic insects, microscopic though some may be, are an integral part of the Earth's ecosystem. If a body of fresh water, large or small, sticks around for a week or so, chances are at least a few aquatic insects will make themselves at home in it. To those keeping watch, these tiny aquatic inhabitants are indicators of ecosystem health, and their presence — or absence — tells a story.

AQUATIC INSECTS

About 751,000 insect species have been identified so far, and many of those at least begin their lives in fresh water. Dragonflies, damselflies and other flies, bugs or beetles deposit their larvae in flowing or standing water. Once they have matured to a certain point, they leave the water and live the rest of their lives on land. Others, such as water beetles, live their entire lives in the water. The larvae of some species are programmed to burrow down into the wet mud if the water dries up, where they remain undeveloped, in a stasis or hibernating-type state, until water appears again.

ECOSYSTEM HEALTH

The absence of aquatic insects in standing or fresh water, or the presence of dead or dying ones, is usually an indicator of the health of the water. Many species of aquatic insects are very sensitive to pollution and other environmental hazards.

Scientists and researchers gain an idea of the health of the water by studying the effect of the water on insects that have little tolerance to pollution, as well as those that are less affected by it. Aquatic insects also perform various roles in ensuring a functioning freshwater ecosystem. Some help keep the water clear by filtering it, others aid in breaking down fallen, dead leaves, while yet others thin the algae, allowing the water to produce more oxygen.

FOOD WEB

The role of aquatic insects in food webs, particularly in wetlands, is of enormous importance. Mouthfuls of these aquatic dwellers, both larvae and adult, provide a nutritious meal to waterfowl, fish, shorebirds and other creatures. This food source is essential to the survival strategy of some fish, amphibians and other animals. Some aquatic insects are predators and in consuming their watery prey they help keep the insect population stable and functioning.

TOXICOLOGICAL DATA

Whether the toxin is unintentional agricultural run-off or the deliberate use of pesticides, aquatic insects provide useful data for scientists and researchers.

This research helps provide information on the toxic effects on non-target species of pesticides aimed at mosquitoes and other pests. Various species of aquatic insects also are used in controlled environments and experiments to measure the effects of pollution.

INVESTIGATION OF AQUATIC INSECTS

Aquatic entomology is the study of aquatic insects. Although it is a discipline in its own right, it often interacts with other disciplines or forms an integral part of them.

For example many aquatic flies are important as disease vectors, hence aquatic entomology interacts with medical entomology and parasitology. Aquatic insects contribute significantly to fresh water ecosystems, one of many groups of organisms that, together, must be considered in the study of aquatic ecology. As such their study may be a significant part of understanding the ecological state of a given ecosystem and in gauging how that ecosystem will respond to stress.

USE OF AQUATIC ENTOMOLOGY IN ENVIRONMENTAL PROTECTION

It would he difficult to overestimate the importance of aquatic insects as food items for other animals, particularly in the food webs associated with wetland environments. Many fishes, amphibians, shorebirds, waterfowl, and other animals forage heavily on both the aquatic and terrestrial stages of aquatic insects, which are essential to their survival.

It is imperative that society fully understand the potential consequences of the uses and alterations of the earth's natural environment. Environmental scientists have come to realise that uses or changes of waterways and lakes may have dire short-term or even irreversible long-term effects, not only on the quality of water itself but also on aquatic ecosystems. As a result, research has been employed to ascertain the effects of perturbations on our water resources and to increase our knowledge about the makeup of natural communities of aquatic organisms and their relation- ships with natural environments.

To this end, aquatic organisms that may be affected by an impending alteration or activity are often surveyed as part of an environmental assessment or impact study. this should be done, tor example, before impounding a river to create a reservoir; before dredging a stream; before constructing power plant cooling facilities, sewage treatment plants, or factories that may deliver potentially toxic effluents into a river or lake; before mining or deforestation activities; or before spraying chemicals over wetlands and forests. Measurements of the richness and diversity of aquatic insect species in relationship to the chemical and physical characters of their environment provide very usable indices for such baseline studies and are commonly made for these purposes.

Aquatic insects and other bottom-dwelling organisms in freshwater systems are also monitored in order to gauge subtle and profound effects that changes in water quality have on aquatic life. Changes in the composition of bottom-dwelling communities, as measured both qualitatively and quantitatively, will reflect, to various degrees, either non-optimum or intolerable quality shifts that result from the addition of pollutants to the water. Sources of such pollutants may be continual, intermittent, or accidental, and they may originate either from precise points or from large areas.

Bio-monitoring of this sort has some decided advantages over chemical and other types of water analysis. Bottom-dwelling insects and other invertebrates, by their very nature, maintain a relatively stable position in the aquatic environment.

Thus the community composition can reflect either previous or long-range shifts in water quality. Analysis of water chemistry or of highly mobile aquatic animals such as fishes tends to reflect the quality of water only at the times that samples are taken. Aquatic insects are also used in bioassay work. Laboratory bioassays are performed to determine the toxicological effects of pesticides on aquatic insects.

The pesticides may be those that are intentionally used in aquatic or adjacent environments for the control of mosquitoes and other pests or they may be those that inadvertently find their way into aquatic ecosystems, for example, via agricultural run-off. Such bioassays are commonly used to ascertain effects on non-target aquatic insect species, and thus have a direct relationship to environmental protection. In addition, several aquatic insects are studied in the laboratory in order to determine the mode of action of pesticides and the relative tolerances that these insects have for pollutants, such as heavy metals, organic enrichment, and heated waters.

Aquatic insects for which sufficient toxicological data are available have some potential for being used in field bioassays. That is, caged stock populations could be implanted at various points in a stream-for example, at the source (or supposed source) of a pollutant, upstream from the source, and downstream from the source. The relative mortality of these implanted populations could then be used as pollution indicators. Certain fish species have been commonly used in such field bioassays. We will use aquatic insect study as part of the overall habitat assessment for the Jemez Mountains, and perhaps for other areas, as an indicator of biological health and diversity. There are many methods for sampling aquatic insect populations: screens, nets, artificial substrate samplers, grabbing devices and dredges. Most of these are used in both fresh and salt water environments. A few of these are shown on the accompanying Aquatic Insect Sampling Devices.

We will use a modification of the substrate sampler in which the substrate is made of leaves ratherthan being artificial. The leaves used are generally those of the gambel oak which provide flatsurfaces like the plate sampler, which have nooks and crannies like the rocks, but which also areorganic and attract insects which eat organic matter.

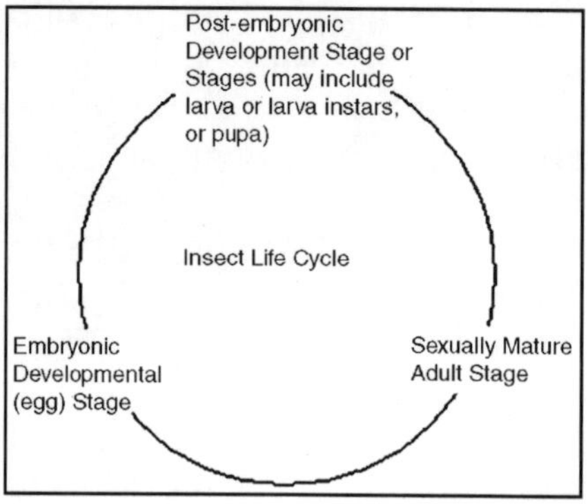

Most aquatic insects are considered benthic because they are found associated with the bottom or with a substrate, but they occupy virtually every possible niche within their habitat.

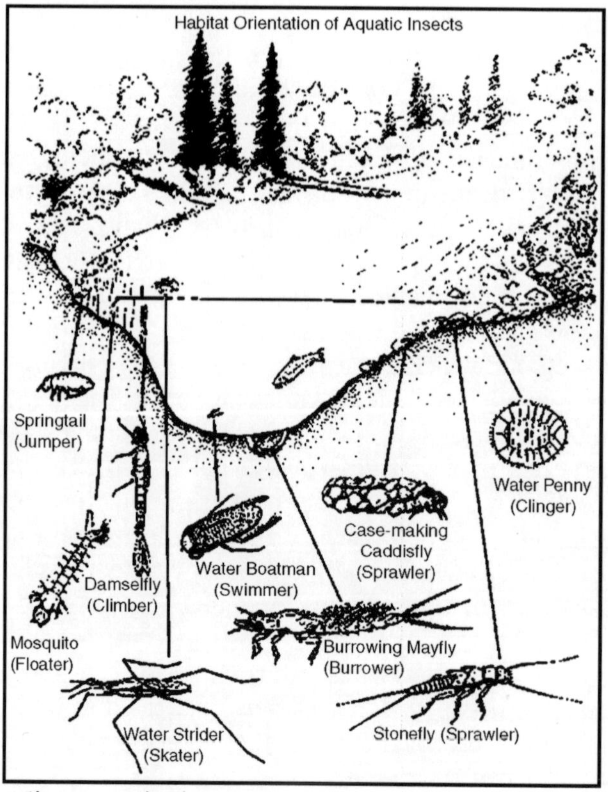

Some are clingers, clinging to the substrate with grasping claws or disks; the sprawlers crawl along the protected surfaces of the substrates; climbers reside on aquatic plant stems and other shoreline substrates; burrowersburrow into the soft bottom; floaters and swimmers are not associated with a substrate for attachment. The water skaters seen on many streams are examples of this type. Aquatic insects, like other insects, go through life cycles which include embryonic (egg) stages as well as post-embryonic stages which may include larvae, pupae etc. leading to the sexually mature adult stage. The term for passing from one stage to the next is metamorphosis.

We will mostly see larvae upon examining the leaf packets left in the streams. These insects are very small and will be examined with a stereomicroscope (a dissecting scope) which renders them up to 45 times their actual size. They will have been fixed at retrieval with Kales Solution, a combination of alcohol, formalin, and acetic acid. This solution is very pungent and irritating to the nose and eyes. Be careful to avoid too much exposure when viewing the specimens.

7

Aquatic Insect Biodiversity and Conservation

The diversity of insects can only be described as amazing. More than half of all known species of living things (microbes, plants, and animals) are insects. There are about 751,000 known species of insects, which is about three-fourths of all species of animals on the planet. While most insects live on land, their diversity also includes many species that are aquatic. In North America, there are more than 8,600 species of insects associated with freshwater environments during some part of their lives.

Just about anywhere you go on the planet, there is some kind of insect that will live in almost any place that stays wet for a week or so. Aquatic insects are important food for fish and waterfowl. They also play important roles in keeping freshwater ecosystems functioning properly.

Not many aquatic insects have been listed as threatened or endangered, but that is only because we have only begun to study their distribution and population numbers. Many species of aquatic insects are very susceptible to pollution or alteration of their habitat. In fact, aquatic insects are the group of living things used most commonly for monitoring the health of aquatic environments.

AQUATIC INSECTS

Insects are invertebrates (animals without a backbone) that are part of the larger group of animals called arthropods. Arthropod means "joint footed." That name was given to these animals because all of the arthropods have legs with joints that are something like our elbows and knees. Some other arthropod relatives of insects are crayfish, crabs, lobsters, millipedes, centipedes, scorpions, spiders, and ticks. Most insects are terrestrial (live on land), and are found in places such as trees, shrubs, flowers, rocks, logs, soil, buildings, and especially our gardens. Everyone is familiar with common terrestrial insects such as butterflies, moths, beetles, ants, bees, wasps, grasshoppers, crickets, cockroaches, and flies. There are also many kinds of insects that live in the water. These are called aquatic insects, and they are often not seen unless you explore places such as puddles, ponds, lakes, ditches, streams, and lakes. There are many different kinds of aquatic insects and almost every type of freshwater environment will have some kind of aquatic insect living in it.

STRUCTURE AND APPEARANCE

Like their other arthropod relatives, insects have their skeleton on the outside of their body (exoskeleton). This outside skeleton is thick, and often hard, so that it can protect the insect, much like our skin protects us. However, an insect's exoskeleton does not grow along with the insect. As an insect grows it must shed its skin and grow a new one in order to get larger. There are several features of insects' bodies that make them different from the other arthropods. The body of an insect is made up of three sections. The head is at the front end of the body. The thorax is in the middle of the body and is usually larger than the head. The abdomen is the rear section of the body and is usually as long, or longer, than the head and thorax together. You can usually see individual segments (up to 10) in the abdomen, but the individual segments in the head and thorax are usually fused together and cannot be easily distinguished.

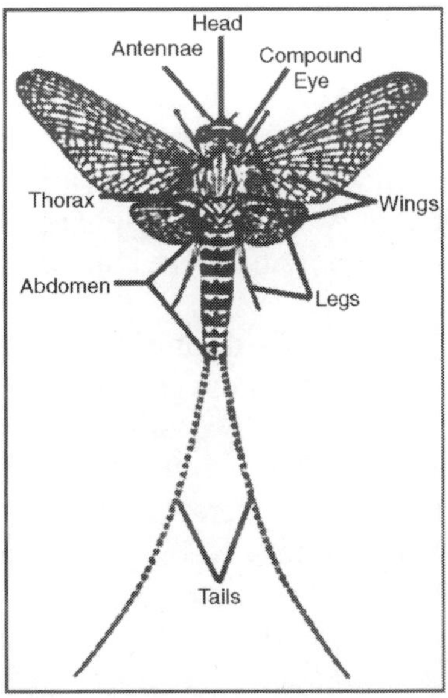

Fig. Major Morphological Features

The head contains structures for eating and sensing the world that an insect lives in. Insects have several different mouthparts that are specialised for tasting, obtaining, and breaking up food. There are two antennae (feelers) on the head, one on each side. There are usually two large compound eyes, which contain thousands of small individual eye cells. The thorax contains the structures that insects use to move around. There are six legs, arranged with three on each side of the body. Some of the legs may be constructed for special movements or purposes, such as running, jumping, digging, or catching food. Most insects have four wings on the thorax. Insects are the only arthropods that can fly, which is the main reason they are more widespread and more diverse than any of the other arthropods.

The abdomen does not have many structures on the outside, except for some structures at the end that are specialised for mating and laying eggs. Many insects have two tails on the rear of the abdomen, which they often use to sense vibrations.

The distinguishing features described above apply to adult insects. For many aquatic insects, only the immature stages (babies) live in the water. Sometimes the immature stages do not have the same features as adult insects. Immature insects never have working wings, and some may not have compound eyes, jointed legs, or a distinguishable thorax section. Some immature insects look more like worms than insects. However, wormlike immature insects will always have an obvious head, or at least some noticeable structures sticking out from the head, such as mouthparts or antennae, while worms never have a head or head structures.

HOW DO THEY BREATHE

Insects that live on land breathe air through holes in their bodies called spiracles. This would not work very well for insects that live in the water, so aquatic insects have special adaptations for breathing in the water without drowning. The most common way for aquatic insects to breathe effectively underwater is to use oxygen that is dissolved in the water rather than oxygen that is a gas in the air.

Many aquatic insects, especially during their immature stages, have gills similar to fish for obtaining dissolved oxygen. The gills of aquatic insects are located on the outside of their body in various locations. Gills come in various shapes, but many are flat oval plates or tufts of small filaments. Other aquatic insects have a soft flexible external skeleton that simply allows dissolved oxygen to pass from the water into their body all over their body surface. Some kinds still use the holes in their bodies to get oxygen from the air. They just keep the holes shut while they are underwater, and only open them when they come to the surface to take in a breath of air. Some kinds take a bubble of air underwater and breathe out of the bubble, which allows them to stay underwater longer.

This is comparable to SCUBA diving. A few kinds of aquatic insects have their spiracles on the end of a long tube at the end of their abdomen. They keep their body underwater and just stick their breathing tube up to the surface to get air, much like snorkeling.

WHAT DO THEY EAT

The foods of aquatic insects are just as diverse as the habitats they live in. Although individual kinds of aquatic insects may only eat one type of food, all organic material in the water, living and dead, is eaten by some kind of aquatic insect. Scientists have found it informative to categorise aquatic insects according to how they obtain their food for studying the ecology of freshwater ecosystems. These categories are called functional feeding groups. Scrapers have special mouthparts that remove algae growing on the surface of rocks or other solid objects.

These mouthparts work like a sharp blade to remove the outermost layer of algae, which is attached very tightly but is very nutritious for those insects equipped to remove it. Collectors acquire small pieces of decaying plant material (detritus). Some kinds use long hairs on their head or legs or silk nets to filter these small particles out of the water. Other kinds of collectors use their mouthparts to gather fine particles lying on the bottom and shove this material into their mouths. Shredders have mouthparts that are designed to nibble off pieces of soft vegetation, such as leaves, flowers, or twigs, and grind up this material.

Most aquatic insects shred pieces of vegetation that have dropped off of plants and are decaying. Most of this material comes from trees and shrubs that grow on land at the edge of the water. Only a few kinds of aquatic insects feed on parts of live plants that grow under the water. Predators feed on other animals that are alive. Predators often have special structures for catching and subduing their prey, such as strong jaws with teeth, a sharp beak, or spiny legs. Predators eat other invertebrates most of the time, but some are large and strong enough to catch small vertebrates, such as fish and tadpoles.

HOW DO THEY GROW

Insects, like all arthropods, must shed their protective external skeleton periodically in order to get larger. Before they shed their old skin, they grow a new one underneath. The new larger skin is soft and folded, so that it can fit inside the old smaller skin.

They also absorb much of the old skin and recycle the materials it contained. To shed the old skin, aquatic insects gulp water to make themselves larger, and they push from the inside on the top of the thorax. The old exoskeleton is thinner there, so it splits and allows the insect to climb out. After getting out of the old skin, the insect continues to gulp water to make itself larger by stretching out the wrinkles in the new soft skin. After a few hours, or in some cases a few days, the new skin hardens into another protective exoskeleton. Different kinds of aquatic insects shed their skin anywhere from three to 45 times. This is a dangerous time in the life of an aquatic insect, and many of them die while they are waiting for their new skin to harden. The new soft exoskeleton is subject to damage, and it does not allow the insect to move and hide very well. As insects grow from eggs into adults, they go through a series of developmental changes that are called metamorphoses. There are two basic types of metamorphosis in aquatic insects. It is important to know about these two types because metamorphosis is used to help identify immature insects and it explains a lot of the biological activities they engage in during their lives. One type of metamorphosis is incomplete (also called hemimetabolous). Aquatic insects with incomplete metamorphosis emerge from the egg looking a lot like miniature versions of the adults, minus wings. They have compound eyes on the head and jointed legs on the thorax. The wings develop in projections on the thorax (wing pads) and get a little larger each time the insect sheds its skin. Scientists have begun to use the term larvae for the immature stages of aquatic insects with incomplete metamorphosis, but for many years they were called nymphs or naiads.

The other type of metamorphosis is complete (also called holometabolous). Immature aquatic insects with complete metamorphosis do not look anything like the adults they turn into. At most they have only a simple eyespot, or a small cluster of eyespots, on the head, and some have no eyes at all. They may or may not have jointed legs on the thorax. The wings develop inside the body, so no wing pads are visible until just before the insect becomes an adult. insects with complete metamorphosis go into an inactive stage before they become adults. During the inactive stage is when compound eyes, jointed legs, and wing pads first show up. The active immature stages are called larvae, and the inactive stage is called a pupa. Some aquatic insects crawl out of the water for the pupa stage, while others spend this stage in the water.

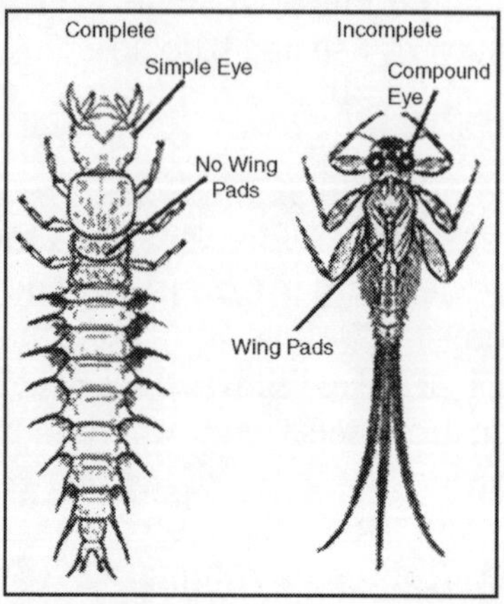

Fig. Two Types of Metamorphosis

HOW DO THEY REPRODUCE

Only adult insects are capable of reproducing, and most aquatic insects spend their adult stage out of the water. After mating on land, females return to the water to deposity her eggs. Eggs are usually stuck on solid objects under water, but a few kinds deposit the eggs on trees or rocks above the water. Eggs usually hatch within a few days or weeks, but some may be programmed to not hatch for many months. A delay in hatching allows an aquatic insect to live in habitats that are too hot, too cold, or dry up during part of the year.

Different kinds of aquatic insects require anywhere from a few weeks to several years to develop into adults. Most grow and develop steadily, but some may go into an inactive state to endure harsh environmental conditions. It is most common for aquatic insects to produce one generation per year, with the adults emerging during the warm months. Some of the larger kinds, or those living in cold environments, require two to three years to develop from egg to adult. Some kinds with short developmental times may produce more than one generation per year. In southern latitudes of the United States, some kinds reproduce continuously throughout the year. Adult aquatic insects usually live a few days to a few weeks. The extremes of adult life span range from a few minutes (some mayflies) to two or three years (some water beetles).

WHERE DO THEY LIVE

One of the most amazing things about aquatic insects is the diversity of habitats where they live. There is no body of water that is too small, too large, too cold, too hot, too muddy, with oxygen too low, with currents too fast, or even with too much pollution for some kind of aquatic insect to live there. About the only restriction to where they live is that they do not usually inhabit the salty water of marine environments, such as oceans and bays. However, there are even a couple of unusual aquatic insects that live on coral reefs and in tide pools of marine environments. Estuaries, where the fresh water of rivers mixes with the salt water of oceans, are home to quite a few kinds of aquatic insects. Anyone who has been to the beach knows about the kinds of mosquitoes that breed in the salt marshes near the beach.

Not all kinds of aquatic insects live in all types of freshwater habitats. The most favourable habitats, and the ones where you can collect the most kinds of aquatic insects, are the edges of ponds and lakes and the sections of streams and rivers where the water is flowing fast enough to splash (riffles). In both standing and flowing freshwater habitats, the most different kinds of aquatic insects will be found in water that is less than three feet deep and can be easily waded. Aquatic insects have a variety of special adaptations for moving around or staying in one place within their habitat. Some are agile swimmers by means of streamlined bodies with long legs or tails, while others climb around on aquatic plants by means of long thin bodies. Some sprawl on top of soft mucky bottom without sinking in because their bodies are flat and their legs extend out from the sides. Others are able to burrow down into soft mucky bottom because they have special structures on their bodies, such as legs that look like shovels or points projecting in front of their heads. Still others can cling to rocks and logs in very swift current because their bodies are very flat and the current just passes over them without knocking them off. Other clingers stay put by using special suckers or by gluing themselves down with sticky silk that they produce. Lastly, many aquatic insects like to crawl around in the tiny spaces among rocks, sticks, and dead leaves.

Because aquatic insects are small and highly specialised, different kinds are often found in small areas with similar features, which are called microhabitats. Examples of microhabitats where you will probably find different aquatic insects are: cobble rocks (about the size of your fist or head), gravel, sand, muck, accumulations of dead leaves and twigs, live plants, and grasses and tree roots that extend into the water from land. Different microhabitats, with different aquatic insects living in them, occur very close together, perhaps within one step of each other.

Aquatic insects even live in temporary habitats, such as small streams or ponds that dry up in the summer. If they are adults, they can simply fly to another place with water. Some immature aquatic insects that cannot yet fly will burrow down into the bottom where it is damp and go into an inactive state, something like animals hibernating over winter. However, most aquatic insects that live in temporary habitats are "programmed" to stay in their eggs, where they are protected, until the time of year when water is present.

COLLECTING AQUATIC INSECTS

Aquatic insects are always easy to collect because they are so numerous and they live in so many different places. The simplest way is just to pick up objects in the water, such as rocks, plants, sticks, or leaves, and examine the material. Sometimes it works best to place the object in a shallow pan with clean water. You can see aquatic insects with your naked eye, but a magnifying glass might help to find some of the smaller ones.

You will probably collect more organisms by using a net with fairly fine mesh. These are available in stores or you can build your own from materials around the house. If there is current, hold the net in the water and move the habitat where you think aquatic insects might live upstream from the net. It works best if you move the habitat with your hands, but it will also work to kick with your feet (it just damages more of the organisms). Either way, aquatic insects and some of the material where they were living will wash right into your net. If you are in standing water, you will need to move the net in the places where the aquatic insects live. Usually short pokes with the net work best. After collecting material in the net, empty the contents into a shallow pan with clean water to find the aquatic insects.

There is no reason to kill aquatic insects unnecessarily, so it is always best to release them back into the water alive after studying them for a while.

However, since most aquatic insects are very abundant, it is acceptable to make a preserved collection of common ones for educational purposes. The best way to preserve aquatic insects is to place them in alcohol in a small clear container with a tight fitting lid. Rubbing alcohol (isopropanol) works fine. You should put a paper label in the container stating what kind of aquatic insect it is, as well as where and when you collected the organism. The labels should be written in pencil or indelible ink.

WHAT GOOD ARE THEY

Aquatic insects are probably best known for their ability to tell us about the water quality in a particular environment. Some of them are very sensitive to pollution, while others are tolerant. If you take a sample of the aquatic insects in a particular place, and analyse the sample in terms of the sensitive kinds versus tolerant kinds, you can get a good measure of the environmental health. Healthy aquatic environments have a lot of different sensitive kinds, while polluted environments have only a few kinds of tolerant aquatic insects. This process is called biological monitoring (or biomonitoring) and is commonly done by government agencies as well as citizen volunteer organisations.

However, the use of aquatic insects for biomonitoring is not the main reason that they are important. Because there are so many different kinds of aquatic insects and their activities are so diverse, they perform essential roles that keep freshwater ecosystems functioning properly. A good analogy would be the various jobs of factory workers on an assembly line that are necessary to make sure that a manufacturing plant turns out plenty of good products.

Some of the aquatic insects are responsible for breaking down the dead leaves and other plant parts that fall into bodies of water from land. This material provides the base of the food chain in some aquatic environments, especially small streams in forests. Some scrape the algae that grow on all firm surfaces in water, such as rocks, logs, and the leaves and stems of live rooted plants. This layer of algae, which produces much oxygen and food for other organisms, is more productive if it is kept thin by the grazing of aquatic insects and other invertebrates.

Other kinds of aquatic insects are specialised for filtering fine particles that are suspended in the water. This is useful because it helps to keep the water clear enough for light to penetrate where algae and other plants are growing on the bottom. Other kinds mix the soft bottom sediments as they burrow in search of food. This makes the bottom healthier for organisms because it puts oxygen from the water into the bottom. Lastly, the aquatic insects that are predators reduce the numbers of other invertebrates and help keep a balance among the different kinds of organisms and the food that is available. Even if aquatic insects did not perform these important jobs in aquatic ecosystems, they would still be useful just because collecting and observing them is so easy and so much fun. Aquatic insects are an excellent way to get people of all ages interested in nature and conservation of natural resources.

AQUATIC INSECT KILLERS THREATS

Hardly any species of aquatic insects have been listed as endangered or threatened. However, the reason for this is that studying the distribution and population numbers of such a diverse group of organisms is an overwhelming task. Also, most people do not understand and appreciate the importance of these small creatures in freshwater ecosystems. It is certain that many species of aquatic insects are threatened and perhaps on the verge of extinction. These are most likely to be species that have narrow ecological requirements, and, thus, live in unique habitats that have not been disturbed by human activities.

In the past, aquatic insects were severely reduced in many bodies of water by discharges of toxic substances, such as those from manufacturing plants and mines. Also, overloaded sewage treatment plants discharged human waste, which used up all of the dissolved oxygen when it decayed. Today in the United States, most of those point source discharges have been greatly reduced by strict enforcement of anti-pollution laws.

However, aquatic insects still face a great threat from non-point source pollution. This widespread problem comes mainly from excessive amounts of nutrients and sediment that get into streams, rivers, ponds, and lakes from so many different sources that it is hard to track them all down.

The human activities that are responsible for most of the non-point source pollution are agriculture, forestry, and urban development. Many kinds of aquatic insects are eliminated because the excess nutrients and sediment cover the surfaces where aquatic insects need to hold on or clog the small spaces where they need to hide. In addition, these pollutants cause the water to have less dissolved oxygen. Other current non-point source problems for aquatic insects include warm water temperature caused by removing the trees that grow along streams and the introduction of toxic substances that wash off of city streets and people's lawns.

MAJOR GROUPS OF AQUATIC INSECTS

There are so many different kinds of aquatic insects, it is difficult to appreciate their biological diversity without considering some of the individual kinds.

The following section provides a brief summary of the eight major groups:

- *Mayflies (Ephemeroptera):* Larvae of mayflies live in a wide variety of flowing and standing waters. Most of them eat plant material, either by scraping algae or collecting small pieces of detritus from the bottom. Larvae breathe dissolved oxygen by means of gills on the abdomen. They have incomplete metamorphosis. Most mayflies are sensitive to pollution, although there are a couple of exceptions. The most unusual feature of mayflies is that the adults only live a few hours and never eat.

- *Dragonflies and Damselflies (Odonata):* Larvae of dragonflies and damselflies are most common in standing or slow-moving waters. All of them are predators. Larvae breathe dissolved oxygen with gills, which are located either inside the rear portion of the abdomen (dragonflies) or on the end of the abdomen (damselflies). They have incomplete metamorphosis. Many kinds are fairly tolerant of pollution, but some kinds only live in unique habitats, such as bogs high in the mountains. The most unusual feature of this group is the way the larvae catch their food with an elbowed lower lip, which they can shoot out in front of the head.
- *Stoneflies (Plecoptera):* Larvae of stoneflies live only in flowing waters, often cool, swift streams with high dissolved oxygen. Some feed on plant material, either by shredding dead leaves and other large pieces of detritus, while others are predators. Larvae breathe dissolved oxygen. Some have gills on their thorax, but others just obtain dissolved oxygen all over their body. They have incomplete metamorphosis. Almost all of the stoneflies are sensitive to pollution. The most unusual feature of this group is that some kinds are programmed to emerge only during the coldest months; hence, they are called the winter stoneflies.
- *True Bugs (Hemiptera):* Most of the true bugs live on land, but the aquatic kinds are most common in the shallow areas around the edge of standing waters. Both the adults and the larvae of the aquatic kinds live in the water. Both stages are usually found on submerged aquatic plants. Almost all of them are predators. They breathe oxygen from the air, either by taking a bubble underwater or by sticking a breathing tube up into the air. They have incomplete metamorphosis. Most kinds are tolerant of pollution. The most unusual feature of this group is the way they kill and eat their prey. True bugs have a sharp beak that they stick into the body of their prey, and then they pump in poison to kill their prey, after which they suck out the body fluids. Some of the larger kinds feed on small fish and tadpoles.
- *Dobsonflies and Alderflies (Megaloptera):* Larvae of different kinds live in flowing or standing waters. They are all predators. They breathe dissolved oxygen by means of gills and their overall body surface. They have complete metamorphosis. Mature larvae leave the water and dig out a protected space under a rock or log for the pupa stage. Different kinds are either sensitive or tolerant to pollution. Larvae of some of the larger kinds are called hellgrammites, which are

popular as live bait for smallmouth bass and other warm-water fish species.

- *Water Beetles (Coleoptera):* There are more species of beetles than any other insects, but most of them live on land. Most of the water beetles are more common in standing or slow-moving waters, but a few kinds are only found in swiftly flowing waters. Both the adults and the larvae of the aquatic kinds live in the water. Water beetles feed in different ways, primarily by preying on other animals, scraping algae, or collecting small particles of detritus from the bottom. All of the adults breathe air by taking a bubble underwater, while most of the larvae breathe dissolved oxygen by a combination of gills and their overall body surface. They have complete metamorphosis and leave the water for the pupa stage. Water beetles range from sensitive to somewhat tolerant of pollution. The most unusual feature of water beetles is that some of the adults live for several years.
- *Caddisflies (Trichoptera):* Larvae of different caddisflies live in a wide variety of flowing and standing waters. They also have a wide range of feeding habits, including scraping algae, collecting fine particles of detritus from the bottom or from the water, shredding dead leaves, and preying on other invertebrates. They breathe dissolved oxygen by means of gills and their overall body surface. Caddisflies have complete metamorphosis and remain in the water for the pupa stage. Most kinds are sensitive to pollution, but a few kinds are somewhat tolerant of moderate levels of pollution. The most distinctive feature of caddisflies is their ability to spin silk out of their lower lip. They use this material to glue together stones or pieces of vegetation into a small house for their protection during the larva and pupa stages. Some also use strands of silk to make a net for filtering particles of food from the water.
- *True Flies (Diptera):* This group has more kinds on land, but there are also many aquatic kinds. They have a wide range of feeding habits, including scraping algae, collecting fine particles of detritus from the bottom or from the water, shredding dead leaves, and preying on other invertebrates. They breathe dissolved oxygen by means of gills and their overall body surface. True flies have complete metamorphosis and remain in the water for the pupa stage. The most distinctive feature of this group is their ecological diversity. Some kinds live in the cleanest habitats (*e.g.*, swift, cool, mountain streams), while others live in some of the harshest natural habitats on the earth (*e.g.*, arctic tundra ponds, geothermal springs, alkaline lakes, mucky swamps). They have equally diverse responses to pollution, with some kinds being exceptionally sensitive, while other kinds endure the

worst imaginable water quality (*e.g.*, raw sewage or acid mine drainage).

AQUATIC INSECTS AND DIET

Aquatic insects are insects that spend at least a portion of their life cycle under or on top of water. Some aquatic insects breathe underwater with gills or breathe from the air at the surface. Some insects, such as dragonflies, mayflies and stoneflies, spend the early part of their life cycle underwater until they molt into an adult with wings. The life cycles vary greatly among aquatic insects. The presence or absence of certain aquatic insects in a lake, stream or river may indicate the overall health and quality of the water and ecosystem. Aquatic insects are also of great interest to trout fishermen, who will use lures that mimic specific insect species.

Fig. Dragonflies are Aquatic as Nymphs.

Aquatic insects are insects that spend at least a portion of their life cycle under or on top of water. Some aquatic insects breathe underwater with gills or breathe from the air at the surface. Some insects, such as dragonflies, mayflies and stoneflies, spend the early part of their life cycle underwater until they molt into an adult with wings. The life cycles vary greatly among aquatic insects. The presence or absence of certain aquatic insects in a lake, stream or river may indicate the overall health and quality of the water and ecosystem. Aquatic insects are also of great interest to trout fishermen, who will use lures that mimic specific insect species. Does this Spark an idea?

INCOMPLETE METAMORPHOSIS

Incomplete metamorphosis is when insects grow slightly larger each time they molt, eventually becoming adults with functional wings. Some common aquatic insects that fall into this category include mayflies, stoneflies, dragonflies and damselflies. These insects spend their lives underwater as nymphs with gills, and leave the water only after their final molt into adulthood. Mayfly and stonefly nymphs are usually found under rocks in streams. Dragonfly and damselfly nymphs can also be found in streams, but can inhabit lakes and ponds as well.

HEMIPTERA

Hemiptera is an order of insects that are often aquatic. They are known as "true bugs" and they may appear beetle-like to some people. These aquatic insects do not breathe underwater with gills. Some, like the water scorpion and giant water bug, have breathing tubes that extend from their abdomen that allow them to essentially snorkel. They also have hooked front legs to capture prey.

The giant water bug can capture prey 20 times larger in size, such as tadpoles, frogs and small fish. Another breathing trick, used by the water boatman and backswimmer, is similar to SCUBA diving. They can carry air bubbles underwater on tiny hairs on their abdomens, and swim with large paddle-like back legs to find prey underwater.

COMPLETE METAMORPHOSIS

Complete metamorphosis refers to the type of insect life cycle that includes larvae, pupae and adult stages. Aquatic insects that undergo this type of life cycle include flies and mosquitoes, caddisflies, dobsonflies, alderflies and beetles. These flies live their larval stage underwater, looking similar to maggots on land. Caddisfly larvae are known for their habit of building homes around themselves out of rocks, sticks and debris, woven together with silk that they produce. Some are carnivorous, while others are scavengers. Dobsonfly larvae are known as helgramites, which are long with their legs towards the front of their bodies. Helgramites have a large set of mandibles which can capture large prey, such as tadpoles. Most aquatic beetles spend both their larval and adult periods underwater. The adults will often carry an air bubble with them underwater, which they can use to breath. They are usually carnivorous during both life stages.

ECOLOGICAL IMPORTANCE

The presence of some aquatic species may indicate the health of a body of water. Some scientists will survey a stream or river by looking for which insects and other invertebrates are present. For example, stoneflies and mayflies require high levels of oxygen, so the disappearance of these species may indicate higher than normal levels of pollution. Some insects thrive in low levels of oxygen or in high levels of biological waste. Changes in the types or numbers of different species of aquatic insects can indicate changes to the aquatic ecosystem because of human influence. Because insects are towards the bottom of the food chain and are easy to capture, they can be used as reliable indicator species.

USES

Individuals who practice the sport of trout fishing have a great interest in aquatic insects. Since trout eat insects, trout fishermen use lures that look like aquatic insects. The practice of fly fishing mimics a flying insect dipping in and out of the surface of water. This motion is mimicking a female insect depositing her eggs into the water, or a terrestrial insect that accidentally fell into the water. To the fisherman, understanding the insect increases their odds of tricking the fish into believing that the lure is a real insect.

WHAT FLIES DO TROUT EAT

Though several types of flies and aquatic insects make up a large portion of a trout's diet, the diet varies depending on the life cycles and availability of aquatic and terrestrial insects and other aquatic creatures. The trout is not fussy when it chooses its food source and will dine on just about any small insect or creature that comes within its hunting range from things as small as plankton and fish eggs to creatures as large as frogs, dragonflies and other fish.

AQUATIC INSECTS

Aquatic insects are those insects that live on or beneath the surface of water or skim along the top. They include springtails, mayflies, dragonflies, stoneflies, dobsonflies, alderflies and even butterflies, moths and beetles. The trout is not particular about which of these insects it catches. It will grab any of them that happen to be over, on or under the water when it is hungry. Mayflies and young dragonflies are most numerous around the fresh water lakes and rivers where trout make their homes but any aquatic insect will do for a hungry trout.

TERRESTRIAL INSECTS

Terrestrial insects are insects that live on dry land. These insects include grasshoppers, crickets, cicadas, leafhoppers, ants and any other insect that lives on land. When they fall into the water by accident, their thrashing movements attract the trout. Though these insects make up only a small portion of a trout's diet, no trout will ignore the thrashing. Most trout will grab anything that comes within their range.

DEVELOPING INSECTS

Another large source of food for many trout is the young of the aquatic insects. Many aquatic insects lay their eggs in or near the water. When the eggs hatch, the larvae, pupae and nymphs spend their development time in or on the water. Brown trout, rainbow and cutthroat will make a feast on these pre-adult insects as well as the unhatched eggs.

OTHER FOOD SOURCES

Though insects make up a large portion of a trout's diet, the fish will not disregard such aquatic creatures as small fish, crayfish and crustaceans. When available, some trout will eat fresh-water shrimp and fish eggs. Some trout that live in lakes even become plankton feeders. In an environment where frogs are populous, trout can make a meal of small frogs and tadpoles as well.

INSECT-TRAPPING AQUATIC PLANTS

According to the International Carnivorous Society, carnivorous plants have three traits that separate them from other plants. They capture and kill prey, they have a way to digest their prey and they meet significant nutritional needs from doing so. These plants grow in all kinds of nutrient-poor environments, including bogs and swamps, and include two genera that are aquatic.

BLADDERWORTS

Bladderworts use small bladderlike traps to capture insects, small invertebrates and tiny fish. The bladderwort's trap works something like a suction bulb. When tiny hairlike projections around the opening of the bladder sense an insect or other creature passing by, they cause the trap to inflate, sucking in water along with their prey, and trapping it inside. Bladderworts live in open water like bog lakes, and can be found throughout the northern hemisphere. They don't have true leaves. Instead, highly branched, leaflike stems can be found floating or submerged underwater. Occasionally, this plant will attach itself loosely to the sediment. Bladderworts have yellow, snapdragonlike flowers with faint purple-brown stripes.

GROWING BLADDERWORTS

Grow aquatic bladderworts outdoors in ponds or fountains. Grow indoors in large aquariums without any fish, because fish will eat your plants. Add peat moss to the bottom of your pond or aquarium to increase the acidity of the water. Don't expose bladderworts to fertilizer or gardening soil. Bladderworts like partial to full sun and a lot of water. If you're growing them outside, make sure the bottom of the fountain or pond is below the frost line because plants sink to the bottom of the pond during their period of winter dormancy. Bladderworts don't produce enough oxygen from photosynthesis and need additional dissolved oxygen in the water where they grow. Aerate their water to provide additional oxygen.

WATERWHEELS

Waterwheel plants are closely related to Venus flytraps, using snap-traps to capture insects and other small prey. These rootless plants float just below the surface of the water. The leaves grow in whorls and end in clamlike traps that snap shut when the long, sensitive hairs are stimulated and triggered by passing prey. Traps close in 1/4 to 1/2 second and reopen in 10 to 20 hours if no prey is captured. Waterwheel plants have small white flowers and grow to be about 6 inches long. They can be found in Australia, Africa, Europe and India and are a threatened species in some areas due to high levels of pollution, illegal collection and destruction of natural habitat.

GROWING WATERWHEELS

Waterwheels are difficult to grow. They don't like algae and do better in acidic water. Grow them in an outdoor bog pool or in a large tank that's at least 30 gallons in size. Plants require a period of dormancy in the winter so, if you're growing them outside, ensure that the pool bottom is about 18 to 24 inches below the frost line. Add 1 inch of peat moss and leaf litter to the water at least one week before planting your waterwheels, to limit the growth of algae and allow tannins to be released into the water. Include companion plants like water lilies, cattail and pickerel weed to help produce food for the waterwheels and reduce the growth of algae. Waterwheels prefer warm, sunny conditions. Since they trap their own food, it's not necessary to feed them.

INSECTS THAT SWARM IN COLUMNS

Just about any insect that hatches in abundance is capable of swarming in columns, but a handful of insects come to mind as commonly forming such columns. Most of the insects known to swarm in this fashion are aquatic and hatch in waves, invoked by the time of year, time of day and weather changes. It is almost always near water: Either fresh, moving water or wetlands. There are several insects known for swarming in columns.

MAYFLIES

Mayflies are of the taxonomic order Ephemeroptera. Their name is derived from "ephemeral," meaning "short-lived." The adult, or mating, stage of their life is so short: one day up to a couple weeks. Mayflies are among trouts' favourite food. They are well-known to fly-fishermen as many flies are patterned after mayflies. In places, hatches of mayflies are sparse. In others, at times they can be so thick they form a dark cloud which is impossible to see through. Mayflies are known to take flight, gain altitude, then flutter back to the ground or water. When mayflies swarm, they can create great columns.

GNATS

Gnats belong to the taxonomic order Diptera — one of the most common insects and among the smallest. Gnats or "midges" can appear in massive quantities. They are an aquatic insect, so they generally appear near wetlands of some kind, but not necessarily moving freshwater like the mayflies prefer.

Like Ephemeroptera, Diptera is a favourite trout food. Subspecies vary, but they they look like small mosquitoes.

NO-SEE-UMS

These little insects with the funny name are a part of the order Diptera. So, they are a kind of gnat. They are, however, a biting gnat. If you encounter a swarming column of gnats and they look smaller than mosquitoes and one of them bites you, you're probably looking at no-see-ums. Alaska and other marshy, Northern areas are notorious for large no-see-um populations. Any marsh-like area that thaws in the spring is ideal habitat. Many consider no-see-ums to be a much more irritating bite than a mosquito's.

CADDISFLIES

Caddisflies are of the order Trichoptera. If a mayfly is a trout's favourite food, a caddisfly is probably its staple. It is among the most common trout food. And like other common trout food, it is a aquatic insect, spending most of its life as a nymph, extracting oxygen from the water. In its short adult life, it becomes terrestrial.

There are many families and geni of caddisflies. They often hatch in small waves. In some areas, at some times, caddisflies hatch in massive, dark clouds of hundreds of thousands of insects, forming dark columns.

INSECTS THAT HATCH IN WATER

Fly-fishermen know the emergence of aquatic insects from water as the "hatch." Fish feed on these insects, which range in colour, size and species. The great challenge for fly-anglers is to "match the hatch"; this involves selecting artificial flies that closely replicate the real species of insects hatching. If anglers are able to "match the hatch," they have a better chance to catch more fish. Fly-fishermen need to understand the life cycle, appearance, habitat and other unique traits of all the different types of flies that trout like to eat: mayflies, stoneflies, caddisflies and many others.

MAYFLIES

Mayflies, also known as dayflies, shadflies, lake flies, fish flies, and jinx flies, are aquatic insects belonging to the order Ephemeroptera. Interestingly enough, these aquatic insects spend a longer time in the immature stage (spent in freshwater) than in the adult stage.

This stage will span anywhere from between a few minutes to one day, depending on the species. The eggs are deposited in lakes or streams, which then sink to the bottom. As such, the naiads (the term used for the immature mayflies) may spend their first few years under rocks or in the sediment. As adults, they are known for their short antennae, large compound eyes (particularly the males), and membranous wings. Mayflies are one of the few insects (if not the only) that possess paired genitalia, with each of the aquatic insects possessing two organs of their gender.

CADDISFLIES

Caddisflies are aquatic insects that belong to the order Trichoptera. Experts believe that these are usually found in clean waters, including streams, lakes, rivers, ponds and vernal pools. Caddisflies closely resemble moths in appearance.

Several of the species belonging to the order Trichoptera spin silk mixed with gravel, sand or other debris found in the area where they live as a protective net. These cases greatly resemble the nets of bagworms but are usually open on both ends. Depending on the species, a caddisfly may build the net as larvae until they are ready to pupate.

STONEFLIES

Stoneflies are aquatic insects belonging to the order Plecoptera. According to the web site Kendall UK, stoneflies are one of the most ancient breeds of insects, with some stonefly fossils found dating back to the Carboniferous geological period. The modern stonefly emerged during the age of dinosaurs (Mesozoic period). The presence of stoneflies along with mayflies and caddisflies is a good indication that the water source has good quality, according to the web site King County Government. While most stoneflies have wings that lie flat on their backs, some species are wingless, as they are aquatic from birth to death.

MIDGES

Midges are like caddisflies as they exhibit a complete metamorphosis, from a larva, pupa, and adult stage. The caddisflies larvae live within stagnant pools, predominantly within water where there are many aquatic weeds. In slow, steady waters like creeks, small streams and some rivers, midges will make up most of the insect food supply for small fish, such as trout that live within waterways.

SPRING FLIES

During early spring, all it takes is a little sunshine for a "big hatch" to occur. With warmer temperatures, adult flies use this time to give birth to their offspring. Fly-fishermen will notice trout rising to snatch flies that land upon the water. To help in selecting the correct artificial fly to use, try to match the colour and size of the flies that land on the water during a hatch. According to Orvis web site, dry flies such as adams parachute, nymphs like the bead-head pheasant tail and the woolly-bugger streamer are top spring baits to catch "game fish."

AQUATIC INSECT RESOURCE MANAGEMENT

Managing natural resources specifically to ensure the continued existence of any aquatic insect has yet to become a reality. Weighted down by the pest stigma, insects have garnered relatively little visibility or clout in the realm of species protection. Consequently, past protective measures and management practices have been directed at a few vertebrate and even fewer invertebrate species. Yet on a number of species basis, 95 percent of the world's fauna is composed of invertebrates, and roughly 70 to 80 percent of this fauna is composed of insects alone.

The species richness and enormous population sizes of invertebrates prompted E. O. Wilson to contend that invertebrates drive our planet's living systems. If we are going to maintain the biological diversity of this world, we must plan for the well-being of invertebrates. Considering that biological diversity is largely a function of insect species richness and that many aquatic insects are imperiled (Morse *et al.*, 1997), work in this area is sadly overdue. In the southeastern United States (Alabama, Arkansas, Florida, Georgia, Kentucky, Louisiana, Mississippi, North Carolina, South Carolina, Tennessee, Virginia), there are more than 4,200 species of aquatic insects known to science. At the regional conference where an oral version of this chapter was presented, the grand total of all species addressed was roughly 6,000. Aquatic invertebrates (insects, decapod crustaceans, and molluscs) accounted for at least 5,000 of these, which was more than 80 percent of the total. Of course, this percentage is an underestimate given that not all invertebrate groups were considered (*e.g.*, non-decapod crustaceans, annelids, bryozoans, nematodes, rotifers). Furthermore, there certainly are aquatic invertebrate species awaiting discovery in the Southeast. Although our knowledge of aquatic insect distributions, population sizes, and habitat requirements is incomplete, it is apparent that some aquatic insect species are imperiled. Morse *et al.* considered 153 species in just four orders of aquatic insects as rare and vulnerable to extirpation. This number is 12.6 percent of the 1,212 species contained in these orders known to occur in the Southeast. If this percentage is applicable to the other orders of aquatic insects (beetles, hemipterans, dipterans, etc.), then of the approximately 4,200 species in the Southeast, over 500 may be imperiled. Morse *et al.* suspect that at least three species (two stoneflies and one caddisfly) may have been extirpated from the southern Appalachian Mountains. However, many insect species occupy small, unusual habitats that are hard to find, and there are few entomologists searching for certain groups. Therefore, we cannot say whether certain species are extinct, rare, in danger of extinction, or whether they still exist as undiscovered populations.

The degradation of our nation's streams is clearly associated with human population density and activity. Where humankind has been most active, watersheds and their associated biota have often suffered. Increasing human populations will mean further demand for space, natural resources, and recreation, and will be accompanied by further degradation of our streams and lakes and the drainage and pollution of our wetlands. Morse *et al.* pointed out that we do not know if aquatic insect species now considered to be in peril are rare because of natural restrictions or because of human activities. However, we can predict that further disturbance to their habitats probably will hasten population declines. The continued existence of some species over the next few decades appears questionable. Recently, even some once-common stream species have drastically declined in numbers. Clearly something must be done to protect insect species and their habitats if we are going to maintain the biological diversity of the southeastern United States.

HABITAT SELECTION AND NICHE CONSERVATISM

Every organism has a place to live in nature, a functional role in that place, and a complex set of adaptations for reproducing its kind. On the surface, this observation might seem to be obvious, even trivial. However, in order to understand our biological world—the biosphere, how it operates and ultimately how to protect it—we need to understand at a deep level how organisms interact with each other and with their physical environment.

We will examine further some of the concepts that ecologists use to organize their thoughts about the ways in which organisms use their environment, relate to each other, and assemble into communities or ecosystems. The most fundamental and perhaps most difficult of these concepts is that of the ecological niche. A niche refers to the way in which an organism fits into an ecological community or ecosystem.

Through the process of natural selection, a niche is the evolutionary result of a species' morphological (morphology refers to an organism's physical structure), physiological, and behavioral adaptations to its surroundings. A habitat is the actual location in the environment where an organism lives and consists of all the physical and biological resources available to a species.

The collection of all the habitat areas of a species constitutes its geographic range. We will examine each of these concepts in turn using a historical approach where appropriate and discuss how these ideas can help us to understand the issues of modern conservation biology.

GEOGRAPHICAL RANGES

One of the first things that people noticed as they started to study and write about the world was that when they visited different parts of the earth they found different species. In fact, most species have limited distributions. Giraffes, for example, are found only in Africa, koalas only in Australia, and lemurs only in Madagascar.

If we choose a particular species, plot a point on a map of the world for every place in which it is found, and then draw a line surrounding all of the points, we will have delimited its geographical range. Range map for three species found in the United States. Some species, such as the Devil's Hole pupfish that only lives in one particular desert spring, have a geographical range of just a few square meters; others (e.g., human beings) range over most of the land area of the earth. Most species, though, have intermediate-sized distributions, as does the bighorn sheep.

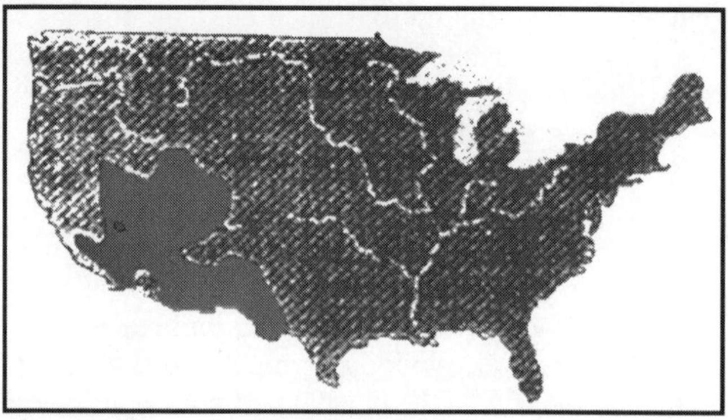

Gray = Geographic range of humans (includes blue area).
BLUE = Geographic range of Desert Bighorn.
Red = Geographic range of Devils Hole pupfish

FACTORS LIMITING THE DISTRIBUTIONS OF ORGANISMS

One factor that limits the distribution of a species is its *dispersal ability*, i.e., how well individuals or their offspring can move from place to place. Obviously in the case of the Devils Hole pupfish, the only route to other waters is overland through the Mojave Desert, so unless they sprout legs and other adaptations they cannot disperse without human aid. Humans, on the other hand, are the ultimate dispersers, being able to transport themselves quickly to virtually any spot on (and even off) the planet. Desert bighorn sheep fall somewhere in between these extremes. Desert bighorn prefer rugged desert terrain and thus inhabit many of the hundreds of small, isolated mountain ranges of the southwestern United States.

Although bighorn are known to move across the flats between close mountain ranges, they don't really like flat open areas and thus are less likely to move between mountain ranges that are far apart. Thus the availability of suitable habitats within a potentially traversable distance affects the distribution of a species. Another factor that limits distributions is a species' *tolerance to different environmental conditions*. Factors such as temperature, moisture, and light have profound effects on species' distributions. Mammals living in hot climates tend to have smaller bodies and longer ears and limbs than mammals in colder climates.

Small body size and long limbs are adaptations that increase the body's surface area in relation to its volume and thus function to help the body lose excess heat. Such mammals would be at an obvious disadvantage if they tried to disperse into an area of cooler climate where it is more important to conserve body heat than to lose it.

Species adapted to the tropical rainforests of Brazil where water is abundant tend to lack water-conserving adaptations and so would have difficulty surviving in the arid deserts of the southwestern United States. Freshwater fish placed in the ocean (with the exception of a few anadromous species like salmon or steelhead) have difficulty maintaining water balance due to the ocean's salinity and actually die of dehydration.

Some plants require days of a certain length in order to flower and reproduce. If seeds of these plants were carried by wind or birds to an area suitable for germination and growth but lacking sufficient day-length for reproduction, the species would never become permanently established.

Interactions with other species can also limit a species' distribution. A species invading a new environment will encounter other species with which it has never had contact. If one of these is a predator that uses unfamiliar tactics, the invading species is likely to be eaten.

Similarly, if one of the species in the new environment uses the same kinds of resources as the invading species and if it is better able to compete for those resources, then the invading species will have trouble gathering enough resources to meet the needs of survival and reproduction. Obviously in both of these cases the invading species will not be very successful in extending its range.

The flip side of this is that native species may be exceptionally vulnerable to an unfamiliar predator or competitor. An example is the decline of most species of large native fish in the Great Lakes due to the introduction of the sea lamprey. The sea lamprey is an eel-like fish that attaches itself to other fishes and sucks out their body fluids, often with fatal results.

When sea lampreys found their way into the Great Lakes, they found an abundance of prey that had no adaptations for avoiding their style of predation. The result was that at least one species of whitefish was driven to extinction and other species were severely depleted in numbers.

IMPLICATIONS FOR CONSERVATION

The concept of geographic ranges has implications for conservation biology. As we have seen, geographic ranges are a function of a species' various morphological and physiological adaptations—characteristics acquired through the process of natural selection. Sometimes these adaptations will also work well in a location different from that in which the species originally evolved.

For a very long time now humans have been moving species around the globe, introducing them into previously unfamiliar habitats. Most often these introductions fail. But as we just saw in the case of the sea lamprey, every so often a species is well-suited to the new habitat and the species proliferates.

This is a major problem for those charged with managing and preserving biodiversity. For example, in the southwestern United States prospectors and miners of the 1800s abandoned their burros (donkeys) in the desert when they no longer needed them. Burros originally evolved in the arid desert regions of Somalia in eastern Africa and so were already well-adapted to the deserts of the American west.

In places like the Grand Canyon and Death Valley they multiplied rapidly, caused changes in the vegetation, and competed with the native desert bighorn sheep for food and water. The burros in Death Valley and the Grand Canyon were removed at great effort and expense and put up for adoption, giving the bighorn a chance to recover. There are many burros in other ecosystems of the Southwest, however, and these have been implicated as important factors in the decline of other bighorn populations.

HABITAT: THE ORGANISM'S HOME

If we examine the geographic distribution of a widely-ranging species, we find that the distributional range consists of both occupied and unoccupied areas. Those areas actually occupied that meet the requirements for a species' survival and reproduction are its habitat. We mentioned earlier that the desert bighorn prefers desert mountain ranges—these mountain ranges, along with all the plants and other animals found there, constitute the bighorn's habitat.

KINDS OF HABITATS

Any place where organisms live is by definition a habitat. Your backyard, an empty lot, an agricultural field, a pristine mountain wilderness—all these are habitats for some group of organisms. Ecologists often find it useful to talk about kinds of habitats or to classify them into more or less general groupings.

Habitats are typically classified on the basis of more or less obvious visual characteristics. Alpine meadows, conifer forests, marshes, lakes, desert scrub, and riparian zones along stream banks are all different habitats that you may be able to visualize from your own experience.

Sometimes habitat classifications have more to do with how the habitat functions than with its visual aspect. Although freshwater and saltwater marshes may appear similar superficially, each supports a different constellation of species. Along the California coastline at the ocean's edge we can distinguish several zones that are functionally different. Above the high tide line there is a splash zone that only the highest ocean waves can reach. Below this is the intertidal zone that is alternately exposed and submerged with the ebb and flow of the tide. Deeper still is the subtidal zone, which is always submerged. Although all of these zones may occur within a few meters of each other, they are fundamentally different habitats for the organisms that live there and they support distinct assemblages of species.

HABITAT FEATURES

Habitats have many features or factors that are important to the organisms living there. Conveniently, we can divide habitat factors into two major groupings, *physical factors* and *biotic factors*. In terrestrial habitats some important physical factors are elevation, steepness, slope direction, soil type, and water availability. Elevation affects air temperature and rainfall—higher elevations are cooler and moister than lower ones. The steepness of a slope will affect the kind of soil that can form there and the amount of water that can soak into the ground after it rains.

Slope aspect (the direction a slope faces) is particularly important. In the northern hemisphere, south-facing slopes get more sun, are warmer and dryer, and thus support different vegetation than north-facing slopes. In aquatic habitats such characteristics as pH, salinity, dissolved oxygen concentration, temperature, and flow rate are important physical factors.

Biotic factors include all the other species that occur in the habitat. For an herbivore such as the desert bighorn sheep many of the grass, shrub, and herb species of the desert mountains constitutes its food source. Trees and shrubs form both hiding cover from predators and thermal the dead trees in which their holes are found but also the other birds (e.g., Pileated woodpeckers) that make the holes in the first place.

Finally, physical and biotic factors may interact to determine the quality of the habitat for a given organism. For example, the nutritional quality of plants available as food for herbivores, such as deer, is determined in large part by the quality of the soils present.

HABITATS AND CONSERVATION

Obviously a species cannot survive without its natural habitat, except perhaps in a zoo. It follows then that the fundamental unit in the conservation of biodiversity is not the species but the habitat. An organism may have more than one habitat. During the summer many of our migratory waterfowl have breeding habitats in the arctic tundra, but they winter on the waterways and marshes of the southern United States.

Mule deer of the Great Basin spend the summer in mountain forests, but winter snows drive them to lower elevation sagebrush zones where they can forage more easily. To ensure that there will always be migratory species such as these we need to protect not only the two habitats at the extremes of their seasonal movements but also the migration routes in between that serve as resting and feeding habitats.

Habitats don't exist in isolation. Many habitats have inputs and outputs. Take Mono Lake, for instance, a spectacular lake on the east side of the Sierra Nevada in California. Its water source is streams fed by winter rains and melting snow in the mountains. In its natural state, water leaves the lake only by evaporation. The balance between the inflowing streams and evaporation created a saline lake with many unique features, including a species of brine shrimp found only in Mono Lake.

As a large, food-rich body of water in a desert area, the lake is a major fueling stop for migratory waterbirds and a major nesting area for other species, such as California gulls. When water from the lake's inflowing streams was diverted to quench the ever-growing thirst of southern California, the lake level dropped drastically. Islands in the lake became connected to the mainland, giving coyotes and other predators access to an easy source of food: nesting California gulls.

With adequate inflowing water, the islands were good nesting habitat; without the water they were unsuitable as nesting habitat. Without adequate inflowing water, the lake also would become too saline to be habitat for the Mono brine shrimp and to be a feeding habitat for migratory waterbirds. Recognition of this fundamental relationship between inflow and habitat for many species was the partial basis of a successful court action that reduced the diversion of water from the inflowing streams.

NICHE: THE ORGANISM'S ROLE

Perhaps the simplest, most general definition of the ecological niche is an organism's "ecological position in the world". Even though this may seem straightforward, determining what constitutes an organism's position in the greater scheme of things is not a trivial pursuit. In fact, the concept of niche is a notoriously difficult one for beginning (and even advanced) students of ecology. Some of the difficulties arise because the word niche has a common meaning to the layperson but a very specific meaning to the ecologist.

HISTORY OF THE NICHE CONCEPT

Webster's Dictionary defines niche as "a place or position suitable or appropriate for a... thing." Biologists and naturalists have long considered each species to have a proper place in nature. It is natural then that the first uses of the word niche in an ecological context had a very strong "place" association. Although not the first ecologist to use the word, Joseph Grinnell, in a series of papers published between 1917 and 1924, is generally credited with being the first to develop the ecological concept of the niche. Grinnell defined niche as the "ultimate distributional unit occupied by just one species or subspecies."

To be more specific, Grinnell was interested in determining which factors governed a species' potential geographical distribution and usually considered these to be physical or climatic factors, as opposed to relationships with other species such as competition or predation. For example, in a 1917 paper entitled "The niche-relationships of the California Thrasher," Grinnell considered the geographical range of thrashers, a common bird of the chaparral, to be limited by temperature since it avoided areas of extreme heat or cold.

Thrashers, Grinnell maintained, were further restricted to chaparral areas because, being shy creatures, they needed dense hiding cover. These factors, because they explained the California thrasher's distribution, constituted the bird's niche. Grinnell's concept of the niche also included two important components: first, animals *evolved* to fill niches and, second, no two species could have exactly the same niche. These two concepts have turned out to be central to the subsequent development of niche theory.

At about the same time, one of the most important ecologists of the early part of this century, Charles Elton, was developing his own concept of niche. Elton's niche concept differed from that of Grinnell in several fundamental ways as evidenced by his definition of niche as an organism's "place in the biotic environment, its relations to food and enemies." When Elton uses the word "place" he really means the organism's role in its community—what it does, how it makes its living—as opposed to the geographical sense used by Grinnell.

In practice, Elton tended to define niches based on an animal's size and feeding habits. For example, birds of similar size that catch insects on the wing were thought to have a similar niche; similarly, all large mammals that eat only grass were thought to have another. Elton's view of the niche concept was simply "to give more accurate and detailed definitions of the food habits of animals" than those afforded by words such as carnivore or herbivore.

A few years later the Russian scientist G. F. Gause combined Elton's view of the niche with the observation that very similar species cannot co-exist within a community. This is because resources such as food generally are in limited supply. Very similar organisms would have to compete with each other for the resource in question and inevitably one species would prove to be the superior competitor.

Although Grinnell and even Darwin had stated much the same thing, Gause based his concept on mathematical reasoning by the Italian mathematician Vito Volterra. This idea, now often referred to as "Gause's principle," or "the competitive-exclusion principle" can be restated succinctly as "no two species in a community may possess the same niche," and has become a central tenet of modern niche theory.

In order to demonstrate this principle, Gause performed what are now considered to be classic experiments. Gause placed two species of *Paramecium*, a single-celled protozoan, into flasks containing a bacterial culture that served as food.

Thus, in this artificial laboratory system both species of Paramecium were forced to have the same niche. Gause counted the numbers of Paramecium each day and found that after a few days one species always became extinct because it apparently was unable to compete with the other species for the single food resource. This process of competitive exclusion has since been demonstrated many times in laboratory and field experiments with many species. However, extinction is not the only possible result of two species having the same niche.

If two competing species can co-exist for a long period of time, then the possibility exists that they will evolve differences to minimize competition; that is, they can evolve different niches.

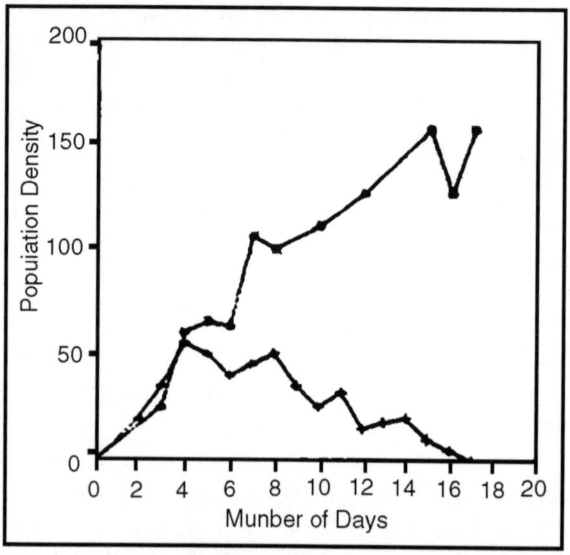

Fig. Results of Competition Between two Species of Paramecium with Similar Requirements.

Two niche concepts have been discussed so far. The first, propounded by Grinnell, is geographically oriented and we can term it a place niche. The second, championed by Elton and Gause, is defined on behavioral considerations and we might call this the functional niche.

Ecologists were relatively happy accepting these views of the niche until the late 1950s when the eminent limnologist and ecologist G. Evelyn Hutchinson devised a rigorous and quantitative concept of niche that, with slight modifications from his original concept, incorporated both place and functional elements and has remained the standard niche model for over thirty years. Prior to Hutchinson the niche was a rather nebulous concept defined only by words; that is, niches could not be measured. Hutchinson's new idea not only allowed a way to measure niches but also a way to compare niches of two or more species. An animal that preys upon other animals will be limited in the range of prey sizes that it can kill. Certain prey will simply be too large to kill; others will be too small to bother with because the amount of energy needed to catch them is greater than the amount of energy to be gained by eating them.

The upper line depicts the range of prey sizes used by the species compared to the total size range available, shown on the lower line. Such a one-dimensional graph is called a niche axis and represents quantitatively how a species uses one of the factors of its habitat. The range of resources utilized along the axis is referred to as niche breadth. A species with wide niche breadth is referred to as a generalist; a species with narrow niche breadth is called a specialist.

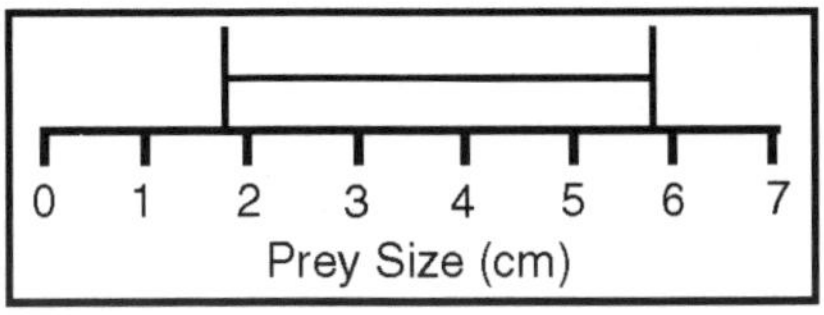

Fig. A Single Axis for a Niche as Described by Hutchinson.

This axis is prey size (0-8 cm) of which the animal is capable only of taking prey of 1.5 to 7 cm.

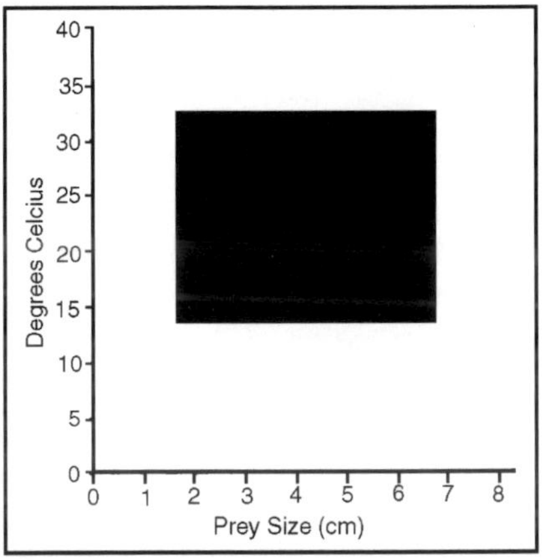

Fig. A Niche of an Organism as Defined by two Axes Temperature Tolerance and Prey Size.

Food is not the only factor that we can graph this way. We add a second axis to indicate the temperature extremes that our hypothetical species can tolerate. Note that this results in a shaded area on the graph indicating the combinations of prey size and temperature that allow the species to exist. It would now be possible to add a third factor such as moisture to our graph making it three-dimensional. The shaded area in Figure would become a rectangular solid and would represent the combinations of prey size, temperature, and moisture that allow our species to survive.

The organism can only live in environments with temperatures ranging from 12 to 33°C that contain prey of 1.5 to 7 cm. Hutchinson's idea was to continue this line of reasoning indefinitely until we had incorporated all of the biotic and physical factors of the habitat to which a species is adapted. Obviously if we have more than three niche axes we are no longer able to draw a graph of the niche, but mathematicians and computers can deal with such hypervolumes with relative ease.

In practice, because it is difficult to determine the responses of a species to all of the factors of the environment, most ecologists limit their investigations of niches to two or three of the most important factors. Such a niche, determined in the absence of relations with other species, is termed the fundamental niche and represents a species' potential to use available resources. Certain interactions between species can affect the breadth of a species' niche along one or several niche axes. For example, the risk of predation could shrink the breadth of a species' food niche axis if searching for certain kinds of food items increased the probability of being eaten. Competition for resources could also reduce the breadth of a species' niche along the resource axis in question.

Thus the fundamental niche is the niche that exists in the absence of predators and competitors (a rare event) and is determined largely by the species morphological and physiological limitations. In the real world, a niche is limited in extent by the presence of interactions with other species and is termed the realized niche. The realized niche of a species may vary from place to place because of the presence of different predators and competitors. In the Eel River of coastal California, for example, rainbow trout live primarily in riffles (shallow, fast-flowing areas) and feed on aquatic insects when predatory pikeminnows are present in the pools; in the absence of the predator, the realized niche of the trout expands to include the pools and more terrestrial insects that fall on the pool surface.

NICHES AND CONSERVATION

Some of the myriad confusions surrounding the niche concept result from the idea of "empty" or "vacant" niches. Some ecologists have suggested that niches actually exist out in the environment and that organisms evolve to "fill" or "occupy" them. Most ecologists consider this to be nonsense.

Hutchinson's niche concept very clearly expressed the idea that a species' niche is the sum total of adaptations to the environment possessed by the species in question. *The niche is just as much an attribute of a species as its colour, size, shape, or physiology.*

In addition to causing confusion, the idea of vacant niches has been misused to justify the presence of introduced organisms. A noted paleontologist has suggested that feral burros in the southwest occupy a niche left vacant by a long extinct small Pleistocene horse and should thus be considered a native species.

Even though we don't know exactly what this horse was, it certainly was not the same species as the burro, and it evolved under very different conditions along with other species that the burro never "knew" in its own evolutionary history in Africa.

Equally important, this argument implies we need to re-introduce a suite of predators to prey on the burros, such African-type lions and grizzly bears. Even if burros were very similar ecologically to this extinct horse, our native southwestern flora and fauna have been without a native horse for a very long time and may no longer be adapted to its presence. For example, grazing by feral burros in Death Valley altered the species' composition of plant communities that formed the bighorn sheep's food source and prevented bighorn from using scarce water supplies. This, in effect, narrowed the bighorn's niche to the point where reproduction became difficult and populations declined drastically. When we humans modify habitats, whether for economic gain, by accident, or even just for the purpose of survival, we further constrain the realized niches of the organisms living there. Very often the altered niche does not permit survival of the species.

There are millions of species on earth and it is probably an impossible task to fully determine the niche of all of them. In fact, it is a difficult task for just a single species. But an understanding of how an organism uses its environment in perhaps two or three important niche axes can help us to predict how a species will react to changes in its habitat. We have discussed in detail some of the concepts that ecologists use to study biodiversity. These concepts are geographic range, habitat, and ecological niche. It should be obvious by now that these are not mutually exclusive topics.

Elements of habitat and niche are subsumed into the determination of geographic ranges. It can also be difficult to say exactly where habitat leaves off and niche begins. Nonetheless, an understanding of these three concepts can help us to effectively plan conservation measures and to appreciate the wonderful complexity of our biosphere.

AQUATIC INSECT MANAGEMENT

One step that has been taken towards protecting aquatic insects is the maintenance of a federal listing of endangered species. In the U.S. Federal Register dated November 15, 1994, 198 species of aquatic insects were proposed for possible addition to the threatened and endangered species list. Of these, 73 species occur in the southeastern United States. The status of most of these candidate species is unsettled, and much time is being spent gathering information necessary to make listing category decisions. Regarding actual listings, 29 insect species are now federally listed as threatened or endangered species. Most of these are terrestrial beetles and butterflies. Only two aquatic species, Ambrysus amargosus, a naucorid bug known only from Nevada, and Somatochlora hineana, a dragonfly which occurs in the Great Lakes region, are on the threatened and endangered (T&E) list. The U.S. Endangered Species Act as it pertains to insects is administered by the U.S. Fish and Wildlife Service. For each species that becomes listed as threatened or endangered, a species recovery plan aimed at de-listing the species is prepared. A suggestion to amend the Act has been made to allow the listing of many species that share a single ecosystem. This would be a step to shift emphasis from management of single species to management of habitat. Although species listing is a positive step that ultimately can afford protection, many species that have not as yet been proposed for listing and which contribute the bulk of biodiversity are declining in numbers due to the widespread disturbance of our running waters.

The actual management of aquatic habitats to benefit aquatic insects is only beginning to be considered, and few examples are available for discussion. Field activities aimed at enhancing habitats for other aquatic taxa, such as stream restoration for fish, probably benefit certain insect groups. However, these efforts do not take into account key differences in the life histories and habitat requirements of various aquatic insects. In general, to enhance stream habitat to maximize insect diversity, a strong emphasis should be placed on providing habitat diversity and bank and substrate stabilisation. Specific protection or enhancement efforts will require knowledge of the target species' life cycle and habitat requirements, both in the immature and adult stages. In providing conditions for a specific insect, care must be taken not to alter the habitat in such a way that would be detrimental to other rare species. Of course, a species approach to aquatic insect management requires two limited resources — time and money. Research on habitat requirements of individual species is beset with difficulties and uncertainties, and is time-consuming. Unfortunately, time may be something endangered species lack. Furthermore, securing sufficient funds to carry out imperiled species projects can be difficult. In contrast, habitat protection or improvement plans can often be relatively more timely and less costly in achieving results. They often can also be more readily accepted by the parties that are affected by them. Because of the biological and sociological complexities associated with managing aquatic insects, we will need to skillfully build upon our successes and learn from our mistakes regarding the methods we use to realise improvement. Below I will briefly discuss several habitat management projects to illustrate some of the types of problems that such projects can pose to insects.

CASE 1. DAMSELFLIES AND THE CONTROLLED BURNING OF AUSTRALIAN MARSHES

Hemiphlebia mirabilis is a small, rare damselfly once thought to be extinct. According to New, it is now listed as a threatened taxon because of population loss and decline associated with agricultural practices and modifications to river flow and flooding regimes. It is the only species in the family Hemiphlebiidae, and it has several interesting ancestral characteristics important in studies of the evolution of the Zygoptera. With its habitat limited to seasonal swamps that are found on sandy heathland, H. mirabilis is a weak flier that depends on emergent aquatic vegetation for cover and as a substrate in which it lays eggs. At one of its major known population sites, controlled burning in a mosaic pattern was used to prevent successional progression of vegetation and promote earlier seral stages, while also preventing a build-up of dead vegetation on the ground which could fuel more intense, unintentional fires. However, it appears that the burning was intended to create food for vertebrates, and was not necessarily intended to provide continued suitable habitat for the damselfly. In 1987, a "controlled" fire spread to the site of the largest colony of H. mirabilis, destroying the vegetation. No damselflies were seen in the area for two years following the burn. A few individuals were observed during the third year after the burn, and recently, numbers have increased nearly to the level present prior to the fire.

Study of this situation reveals several critical points. First, careful burning of patches of the surrounding habitat can be an important management tool regarding species like H. mirabilis because fires can prevent the maturation of the shrubby trees which can be responsible for the long-term drying of swamps. Secondly, H. mirabilis can withstand even drastic temporary alterations of its habitat if there are nearby locations from which individuals can immigrate when habitat conditions again become favourable. Of course the widescale reduction in the habitat of H. mirabilis makes the last mentioned possibility extremely tenuous, and underlines the vulnerability of this species. Together, the above experiences illustrate the need to carefully monitor habitat alteration projects and to consider the long-term effects they might cause.

CASE 2. RE-ESTABLISHMENT OF PRAIRIE HABITAT IN ILLINOIS

In a small area of northeastern Illinois, there is a particular site being managed to restore mid-western prairie habitat. This is accomplished mainly by periodic controlled burning and removal of trees and shrubs. Within this area, a rare dragonfly (Hine's emerald, Somatochlora hineana) inhabits an unusual microhabitat, namely shallow, spring-fed sheet flow, and narrow streamlets through calcareous (dolomitic) cattail marsh. This dragonfly was once thought to be extinct, but a few isolated colonies have been found recently in northeastern Illinois and eastern Wisconsin. Somatochlora hineana was recently placed on the T&E list, partly because of its apparent extirpation from previously known localities in Ohio and Indiana. For two main reasons, the cut-and-burn management practice at the Illinois location creates concern regarding whether the future survival of S. hineana is being considered. First, burning of the wetlands may directly destroy dragonfly individuals if aestivating nymphs occur in dry peat/muck or if the shallow habitat is overheated. Furthermore, this species' distribution may be limited in part by water chemistry, and an additional problem might be created if burning temporarily changes the water chemistry where eggs are laid and nymphs develop. Secondly, removal of nearby trees and shrubs denies adult dragonflies access to preferred perching sites, as well as to mating and foraging areas. This last issue underscores the importance of the terrestrial environment to an aquatic insect species.

The most pressing need regarding S. hineana is the immediate protection of the wetlands area, as well as research aimed at developing management strategies that will ensure continued existence of the species' required habitat. Close communication and cooperation among all concerned parties along with a monitoring programme are needed to manage the site properly for the continued existence of the variety of organisms now present.

CASE 3. WEIR PROJECT, SOUTH FORK OF THE HOLSTON RIVER, TENNESSEE

In an effort to improve levels of dissolved oxygen (DO) and minimum flow in the South Fork Holston River in northeastern Tennessee, the Tennessee Valley Authority (TVA) constructed an aerating labyrinth weir (completed in December, 1991) 2 km (about 1.2 miles) downstream from the South Holston Dam. This was the first project under TVA's Lake Improvement Plan (Hauser, 1993). DO levels have improved from 1.5-3.0 mg per l upstream of the weir and from 6.0-8.7 mg per l downstream, and a minimum flow of 2.55 cms is usually maintained between electricity generating periods. Rainbow trout have been stocked below the weir, and the monitoring of benthic organisms downstream of the South Holston weir has been conducted.

The impoverished macroinvertebrate fauna in the tailwaters below the South Holston Dam, characterised by low taxonomic diversity and a predominance of stress-tolerant species, is characteristic of hypolimnetic release areas. The minimum flow regime established now provides a more constant habitat for benthic insects. Eleven aquatic insect taxa have been collected immediately downstream of the weir and about twice that number in the 1.6 km (one mile) stretch further downstream (Yeager *et al.*, 1993). However, only two EPT taxal were found. Despite a slight increase in biodiversity below the weir compared to the area immediately upstream, there was still a much lower diversity of aquatic insects than would be expected in a river of this size in eastern Tennessee. Moreover, blackfly (Diptera: Simuliidae) production below the weir was extremely high, and was undoubtedly a response to continuous flow combined with the high organic productivity of the upstream lake water.

This example illustrates the need for monitoring the response of aquatic life to supposed habitat improvements. It also raises the question, is altering one or two physical/chemical parameters adequate to improve habitat and enhance biodiversity? Other questions follow. What can be done about nutrient overloading in rivers, especially when conditions are optimum for large populations of pest insects to be produced? What factors other than low DO, low flows, and nutrient loading are responsible for depressing desirable aquatic insects in tailwaters of large dams? Solving these problems would help reinstate well-balanced communities in rivers below reservoirs, of which there are many river-miles in the southeastern United States. 1EPT stands for the three aquatic insect orders Ephemeroptera, Plecoptera, and Trichoptera. EPT counts are generally used as a measure of taxa richness and hence stream quality. An EPT value less than 10 generally indicates a stream in poor biotic condition.

CASE 4. WALKER BRANCH STATE NATURAL AREA, TENNESSEE

Walker Branch is a tributary of Mud Creek, flowing through an ecotone between an upland area and big river bottomland in Hardin County, Tennessee. The area is a mosaic of aquatic habitat types, including wooded hillside seeps, first-order runs, shaded and partly open second-order streams, larger slow-flowing and pooled streams, and swampy wetlands with bald cypress and Tupelo gum.

Disturbance from past farming and logging activities is still evident, although the area has experienced considerable regrowth and habitat recovery. The diversity of Odonata I found while collecting there between 1979 and 1982 prompted me to suggest that the area be looked at by other biologists and perhaps be considered for some type of protection. Besides the great number of Odonata species I found (nearly 40), the combination of rare and unusual species at this site is not easily found elsewhere. In addition to rare insects, several state-listed plants were also found, including Iris brevicaulis, Carex lacustris, and Malanthium virginicum.

The area in which the seeps and cypress wetlands lie is designated to be purchased by the Tennessee Wildlife Resources Agency and registered as a state natural area known as Walker Branch Hills. This positive action is mainly being realised due to efforts by staff at the Tennessee Department of Environment and Conservation. No development or publicity of the area is proposed, in hope of keeping human visitation to a minimum. The seeps I surveyed constitute only a small part of the area, 0.81 ha (roughly two acres). They are vulnerable to trampling and are still recovering from past disturbance. However, prospects are now good for the continued existence of this rich community of dragonflies and other species. This case is a rare example of an important natural area being discovered and eventually protected because of its aquatic insect fauna.

CASE 5 GOMPHUS SANDRIUS —THE TENNESSEE CLUBTAIL

The clubtail dragonfly, Gomphus sandrius, is unusual in that it has the most restricted distribution known within the entire Gomphus species complex. It occurs in only five tributary streams of the Duck River in a 150 km2 (about 60 square miles) area of the Central Basin south of Nashville, Tennessee. Population sizes are small, estimated to be one to six individuals per 30 m (about 100 feet) depending on the stream and locality. The streams this clubtail inhabits are shallow and slow-flowing, with bedrock and gravel bottoms. Nymphs of G. sandrius occupy gravel bars consisting of mixed particle sizes that accumulate downstream of rock fissures and small islands of water willows.

The surrounding lands are heavily used as pastureland for cattle, with some fields cropped for hay. Cattle have access to the streams and surface run-off contributes organic waste and perhaps agricultural chemicals to the streams. Removal of much of the riparian vegetation undoubtedly has raised water temperatures and increased algal growth and sediment input.

Although G. sandrius is on the federal candidate T&E list and a status survey for this species has been conducted, a decision to list it has not been made. Populations at several of the streams appear to have declined since discovery of the species, but because aquatic insect population size is difficult to estimate, longer periods of time often are needed to discover trends. It is possible that T&E status would not benefit G. sandrius, because the species might be faced with a range of adversities that place it close to extinction. Furthermore, forcing regulations on private landowners often generates entirely new sets of problems. An alternative and perhaps more timely approach to prevent further damage to these streams is to form cooperative agreements with the local citizenry. By showing landowners the benefits of improved water quality and stream habitat, and involving them in decisions on what can be done to lessen agricultural impacts, their help can be enlisted for the benefit of all forms of aquatic life, including the Tennessee clubtail.

CONFLICTING LEGISLATION

Roberts pointed out that the mandates of the U.S. Endangered Species Act (ESA) conflict with those of the U.S. Mosquito Abatement Act (MAA).

He illustrated this by applying a strict interpretation of these acts to two situations as follows:

- *Example:* A wildlife biologist trying to manage a wetland to support an endangered aquatic or semi-aquatic species inevitably might create conditions that produce mosquitoes. This would be in violation of the MAA, which states that aquatic habitats that support mosquito production are public nuisances to be legally abated.
- *Example:* A vector biologist trying to control mosquitoes in a wetland could negatively affect an aquatic or semi-aquatic endangered species. This would be in violation of the ESA.

Such dilemmas result from various specialised disciplines considering environmental management independently rather than working together towards a more harmonious approach. Ecosystems are complex, integrated systems, and Roberts warned that attempts at piecemeal solutions to many environmental problems are bound to fail. The ESA requires federal agencies to consult with the U.S. Fish and Wildlife Service concerning any activity which may adversely affect listed species, and this at least provides a gateway for interagency cooperation. Cooperation between biologists and resource managers during the initial planning stages of projects is especially critical.

Of course, the issue of public health and human rights versus animal protection and animal rights is often at the heart of conflicting legislation, and taking inflexible sides on such matters sets up confrontation scenarios which can thwart lasting achievements. However, through cooperation aimed at trying to resolve or otherwise minimize apparent differences, rewards can often be realised which benefit the greatest number of involved parties. As an example, while no insect species that has been implicated in the transmission of human disease is currently considered endangered or threatened, if such an instance is identified, the insect would not be afforded protection under the ESA.

CLOSING THOUGHTS ON THE MANAGEMENT OF INSECT RESOURCES

In contrast to the present gloomy state of affairs regarding the protection of imperiled aquatic insects, the future can bring brightness through education and change. Our industrial, capitalistic society, which put a man on the moon before adopting a national environmental policy, has become conscious of its environment, and a large faction cares for or at least appreciates the other living things around them. The term "biodiversity" is being used outside biological circles by politicians and the general public, and understanding of its importance is being commonly gained.

We are in the process of looking at aquatic species and their habitats in hopes of identifying what is in danger of being lost forever. The federal government is concerned with endangered species, including candidate aquatic insects, and is attempting to rectify some of its own past assaults on our rivers. Private citizens are becoming involved in clean-up and restoration activities, and a majority of people express the desire for a "healthy" environment. On the other hand, the world's human population continues to increase at an alarming rate. Globally, the human population is increasing at about 10,000 people per hour, or 93 million people per year. In the United States, the 1994 census was 261 million, and this number is projected to increase by 25 percent to 326 million by the year 2020. In the 11 southeastern states considered in this chapter, the population will increase from approximately 60.9 to 78.2 million by the year 2020, a 28 percent jump in just 25 years. The Southeast will remain the most densely populated region in the country. Florida currently has the highest population within the Southeast (about 14 million people) and the greatest projected rate of increase.

More people will place greater demands on natural resources, increase demands for living space, and increase encroachment on natural habitats. Many people desire lakefront or streamside property and most seek high standards of living. These demands often result in water quality reductions and habitat degradation. Protecting species requires a constant vigil, because losses are irrecoverable. Increased human habitation of remote areas can result in a greater number of species whose ranges are fragmented or otherwise diminished in size. For many, it is extremely disturbing and ironic that although we have instituted more policies to prevent or regulate the degradation of our waters over the last 20 years than ever before, more damage seemingly has occurred during this period than ever before. How do we satisfy the needs of our rapidly growing human population and still keep aquatic habitats capable of supporting other life forms? The solution lies in getting appropriate agencies and affected organisations and individuals in watersheds concerned with local projects and to work through cooperative efforts. Where regulation might fail, communication, education, and recruiting stakeholders can often succeed. We need to instill and foster an appreciation of all forms of life at an early age through education, both formal and by example.

The belief that other living things were put here for our use must be challenged. Furthermore, we need to educate everyone of the benefits of protecting habitats and maintaining the levels of biodiversity that are still present. By stressing the importance of habitat protection and its benefits to society as well as to the natural world, even skeptical parties might be persuaded to become good stewards of nature. The integrated ecosystem approach to environmental management was recommended by the U. S. National Research Council in a proposal to embark on a major national aquatic ecosystem restoration programme.

I suggest that we might benefit from a more proper term than "resource management" for trying to improve aquatic habitats. To manage means to handle, direct, govern, or control in action or use. This concept is difficult to apply to human tinkering with natural systems. Furthermore, some might question the goodness of human control regarding a workable philosophy in dealing with ecosystems. The term resource management also implies knowledge of how to change things to achieve particular goals. Lastly, the term "resource" carries a connotation that there is something in it for people. A term that is more neutral and also connotes resources for all living things is "habitat." Terms such as "habitat manipulation" and "habitat alteration" do not imply that we can control the outcome of our actions. After all, a lack of knowledge concerning an organism's requirements and responses to change reduces our management efforts to trying to manipulate one or two environmental variables and then observing the outcome. The term "restoration" also seems inappropriate, as it means to bring back to a former, original, or normal condition. We have changed many of our larger streams so drastically (by damming, dredging, etc.) that they arguably cannot be restored to their original conditions.

Strictly adhered to, habitat restoration of a large system where a reservoir is now in place would dictate that the dam be removed, allowing the river to resume its natural dynamic state. Of course, many problems would result from removal of large dams (*e.g.*, flooding, bank cutting and other forms of erosion) such that this action might lead to new forms of environmental disaster. It might be more prudent to try to improve existing reservoirs and remaining river stretches by re-establishing riparian vegetation and conducting other bank stabilisation projects, by reviewing dredging proposals, and by preventing livestock, organic wastes, and toxicants from entering water courses. Monitoring programmes should be initiated to document that natural processes are being maintained in our rivers.

The next ten years will probably be critical regarding the overall fate of our aquatic insect fauna in the Southeast. A heightened public awareness of stream biodiversity and its importance will be needed to abate the negative pressures impacting our streams. People from all disciplines will be needed to work together using integrated approaches to maintain watersheds as natural functioning systems. How will we be able to measure our success in these matters? Some will count the numbers of species being removed from the federal T&E list. Some may tally the number of river kilometers improved. I hope to go to a stream, pick up some submerged rocks, and see many various mayflies, stoneflies, caddisflies, and dragonflies crawling on them.

CLASSIFICATION AND IDENTIFICATION

When people formally identify a plant or animal or for the examples below, aquatic insects, they are going beyond just giving it a name. In reality, the specimen is being placed in an hierarchical classification of living things. Humans classify everything in the world. For example, books can be divided into fiction, non-fiction, etc. Each of these groups can in turn be further subdivided (romance, comedy, self-help and so on). The same has be done for living organisms. Initially classifications were for practical purposes: edible, inedible, useful, dangerous, medicinal, innocuous, etc. As biology developed similar looking organisms were grouped together. With the synthesis of ideas in the 18th and 19th centuries by workers such as Linnaeus, Lamarck, Wallace, and Darwin, classification of organisms began grouping organisms together that shared a common evolutionary heritage (common ancestry). The process of developing this "natural" classification continues to evolve as new methods such as DNA analysis are developed to supplement morphological, anatomical, behavioural and ecological information. Organisms are usually referred to by their genus and species name together. Always written with the genus capitalised, the species name lower case and in italics, Aeshna eremita. This is unique to that type of organism. Often included with the genus species name are the author, the person who is credited with first describing the organism and the publication date for the description. For example the full citation for the Lake darner would be, Aeshna eremita Scudder 1866. If there are parentheses around the author's name this indicates it was originally described as belonging to another genus and a subsequent worker assigned it to its current place in the classification scheme. There are very strict rules for naming new organisms and assigning organisms to new groups.

For the general public naming aquatic insects does not go much beyond common names (dragonflies, mosquitoes, beetles, blackflies, etc.). These names may refer only to groups of insects and vary in their precision. For example, dragonflies refer to the suborder Anisoptera of the order Odonata (Dragonflies and Damselflies). There are about 45 different species of dragonflies reported from Saskatchewan. Mosquitoes refer to the family Culicidae with 47 types present in the province. The term beetle refers to the order Coleoptera which has over 230 aquatic representatives in Saskatchewan. Blackflies refer to the family Simuliidae, which has 32 species recorded from the province. And so on....

"The species is the thing": Typical generalised identifications, such as those above, carry little real information regarding life history, ecology, and environmental requirements, as these vary, sometimes significantly, among the species of the "group". It is the species level identification that carries the detailed information regarding life cycle and biology that researchers use to communicate, develop experiments and test hypotheses. Accurate species identifications and ecological information is especially critical in the case of insects that carry diseases. For example, not all mosquitoes feed on humans and mammals, some feed on birds, others will feed on both birds and mammals and still others feed on amphibians and reptiles.

Only certain species carry West Nile Virus and are able to transmit it to humans. To monitor the potential of disease spread and develop effective control measures it is vital to be able identify the species, know its life history, its preferred breeding habitats, feeding behaviours and its general ecology. Furthermore, biodiversity studies require species level work in order to accurately document the faunal (and floral) richness of areas or regions of concern. And, the monitoring of invasive species, as the name implies, requires being able to identify the specimens in question to the species level in order to know its detailed biology so the invasion may be prevented, or if the invasion as already occurred, controlled effectively.

The species category is commonly the most detailed level of classification used in Biology. Species information in itself is a generalisation of life history, biology, ecology and distribution data of populations that make up a potentially interbreeding group. It is the basis for all generalisations of higher levels of classification. In many instances species within the same genus have significantly different life histories, biologies, and environmental requirements. Without first determining the species in a study it is impossible to assess the accuracy of genus or family level reporting. Therefore, studies at the supraspecific level have their accuracy fundamentally compromised especially in cases that involve a number of multispecies genera. Unfortunately, genus and family level analysis without supporting species level data seems to be the current vogue of aquatic ecology and impact monitoring even though the value of genus/species level taxonomy can provide significant additional information (Lenat and Resh 2001). Supraspecific research (family and generic level studies) not only affects the quality of data available to researchers and decision makers but has also curtailed the development of aquatic insect systematics and species level biodiversity research.

To identify some aquatic insects to species can be as simple as comparing pictures in a guidebook similar to "birding". If the specimen looks like the "Giant Water Bug" and it was collected in Saskatchewan then you can be reasonably sure it is Lethocerus americanus as giant water bugs are so distinctive in shape and there is only one species found in Saskatchewan at the present time.

Unfortunately, easy species identifications like the above are rare. In most cases the process of identifying an aquatic insect to species is an arduous task that requires a great deal of knowledge and years of experience to do correctly and effectively. As E.O. Wilson wrote in his book "Naturalist"," [Taxonomy] is a craft and a body of knowledge that builds in the head of a biologist only through years of monkish labour..... If a biologist does not have the name of the species, he is lost. As the Chinese say, the first step to wisdom is getting the right name."

The typical process of identification is to "run" a specimen through a dichotomous key that uses choices of character states to work through a series of couplet statements to reach an identification. For Saskatchewan two very good taxonomic keys are Merritt, Cummins and Berg and Clifford. Both are excellent introductions to the family and genus level. Unfortunately, Clifford's book has become extremely dated and is in need of revision for many groups to incorporate taxonomic and ecological advances. Both texts are also not designed for Saskatchewan exclusively so there are many families and genera included that do not occur in the province and there are a significant number of insects found in Saskatchewan that are not covered in Clifford's Alberta text.

This last point is significant as regional keys often are able to use shortcuts because certain taxa are not found in the particular region and therefore specimens worked through the key will be incorrectly identified. Many workers do not fully appreciate that rather than these texts being the end of the identification process they are actually only a good starting point.

Both texts only provide taxonomic keys to genus and lack discussions regarding variations in characters within a genus that are often required for positive identifications. And, species level identifications are beyond the scope of these books. More specialised books and papers, such as generic revisions, must be consulted to ensure the genus level identification is correct and to continue to the species level. Even with the proper literature assembled identifications can still be difficult and frustrating because more often than not the specimen is damaged and critical parts are missing. Another problem is the vast majority of taxonomic keys require certain stages. Most larval keys are designed for final instar larvae and will not work if the specimen is an early instar. This is particularly important when lengths of the body, or a body part, or a ratio of one part to another, are used as the distinguishing character. Often the adult stage is required for species level identifications. For many groups the adults, if not aquatic, are not collected in regular aquatic sampling and special effort and techniques must be used to collect them. Furthermore, without positively associating the larval stage with the adult stage it can be impossible to identify the larvae to species by itself.

The above information is primarily related to the traditional morphological methods of identification where structural differences are used to distinguish taxa. However, in many cases, interspecific morphological differences may be very subtle in closely related species to the point of two (or more) species being essentially indistinguishable in all stages. The only suspicion of the presence of multiple species may be from distributional, ecological or life cycle differences and "gut feelings" of taxonomists. The use of DNA barcoding has proven to be quite promising in splitting such species groups and discovering unknown species within these groups. However, barcoding does have its disadvantages as it requires much more than just a microscope. So at present barcoding is essentially only really practical for taxonomic research, although preliminary attempts are starting to explore its use in ecological and monitoring researches.

8

Aquatic Plant Ecosystem and Soil Analysis

INTRODUCTION

Plant ecology is a branch of the scientific field of ecology which focuses specifically on plant populations. There are a number of applications for plant ecology, ranging from helping people develop low water gardens to studying endangered ecosystems to learn about how they can be protected. Researchers from this field tend to come from an interdisciplinary background which can feature training in a wide variety of scientific pursuits, including plant anatomy, general ecology, biology, and so forth.

The field of plant ecology includes the study of plants and their environment. Rather than just looking at plants in a vacuum, researchers consider how they interact with each other and their environment to create an interconnected system. Plant ecology can include the study of entire ecosystems, such as the rainforest or plateau, or the study of specific areas of interest, like plant populations which manage to survive next to a polluted stream. Plant ecologists also look at animals, soil conditions, and other influences on a plant's environment.

Ecology is a complex and vast field of study which can encompass everything from understanding how natural environments function to how humans interact with the natural world, and how various behaviours can fundamentally alter the natural environment. In plant ecology, people can focus on topics like climate change and its effect on plants, plant evolution, how plants disseminate themselves in nature, symbiotic relationships between plant species, plant diseases, and so forth.

A great deal of field work is involved in plant ecology, as researchers like to see their subjects in nature so that they can learn in context. A single sample of a plant can provide interesting information and data, but actually seeing that plant growing can provide a researcher with a great deal more data. For example, looking at a plant alone, a researcher might not understand why its leaves are shaped the way they are, but when the researcher sees the plant in nature, he or she might realize that the leaves conferred some sort of benefit on the plant or the surrounding environment, ranging from signaling the presence of the plant to pollinators to providing shelter for seedlings so that they can grow up.

Plants make up a vital part of the natural environment, and plant ecologists are well aware of this. In a healthy ecosystem, plants provide food and shelter for animals, secure the soil to prevent erosion, cast shade to create microclimates, conserve water to keep it in the ecosystem instead of allowing it to be lost, and participate in the breakdown and recycling of organic material to keep the ecosystem thriving. Plants are also of critical interest because they produce oxygen, and plants have been heavily implicated in the creation of the Earth's currently oxygen-rich atmosphere.

The relation of organisms and their processes individually to the physical conditions of their internal and external environments is the especial province of the plant physiologist but that of the plant ecologist is to study the integrated effects of the interaction of these physical conditions and the organisms they support, having due regard to the important part played by what we comprehensively term competition. It is the purpose of this article to present some of the salient trends in the study of the causal factors which influence the character and composition of plant communities.

Often in the past it has been found convenient to consider the habitat factors in separate categories, the historic, the climatic, the edaphic and the biotic, but there is an inevitable danger in such abstraction that we underrate the degree to which these interact to affect their influence on plant life. Those influences which have operated in the past to produce the conditions of the present can be classed as historic factors and these, it is manifest, must often have been secular in character.

The replacement of the prevailingly Pteridophytic and Pteridospermous flora of the Coalmeasure Period by the predominantly Gymnospermous vegetation of the succeeding epochs, must have depended upon prolonged changes in the environment, both physical and biotic, as also, no doubt, the transition to the predominantly Phanerogamic vegetation of the present era. The persistence or survival of any one group that marked the facies of plant communities may well reflect, not so much the inherent capacity of the gene complexes involved, to provide organisms of survival value, as their capacity, relative to those of other organisms, to change with sufficient rapidity.

The distinction is important since the one implies an intrinsic lack of physiological plasticity, the other a genetic complex that is merely resistant to adjustment. The one would lead to complete extinction under changing conditions, as with the Pteridosperms, the other to a rare persistence as exemplified by the Royal Fern *(Osmunda)* and the Quill-wort *(Isoetes)* alike survivals of very ancient lineage. Though very remote in the time of their initiation these secular changes are probably still operative in their effects. Indeed the vegetation of the Earth's surface today has, we all recognize, been profoundly modified by the age of steam and the internal combustion engine which alike depend for their source of power on our heritage of fossil fuel from the extinct floras and faunas of the past; their exploitation has accelerated the progressive replacement of natural plant communities by the semi-natural and the quasi-artificial, a transition that the geometrically increasing population of the world is continually augmenting.

If the effects of the remote, and even of the immediate, past are often difficult to evaluate that in no degree denigrates their significance. In our own time man's influence in changing the protective cover of vegetation has disastrously accelerated erosion in many areas. By regulating waterways and creating hard road surfaces he has speeded up drainage and lowered water tables in an ever increasing area, whilst through forest destruction on catchment areas the regulatory influence of the organic sponge has been reduced so that floods are liable to alternate with drought and bring changes far beyond the extent of obvious interference. Anti-erosion measures have recognized the important role that vegetation has to play and the time must one day come when foresters, agriculturists, water engineers and others will recognize a common interest and responsibility in initiating changes in the ecology of the plant cover which they severally exploit.

From the nature of evidences of the past and their often fragmentary character it is only in relation to the most recent events that it is possible to do more than note the contemporaneous occurrence of vegetation changes and physical alterations so that historical ecology is more often descriptive than causal and so outside the scope of this present consideration.

The more obvious climatic influences of wind, precipitation, humidity, temperature, and illumination were, in the earlier ecological studies, considered far too much in isolation from the point of view of the average conditions and with little regard either to their interaction or their indirect effects. Furthermore, it was not till attempts were made to correlate vegetation with the climatic pattern that it was realized how complicated this pattern might be.

EFFECT OF OROGRAPHIC FACTORS IN MODIFYING THE LOCAL CLIMATE

The effect of orographic factors in modifying the local climate was soon appreciated and the numerous studies in montane areas, such as Switzerland, showed clearly the importance of aspect and of the general level of the mountain massif on altitudinal limits of communities and species. This rapidly led to the recognition of what has been termed cold-air drainage and the existence of frost pockets which could result in temperature inversions and accompanying inversions of altitudinal limits, features that have been closely studied by fruit growers for obvious economic reasons. Our knowledge of the more detailed pattern of climatic features as they affect plant life is, in many areas, inadequate despite the broad correlation between vegetation types and saturation deficits that may obtain in some regions or with length of frost-free periods in others. Moreover our knowledge of the microclimate of the various strata of plant communities and its seasonal variations is generally meagre whilst any concept of the microclimate within the soil is commonly wanting. So, even though we may know the climatic environment of the shoots with some degree of accuracy, we are often quite ignorant as to the significant environment of the root systems.

During prolonged cold spells the depth to which the soil is frozen solid, for example, can vary widely in different plant communities growing on the same type of sub-soil. In one instance bare soil was frozen to a depth 2 to 3½ in., under rough grass to between 1 and 1½ in. and in a hazel copse to only W! in., or not at all, according to the depth of the leaf litter.

OPERATION OF CLIMATIC FACTORS

The long term operation of climatic factors may clearly limit the capacity of a species to flourish and compete with others or it may even impair or inhibit its power of reproduction. The detailed studies on frost resistance of Apple varieties have, for instance, revealed the diverse times at which temperature-sensitive crises may supervene in the life-history so that if this obtains also in wild populations climatic differentiation in the occurrence of strains as well as species may obtain.

But the elucidation of such relations demands a knowledge, for each species, of these critical periods, such as flower initiation, pollen-tube development, fertilization, etc., and the temperature tolerances of each. The success of any type of plant in the wild state will depend upon its capacity to flourish and reproduce under the local climate which the plant itself in part creates.

The determining conditions are not, however, the average but the extremes to which it is subjected and hence the need for continuous records of the climatic factors within the community itself not forgetting the steep vertical gradients that may obtain between different levels, of the same vegetation unit, and the interference effects of the different species with respect to the penetration of light and rainfall.

The climatic pattern is often conditioned by the structure of the community and the physiognomic character of its constituent members no less than these are themselves determined by the climatic pattern which they can modify and endure. We have referred to the fundamental principle that it is the most severe conditions of habitat, rather than the average, that mainly determine the characteristics of communities. But this, which we may term "the *Principle of Extremes,*" must not be over-simplified and regarded as being in the nature of so many sieves of lethal conditions. Species that will survive a low temperature during a single winter may succumb to two successive winters of no greater severity. Again, it is commonly the extreme conditions, that are not lethal, but sub-lethal, operating from season to season that undermine the vigour of a species and give advantage to the more tolerant vegetation over the less tolerant. The crude early concepts as to the relationship of plants to low temperature gave place when subjected to experimental tests to the recognition that resistance of a plant or its parts was largely conditioned by its water content and that the capacity of seeds to survive storage in liquid air was only true so long as their water content was negligible.

The degree of toleration of low temperature is well known as a specific character that limits the range of many plants but the complexity of the phenomenon of frost resistance has only been appreciated in the light of recent observations and experiments which have shown that the low temperature toleration of a species varies seasonally and that capacity to endure a certain amount of frost is no guarantee that the same species will not succumb to far less severe conditions if the temperatures be fluctuating.

Similarly with respect to drought tolerance the capacity to endure desiccation is one that is partly inherent, partly dependent upon the time of the year and partly on the manner in which the desiccation comes about. Both with respect to desiccation by drought and by low temperature, the mode and rate of recovery are operative factors in determining survival since rapid expansion of a contracted protoplast, unsupported by the cell wall from which it has shrunk, may involve lethal rupture. A detailed consideration of the general topic of plant hardiness has been furnished by J. Levitt which emphasizes the diverse qualities that render the protoplasm resistant to disorganization but the particular aspect we would here stress is that the precise pattern of climatic change in a particular habitat or partial habitat, often referred to as the microclimate, can alone provide a clue to the climatic factors as they affect the constituents of a natural community.

Thus the seasonal incidence of precipitation if confined to a few thunderstorms in the hottest months may scarcely support a desert type of vegetation whilst the same average rainfall spread evenly over the cooler period may suffice for a plant community which evades the waterless period by shedding its leaves. We must not lose sight of the fact that the precipitation itself may vary to a significant degree in kind as well as in amount.

For example, rain contains chloride corresponding in amount to the proximity of an area to the sea. Dr. Miller of Rothamsted, half a century ago recorded the chloride content as ranging from 218 p.p.m., near the coast, to between 2 and 3.4 p.p.m. fifty miles inland, whilst for sulphur a range of from 0.6 to over ten times that amount has been determined in open country and much higher values in the polluted atmosphere near industrial areas.

LNOWLEDGE OF CLIMATIC CONDITIONS

Calcium would appear to be present in amounts ranging from one to six parts per million and though such amounts be small their repeated addition to the soil whenever there is a fall of rain does not permit their being ignored. It would, however, seem that variations in the amounts of micronutrients present in rainfall is likely to have but a negligible influence upon plant distribution.

The advances in our knowledge of climatic conditions have emphasized the importance of detailed studies, both spatial and temporal, that take cognizance of all types of humidity and precipitation both above and within the soil. The clue to the presence of vegetation in so-called rainless desert! has been found to rest in the dew formation that even in temperate communities may play an important role.

Quite recently the probable importance of high humidity under exceptional environmental conditions has been further emphasized through the experimental proof by Breazeale*et al.* that plants can absorb water through their leaves from a saturated atmosphere and that this together with contained solutes can exude through the roots into the soil which may thereby attain field capacity.

To descriptive ecology we owe the recognition of a number of major vegetation types that are associated with particular climatic conditions, as for example the occurrence of the deciduous forests in areas where a favourable climate alternates with an unfavourable, due to low temperature or inadequate humidity. Another example is the association of evergreen coniferous forests with areas where the length of the growing season is restricted.

RELATIONS OF PLANT COMMUNITIES

But now that the broader features of the relations of plant communities with climatic types have come to be recognized the unexplored areas of the world's surface, though likely to provide further examples and interesting variants, are unlikely to contribute materially to our real understanding of the causal factors of distribution unless investigated in considerable detail.

Further progress demands a detailed analysis of climatic and other habitat conditions throughout the seasonal cycle and experimental studies of individual species to determine their tolerances and the critical phases in their development which climatic conditions may promote or inhibit and the temperature and humidity limits these demand.

Other features of the climatic complex that call for more detailed analysis are the responses of individual species to direct and indirect effects of wind action and the influence of radiation of different intensities and wave lengths. The comprehensive work edited by Professor Duggar constitutes a valuable summary of the position attained at that period, since when the amount of research, especially upon photoperiodism, to which we shall revert, has been both extensive and significant.

In the competition for radiant energy potential heights and spread are morphological features that determine the degree to which one kind of plant or individual can become a sunshade to another but clearly the time factor is here significant and in a population of seedlings, for example, those individuals that germinate first will have an advantage over those germinating later which emphasizes the survival value of what has been termed germination energy.

The more continuous the light screen provided by the foliage of a species the more effective will be its suppression of the plants in its shade though the mechanical vulnerability to wind imposes on leaf area the compromise presented by the compound character of most large leaves, although the rosette-leaves of many biennials which have the support of the soil or other vegetation beneath are immune from this type of injury and are often large and quite entire.

We can, indeed, in many morphological features discern indications of this struggle for light. The dominance of the tree habit in the climax phase where habitat conditions can support forest is but an expression of the fact that in the struggle for light the tallest members tend to survive. The wane of the arboreal vegetation is probably an accompaniment of man's increasing interference with nature making ever-augmented inroads upon woody vegetation so that today, almost throughout the world, herbaceous communities are becoming more conspicuous.

It is, perhaps, due to the importance of the time factor that the more successful species are in general those that have the potentiality of large seed-output and the marked association of average seed size with the shadiness of the plant community that a species normally colonizes is a further emphasis on the importance of the light factor in competition at the juvenile phases of developments. In the competition for nutrients the extent and rate of development of the root system to which we have already referred, is manifestly correlated with the success of the assimilating organs and the mode of spread of a species whether by seed or vegetatively will mainly determine the habitat conditions that it can invade.

From experiments with Barley growing in competition with *Holcus mollis* H. H. Mann and T. W. Barnes concluded that though the root space and supply of plant food was apparently ample for both they reduced the growth of each other and the dominant appeared to be determined by the *time of development* as well as by the density. One could multiply examples but it is clear that temporal factors play an appreciable part in the competition both for light and nutrients and emphasize the need for a much more detailed knowledge than we at present possess concerning the life histories of the chief species contributing to plant communities and the conditions that modify them.

An interesting example of the differential effect of species upon one another was exhibited by some experiments carried out with *Juncus effusus.* A. Lazenby found that the number of seeds of this Rush which germinated and the subsequent growth differed strikingly according to the other species of seeds sown with them. Thus when sown with Bent *(Agrostis tenuis)* and White Clover *(Trifolium repens)* only about one-tenth of the population of Rush plants was obtained as when the accompanying species were Ryegrass *(Lolium perenne)* and White Clover.

It is only natural that the visible, overground, parts of the plant should have received major attention from both morphologists and ecologists though it is perhaps remarkable, having regard to the early recognition of the importance of the root-system, from the nutritional as well as the mechanical aspect, that its nature, extent and mode of development should have so long been neglected. Such neglect is perhaps partially to be explained by the difficulty of exhuming the entire root system, in most soils, as a consequence of which their extent has often been underestimated.

The valuable researches of Weaver and his associates in America have served to call attention to this neglected field of study. From these the considerable labour involved in such studies can be appreciated but also the great diversity and significant distinctions that obtain. Sandy soils afford an especially favourable medium for such investigations since by blowing away the sand from a working face as it dries entire root systems can be excavated without risk of losing the finer ramifications and by appropriate means these can be supported in their original positions as they emerge from the matrix. The present writer's own studies on the root systems of plants characteristic of various phases of dune development have shown how varied these can be not merely between different species but also between different individuals of the same species when growing in conditions of diverse water supply and aeration. Root studies have emphasized the distinction between those that exhibit a restricted but richly branched system, which exploit a relatively small volume of soil intensively, and those in which the root system is extensive but the large soil volume is much less completely interpenetrated.

Modern studies of economic plants have shown what marked differences in this respect may obtain between cultivars of the same species as, for example, the root systems of the "Wealthy" and "Doucin" apples or the various East Malling Apple stocks, and there is no reason to suppose that similar differences both interspecific and intraspecific, do not occur with respect to the root systems of wild plants and that they play an important role in their competitive equipment. The recognition of the fundamental importance of a number of trace elements and their frequent presence in inadequate amounts, or in more or less non-available condition, has brought into prominence the great significance that may wcll attach to thc type and degree of root exploitation.

Furthermore it has been shown that the availability of some trace elements, such as zinc, in the early stage of development of a species, can be decisive for the subsequent healthy growth and its early supply will differentiate between failure and success. The importance therefore of the *rate* of development of the root systems of different kinds of plants in what we may perhaps be permitted to describe as their race for the necessary nutrients which are in short supply, scarcely requires emphasis. The great success of the grasses may be in no small degree due to the extent and rapidity of their root development. Weaver showed that the latter is frequently more than 12 mm in a day whilst the root of a maize plant may extend more than 50-60 mm in the same period.

With uniform genetic material the extent of the root system is very susceptible to the influence of water soluble substances, for example the calcium ion is known markedly to affect root growth. The extent of the root surface in the presence of calcium was several times that in its absence and the volume of soil that can be exploited is probably proportionately increased. The extent of the root system is also affected by the soil texture and indeed one aspect of cultivation is that it reduces the resistance that the growing root has to overcome in its penetration of the soil. Again, optimum moisture content and above all good aeration promote not merely the extent of the root system but in particular the development of root hairs.

One may note that root hairs can augment the absorbtive surface sevenfold, so that factors affecting their production and growth may be of great significance. All these conditions affecting the size and efficiency of the absorbing structures stress the importance of the morphological-physiological relations and one may expect that future ecological studies will tend to integrate still further the two angles of approach, but if this is to be achieved it is essential that the physiologist should be apprehensive of the extent to which field competition is likely to modify the laboratory indications and the ecologist must be alert to appreciate the field significance of the controlled experiments which, however oversimplified they may be from the point of view of their translation to the plant community, can rarely fail to make their contribution to our understanding of the more complex conditions.

In 1929 I called attention to the striking reduction in the extent of the root system of a species when subjected to the competitive influence of the root systems of other plants. Eight years later Pavlychenko furnished remarkable quantitative data respecting the magnitude of this effect as exhibited by cereals in competition with weeds and with one another. The importance of the competition factors in natural communities cannot be over-emphasized and such may well be operative from the earliest phases of plant development since seedling mortality, which is very high, appears to be influenced not only by the physical conditions but by the proportions in which the seedlings of different species are present.

Statistical studies have confirmed what general observation had suggested that individual plants are not normally dispersed in a random manner. This was for instance clearly brought out in regard to grassland communities by G.

E. Blackman, in 1935, but it yet remains to be ascertained to what extent such lack of uniformity in dispersion reflects the lack of uniformity in the habitat conditions. For example, the marked tendency towards aggregation of the individuals of the herbaceous species of *Salicornia* is clearly to be expected from the crozier-like hairs upon the testa which anchor the seeds to suitable surfaces, such as the wefts of salt-marsh algae, and to one another, when they are dispersed by the rising and falling tides.

So, too, the classical examples of the hygroscopic arms of the dissected spore-coats of *Equisetum* which ensure some measure of aggregation of the unisexual gametophytes. These and other similar morphological determinants may clearly be important for survival but unrelated to any permanent feature of the habitats concerned. We still do not know why it is that the annual vegetation of a desert, superficially uniform, may present a Joseph's coat with large areas each dominated by a different species of *Helipterum.*

Are such phenomena visual expressions of vagaries of dispersal, or evidences of survival from selection accompanying real though obscure habitat conditions? Attention has more than once been drawn to the phenomenon of certain species that occur with great abundance in an area and produce copious seed only to disappear for a long period subsequently or, if not completely absent, they are rare. A striking example that has been observed both in this country and on the Continent is afforded by the Yellow-wort *(Blackstonia perfoliata),* which is very pertinent to our present context because its root system is of comparatively small extent.

The output of its small, almost dust-like, seeds which exhibit a high per centage germination is large. So it is not unreasonable to assume that these are in fact broadcast over the entire area. Hence, one may ask, is it that such species require some nutrient condition, perhaps a trace element in short supply and that this becomes depleted? The discontinuity of the densest population of *Blackstonia* would not warrant suggesting that its roots completely exploit the entire surface soil but it may well be that, from a vast population of seedlings, the mature plants we see are in fact survivors where the requisite nutritional conditions obtain and that the discontinuity of occurrence does indeed reflect the heterogeneity in this respect of a habitat otherwise apparently uniform.

PROCESSES OF PEDOGENESIS

The role of vegetation in the processes of pedogenesis has been studied under conditions where the distribution of the phases of plant development in space corresponds to their sequence in time and thus provides an unequivocal chronological succession over a prolonged period with respect alike to soils and organisms.

For sand dunes it has been shown that the soil development is dependent upon a succession of species that incorporate organic material beneath the surface as well as providing deposition upon it and augment water retention and stability so that less and less specialized types can effect colonization ultimately ousting the pioneer specialists.

The original inhospitable soil conditions with an open community of perennial and annual species, characterized by features that enable them to endure the rigourous conditions, is succeeded by conditions favourable to an increasing number of species, far more diverse in their biological equipment, that ultimately constitute a closed community and this in turn may finally give place to scrub and woodland occupying soil that is now favourable to a great diversity of plants but where in fact the number of species has again diminished owing to the dominance of trees and shrubs.

The initial edaphic specialization has thus, in the course of perhaps a century, given place to a biotic specialization that in its own way may be almost equally severe. The studies of R. L. Crocker and his associates on the recessional moraines deal with a comparable sequence from an open community of pioneers that in the course of some 120 years has developed to spruce forest and, as in the dune succession, there is an increasing organic content and transition from an initially alkaline soil to one of appreciable acidity with an augmenting gradient in the soil profile. The recognition of the major role in water retention by the soil played by the organic material gives added significance to its vertical distribution and it thus came to be realized that, in the naturally stratified soil the surface layer might be of sufficient thickness and its organic content so high that it constituted a sponge which could starve the lower layers of water except perhaps in the heaviest rainfall. Thus is indicated one of the ways in which during the course of years a tall deeply rooted layer of dominant species continually adding its quota to the superficial organic accumulations can gradually contribute to its own destruction and an acidophile shallow-rooted community replaces the erstwhile dominants.

Each species has its own complex of conditions in which it can best develop. If we grow any one in conditions that vary with respect to one factor, such for instance as soil reaction, we find there is a range over which the plants exhibit maximum growth on either side of which increasing acidity or increasing alkalinity is accompanied by diminishing vigour. In the absence of competition the range of toleration is often wide but in the presence of other species this may become very restricted.

This fact emphasizes the fundamental principle of the interaction of habitat factors. For whether it be soil reaction or any other there is usually no absolute optimum but a relativity that is the more complex because any single one may operate in so many different ways. Soil acidity, for example, may be beneficial as enabling certain nutrient ions to be readily available but it may also be directly harmful by reason of toxicity of the hydrogen ions themselves, or indirectly through excess of ions of alumina or manganese in the soil solution, or again it may adversely affect beneficial soil organisms such as the nitrogen fixing bacteria and encourage harmful ones such as *Plasmodiophora brassicae.*

Above all, adverse or beneficial factors cannot be considered in isolation since plants grow where they must and not where they will so that what, in the laboratory experiment, may seem harmful appears in the field as a benefit because the species concerned is more tolerant of this condition than its normal associates.

For instance *Viscaria alpina* can apparently tolerate a copper content in the soil toxic to most species. Again what favours vegetative growth is often detrimental to seed production and which of these is the more important may vary from time to time in one and the same plant community. Experiments have repeatedly shown that the optimum conditions with respect to various factors alter with the stage of development of the individual and also its susceptibility.

"The field for the problem and the laboratory for the solution" is only very partially true for ecological studies, for laboratory findings must be transmuted into the far more complex context of so many other variables that experience alone can guide as to what may be applicable and to what degree. But field studies, both observational and experimental, should serve as the means for testing hypotheses based upon what the plant physiologist has ascertained. Ecology is essentially applied plant physiology with the competition factor as an important modifying influence in a physical environment that represents the interaction of climatic, edaphic, and biotic factors that are themselves abstractions from a sequence of continual change.

The emphasis on physiognomy that has formed the basis for some systems of classification of plant communities can be regarded as an attempt to utilize the plants themselves as integrators of the physical conditions and the concept of the so-called indicator species is but another aspect of the same theme.

When extreme conditions are under consideration the correlation of physiognomy with habitat conditions is sometimes surprisingly high as is well exhibited, for example, on some of the sandy and areas in South-West Australia, where communities are encountered consisting of species of the most diverse genera and families yet resembling one another so closely in their general physiognomy as to be difficult to distinguish except when in flower or fruit.

In less differentiated conditions, however, as for instance in the temperate woodland, the same stratum, where there is relative uniformity of environment, often presents a great range of morphological diversity that suggests either that physiognomy is an unsatisfactory guide or that our knowledge must be greatly increased by experiments, in the field of environmental morphogenesis, before we are in a position to assess what morphological characteristics are significant and what are irrelevant.

STRATIFICATION AND SEASONAL DEVELOPMENT OF AERIAL SHOOTS

The stratification and seasonal development of aerial shoots, to which attention was drawn by Kerner Von Marilau nearly a century ago, and of the subterranean organs, which the late Dr. Woodhead was one of the first to investigate, both affect the competitive efficiency of the species whilst the periodicity of the important phases of the life history and their modifications by the environment are no less significant though too little investigated.

These latter, for economic reasons, are better known with respect to horticultural plants. The gardener is familiar with the fact that most plants are best moved in the dormant state, yet most species of *Helleborus* thrive best if transplanted when in full flower, a difference probably related to the phases of root development. Or again, careful investigations of bulbous species have shown how temperature fluctuations can affect differently diverse cultivars of the same species apparently by reason of their individual rhythms in the production of flower initials.

What is true of cultivated plants is probably equally true of wild species and emphasizes for us that the understanding of the communal life of plants is dependent upon a knowledge of the idiosyncrasies of its chief constituents. The structure and chemical characteristics of the soil as they affected plant life gained no small impetus from the work of Lawes and Gilbert, initiated in the early forties of last century, but the classical work of E. J. Russell on *Soil Conditions and Plant Growth,* which first appeared in 1912 and which passed through eight editions and has been translated into various languages, probably did more than any other single publication to bring before the minds of botanists the complexity as well as the importance of edaphic conditions upon vegetation.

The soils that the agriculturist had studied were, however, to a very considerable degree physically homogenized by the processes of cultivation and chemically altered by their manurial treatment. It was not until ecologists began to pay attention to the organization that is exhibited by natural soils that the degree of applicability of the findings of the agriculturists could begin to be assessed. One of the earliest recognitions of the structure of natural soils and its relation to the vegetation which they bear was the study by Gesser and Siegrist of the profiles in the soils of the Aare communities in Switzerland. In this country the present writer called attention to the striking vertical gradients with respect to organic content, water content, soil reaction, and other features exhibited by woodland and sand-dune soils which provide a gradient of conditions for the root system.

These vertical changes not only contribute to the establishment and maintenance of the complementary communities of species but, as soils derived from a calcareous substratum strikingly show, can provide one and the same root system with a diversity of conditions favourable for the easy acquisition of the whole range of nutrient requirement from an acid surface layer from which the ions of iron and manganese may be readily obtained to an alkaline subsoil from which molybdenum is readily available.

PLANT NUTRITION

Complete nutrition results in superior plant growth. Why choose anything less for your plants? There are 20 elements necessary for optimum plant growth. Air and water supply carbon, hydrogen and oxygen. Macronutrients are required by plants in large amounts.

Micronutrients are required in smaller amounts. Eliminate any of these elements,and plants will display abnormal growth, deficiencies or may not reproduce. The following is a brief guide to the role played by each of these essential nutrient elements.

MACRONUTRIENTS

- *(N)-Nitrogen*: Component of proteins, hormones, chlorophyll, vitamins and enzymes. Promotes stem and leaf growth. The ammoniacal and nitrate forms are used directly by plants for stem and leaf growth. The urea form of nitrogen must be broken down by soil borne microorganisms or urease before it can be utilized by the plant. Urea can cause leaf tip and root burn. Deficiency (Def.): reduced yields, yellowing of leaves, stunted growth. Excess nitrogen can delay fruiting and flowering.
- *(P)-Phosphorus*: Seed germination, photosynthesis, protein formation, overall growth and metabolism, flower and fruit formation. Def.: purple stems and leaves, retarded growth and maturity, poor flowering and fruiting. Large amounts without zinc cause zinc deficiency. Low pH (<4) ties up phosphates in organic soils.
- *(K)-Potassium*: Formation of sugars, carbohydrates, proteins, cell division. Adjusts water balance; improves stem rigidity and cold hardiness; enhances flavour, colour and oil content of fruits; important for leafy crops. Def.: spotted, curled or burned look to leaves; lower yields.
- *(Ca)-Calcium*: Activates enzymes; structural part of cell walls; influences water movement, cell growth and division Required for uptake of nitrogen and other minerals. Leached from soil by watering. Immobile-requires a constant supply for growth. Def.: stunting of new growth in stems, flowers, roots; black spots on leaves and fruit; yellow leaf margins.
- *(Mg)-Magnesium*: Critical component of chlorophyll; needed for functioning of enzymes for carbohydrates, sugars and fats; fruit and nut formation; germination of seeds.

- Def.: yellowing between veins of older leaves; chlorosis; leaf droop. Leached by watering. Foliar spray to correct deficiencies.
- (*S*)-*Sulfur*: Component of amino acids, proteins, vitamins, enzymes. Essential for chlorophyll. Imparts flavour to many vegetables. Def.: light green leaves. Water supply may contain sulfur. Leached by watering.

MICRONUTRIENTS

- (*B*)-*Boron*: Affects at least 16 functions: flowering, pollen germination, fruiting, cell division, water relationships, movement of hormones, cell wall formation, membrane integrity, calcium uptake, movement of sugars. Immobile; easily leached. Def.: terminal bud die back causes rosette of thick, curled, brittle leaves or brown, discoloured, cracked fruits, tubers and roots.
- (*Cl*)-*Chlorine*: Involved in osmosis (movement of water or solutes in cells), ionic balance necessary to take up mineral elements and photosynthesis. Def.: wilting, stubby roots, yellowing, bronzing. Scents in some plants may be decreased. Leached by watering.
- (*Co*)-*Cobalt*: Required by nitrogen fixing bacteria; formation of B12 vitamin; formation of DNA. Will extend life of cut flowers such as roses. Def.: may result in nitrogen deficiency.
- (*Cu*)-*Copper*: Necessary for nitrogen metabolism; component of enzymes - may be part of enzyme systems that use carbohydrates and proteins. Bound tightly in organic matter. May be deficient in highly organic soils. Not readily lost from soil but may be unavailable. Def.: die back of shoot tips; terminal leaves develop brown spots. Excess is toxic.
- (*Fe*)-*Iron*: Enzyme functions; catalyst for synthesis of chlorophyll; essential for new growth. Def.: pale leaves, yellowing of leaves and veins. Leached by water and held in lower parts of soil. High pH soils may have iron present but unavailable to plants.
- (*Mn*)-*Manganese*: Enzyme activity for photosynthesis, respiration and nitrogen metabolism. Def.: young leaves are pale with green veins similar to iron deficiency; advanced stages-leaves are white and drop; brown, black or gray spots may appear next to veins. Plants in neutral or alkaline soils often show def. Acid soils may increase uptake causing toxicity.
- (*Mo*)-*Molybdenum*: Structural part of enzymes that reduce nitrates to ammonia for amino acid development essential to protein formation; required by nitrogen fixing bacteria. Def.: pale leaves with rolled, cupped margins. Seeds may not form. Nitrogen deficiency may occur if plants are lacking Mo.

- *(Ni)-Nickel*: Recently recognized as essential. Required for the urease enzyme to break down urea into usable nitrogen and for iron uptake. Seeds require nickel to germinate.
- *(Na)-Sodium*: Improves nitrogen metabolism in many plants, involved in osmotic (water movement) and ionic balance in plants. Def.: yellowing of leaves and leaf tip burn; may inhibit flower formation.
- *(Si)-Silicon*: Component of cell walls; enhances resistance to sucking insects and fungi.Foliar sprays reduce populations of aphids on some plants. Enhances leaf presentation; improves heat, drought and cold tolerance; improves photosynthesis; extends bloom life. Def.: wilting, poor fruit and flower set, increased susceptibility to insects and disease. Disease resistance is enhanced by regular foliar feeding.
- *(Zn)-Zinc*: Functional part of enzymes including auxin (growth hormone) synthesis, carbohydrate metabolism, protein synthesis, stem growth. Def.: mottled leaves, irregular yellow areas. Zinc deficiency leads to iron deficiency. Occurs in eroded soils; least available at pH of 5.5-7.0. Lower pH can cause availability to the point of toxicity.

NUTRIENT BALANCE AND NUTRIENT DEFICIENCIES

Plants (like people) need a 'balanced diet'. They need all 13 nutrients to remain healthy. If one is missing, the plant will not grow well. Poor plant nutrition causes plants to grow slowly in the nursery and in the field, and to be more susceptible to diseases. Many people confuse the symptoms of nutrient deficiencies with those of too much or not enough shade or water.

In fact, all three factors, shade, water and nutrients affect plant growth, and interact to produce healthy plants. A plant that grows in full light with abundant moisture and receives all the 13 nutrients will grow fast and have a dark green colour in its leaves. A plant that grows slowly in the shade may also have dark green leaves, but when exposed gradually to the sun, the leaves may turn yellow.This does not mean that plants do not like full sun—it might indicate a nutrient deficiency which did not show up in the shade because the plant did not have enough light to stimulate fast growth. Together, water, shade and nutrients must be monitored and adjusted to produce quality seedlings. It takes practice to learn the signs identifying a missing nutrient or nutrients, but you can learn to do so, and some of the signs are common to many plants. A *good nursery practice* is to carefully monitor the leaves of your plants for signs of nutrient deficiency, and correct them with a better substrate or with fertilizer.

INORGANIC FERTILIZERS

Inorganic fertilizers are mined from the soil, or produced during complicated chemical reactions. A *good nursery practice* is to read the fertilizer labels. This allows you to apply what the plants need without wasting nursery resources. Fertilizers contain only plant nutrients; they are not used to combat plant diseases or insects. Inorganic fertilizers do not improve the substrate physical properties, whereas organic material such as compost does. Inorganic fertilizers are also expensive and not always available in the stores. Nursery managers should carefully consider the cost and benefit of buying these products.

Granular fertilizers are commonly given names like "17-17-17", or "10-30-10". What do the numbers mean? They represent the percentages of nitrogen (N), phosphorus (P), and potassium (K) in the fertilizer—17 per cent N, 17 per cent P, 17 per cent K. In this case, 51 per cent of the mixture is made up of N-P-K, and the rest is inactive material used to help spread the fertilizer evenly. Urea contains only N, and is labelled as 46-0-0. Urea is very strong and can easily burn the plants if too much is applied.

GRANULAR FERTILIZERS

Granular fertilizers can be mixed into the substrate or into the irrigation water, or be applied to older plants on the soil surface. It is better to mix the fertilizer directly into the substrate before planting the seed because the roots can avoid or seek the fertilizer as they need it. Use only small quantities such as 2 or 4 grams (1/2 teaspoon) per 1 kg of soil. It is better to add too little then too much.

You need to experiment with different levels. Plants should respond within two weeks. When dissolving fertilizer in warm water, carefully note whether it is thoroughly dissolved. If not, it is probably the phosphorus that remains. It may be better to apply fertilizer in granular form if it does not dissolve thoroughly. Apply liquid fertilizer to the soil, not to the leaves which are easily burnt if fertilizer remains on them. Be extremely cautious when applying fertilizer to young plants.

FOLIAR FERTILIZERS

Foliar fertilizers are used in order to get the nutrients to the plants quickly. They are specially formulated to apply directly on the leaves. Foliar fertilizers are absorbed by the leaves, not by the roots. When plants are acutely deficient in nutrients, foliar fertilizers often help 'green them up'. Frequently, foliar fertilizers only contain the micronutrients, since it is assumed that the macronutrients are available in the substrate. However, some such as 'GroGreen'® contain both micronutrients and 20-30-10 of N-P-K. Often an adhering agent such as 'Da-Plus' is used to help the fertilizer stay on the leaves so that it is not washed off in the rain. Because foliar fertilizers are expensive, and they may not encourage strong root growth, they should not be used as a long-term solution for plant nutrients.

COMMON NUTRIENT DEFICIENCY SYMPTOMS OF MACRONUTRIENTS

- *Nitrogen*: This is a mobile nutrient, which means that when nitrogen is deficient, plants move it from the older foliage to the younger, actively growing leaves. The older leaves (the ones lower on the stem of the tree) become yellow first, while the new leaves remain green.
- *Phosphorus*: The entire seedling is stunted, especially during early growth. Depending on the species, the leaves may become dull green, yellow or purpletinged. The purpling of leaves is a classic symptom, but sometimes there are no colour differences in leaves, so visual diagnosis is not always reliable. The purple colour should not be confused with new leaves that often appear purple or red when they first flush out.
- *Potassium*: Symptoms appear in older leaves first. These start to yellow at the edges, and have some green at the base. Later, leaf edges turn brown and may crinkle or curl and small necrotic (dead) spots may appear. Plants may wilt, even though sufficient water is available in the substrate. When deficiencies are severe, leaves will die. Calcium: This is difficult to detect because signs include slow growth, and dieback of bud or root tips. Seedlings will have stubby little roots with brownish discoloration. The problem is most common in very acidic soils. A well-developed root system with many fine root hairs is important for calcium uptake.
- *Magnesium*: This nutrient is commonly deficient in coarse-structured soils and in acidic soils. Uptake may be blocked if there is too much potassium in the soil. Like nitrogen, magnesium is a mobile nutrient, so deficiency symptoms show up in the older leaves first. These leaves show a very characteristic yellowing between the veins or ribs, and they appear streaked.
- *Sulphur*: Plants will be slightly stunted. This is not a mobile nutrient, so the symptoms show up on younger leaves which are initially light green, but eventually develop scorched and curled margins. Dry areas can form along the margins and then spread inward to the leaf midrib.

ESSENTIAL MINERAL NUTRIENTS

Solution culture experiments have established a number of facts basic to plant growth on soils—that plant roots must have a supply of oxygen, that solid soil particles and micro-organisms, while sometimes benecial, are not essential, that in addition to C, H and O some 14 or 15 chemical elements are essential for plant growth, and that all of these essential elements may be supplied to plant roots as simple ions of inorganic salts in solution and must be supplied in adequate but non-toxic amounts.

A chemical element is regarded as essential if, in its absence, a plant cannot complete its life cycle. An alternative criterion has also been suggested—that the element is part of an essential plant constituent or molecule. So far, only the rst criterion has been used to establish essentiality although the latter has proved useful as a guide to further research, as, for example, in the association of Mn with laccase activity and Ni with urease activity.

'Essential mineral nutrients' or simply 'essential nutrients' include all those chemical elements which are normally absorbed from the soil solution by higher plants. They exclude C, H, and O which comprise more than 99 per cent of metabolically active leaves and some 90–95 per cent of their dry matter, but, paradoxically, they include N which does not occur in any soil mineral and is supplied to some plants from air.

Essential mineral nutrients comprise less than 1 per cent of fresh mass in active leaves and 5–10 per cent of their dry matter; yet within this small proportion, all 14 or 15 elements must be present in adequate amounts for healthy growth and effective reproduction.

FUNCTIONS OF NUTRIENTS

A Summary of notional values for concentrations of nutrient elements (on a Dry Mass Basis) considered as just adequate for maximum palnt growth, and accompanying data on the relative number of atoms of those elements witnin a plant (referenced to Mo=1).

A broad distribution between plant requirements for micronutrients (trace elements) and micronutrients is than evident. Mobility form leaves is meant to reflect overall impressions chemical analaysis of nutrient budgets for whole plants and organs.

Historically, essential nutrients have been classied in one of two groups based on amounts required by plants, namely macronutrients or micronutrients, and that convention has been adopted for convenience in Table. This distinction reflects historical sequence and experimental difûculties in discovering essential nutrients. Following advances in nine-teenth century chemistry, essentiality of macronutrients was relatively easy to prove, but essentiality of micronutrients (except Fe) was elusive, requiring great care in eliminating contamination from macronutrient salts, from water and from other environmental sources.

The classication remains in common use, and was therefore used in Table but the distinction between macro and micro is somewhat arbitrary. In terms of chemistry and function, an alternative grouping is outlined below.

GROUP 1: N, S

N and S are covalently bound in organic compounds in reduced states. They are essential components of proteins providing reactive groups for interaction with metabolites and other nutrients. They are generally absorbed as oxyanions which, except for small amounts of sulphate in a few organic com-pounds, must be reduced before use.

GROUP 2: P, B

P and B are covalently bound in organic compounds in their fully oxidised states. P is present in phospholipids of cell membranes, nucleic acids of chromosomes and a large number of intermediary metabolites including many in which it plays a crucial role in bioenergetics. P is always present as mono- or polyorthophosphate which behaves as a weak acid. B, like P, is thought to function only in its fully oxidised form; B oxyacid is much weaker and is largely undissociated at the pH of cells. B complexes strongly with hydroxyls in adjacent *cis*-diol conguration in sugars and their derivatives such as those in the hemicellulose of cell walls, but key metabolic or structural functions still remain unknown.

GROUP 3: K, CA, MG, NA, CL

K, Ca, Mg, Na and Cl are present primarily in ionic form as either free ions in solution or as entities reversibly adsorbed by electrostatic forces on charged sites. K, Ca and Mg are required in relatively large amounts while Na and Cl, though often present in large amounts, are required only in trace amounts. Group 3 nutrients function in osmotic adjustment of cell activities and in enzyme activation, probably by modifying the shape and orientation of substrates and enzymes.

Ca and to a lesser extent Mg also stabilise cell structures such as membranes and cell walls, probably via their divalent positive charge which forms cross-links with negative charges on cell structures. Some Ca binds strongly and reversibly to the protein calmodulin which in turn activates several enzymes and controls Ca transport. Some Mg is structurally bound in chlorophyll (a strict 1:1 stoichiometry of Mg atoms to chlorophyll molecules is universal)

GROUP 4: MN, FE, CO, NI, CU, ZN, MO

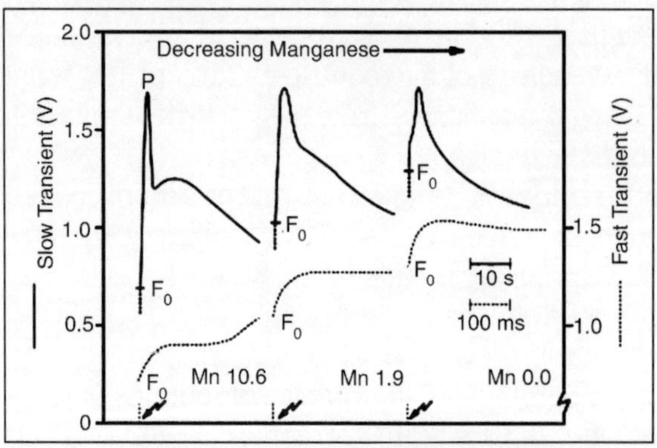

Fig. Induction Kinetics for *in Vivo* Chlorophyll *a* Fluorescence Change Dramatically According to Leaf Mn Status of Wheat Leaves.

Constnat yield fluorescence (F_0) increases but variable fluorescence (F_v; where $F_v = P-F_0$) decreases with decrease in leaf Mn. A ratio of F_0/F_v increases abruptly as leaf Mn drops to a critical level, and provides an early indication of impending Mn deficiency.

Preplanting soil dressings corresponding to transients on leaf samples were, left to right, 10.6, 1.9 and 0.0 kg $MnSO_4$ ha^{-1}. Fe, Co, Ni, Cu, Zn, Mo and, to some extent, Mn are tightly bound in proteins as metalloproteins or, in the case of Co, which is only required in plants xing N_2, in a coenzyme. All are transition series elements, and with the exception of Zn all show variation in valency state with attendant variation in physiological effect. Group 4 elements govern a wide range of reactions including many oxidation–reductions in which all except Zn can act as electron carriers by undergoing reversible oxidation–reduction.

All are micronutrients, and all except Mn form strong complexes with organic and amino acids. Mn is exceptional in that it resembles Group 3 nutrients in being present largely as a free divalent ion which can replace Mg in the activation of a number of enzymes. But some Mn is tightly bound within the water-splitting apparatus of photo-system II and in superoxide dismutase, an enzyme system that helps dissipate harmful effects of singlet oxygen formation. Mn deciency results in highly charac-teristic induction kinetics for *in vivo* chlorophyll *a* fluorescence and can be used as a diagnostic tool in combination with chemical assay.

FERTILIZER MANAGEMENT PRACTICES IN PLANT

RATES

Nitrogen fertilizer rates are determined by the crop to be grown, yield goal, and quantity of nitrogen that might be provided by the soil. Rates needed to achieve different yields with different crops vary by region, and such decisions are usually based on local recommendations and experience. The quantity of nitrogen supplied by the soil is determined by the quantity of nitrogen released from the soil organic matter, that released by decomposition of residues of the previous crop, any nitrogen supplied by previous applications of organic waste, and any nitrogen carried over from previous fertilizer applications.

Such contributions can be determined by taking nitrogen credits (expressed in lb/acre) for these variables. For example, corn following alfalfa usually requires less additional nitrogen than corn following corn, and less nitrogen fertilizer is needed to reach a given yield goal when manure is applied. As with rates, credits are usually based on local conditions.

Soil testing is being suggested more often as an alternative to taking nitrogen credits. Testing soils for nitrogen has been a useful practice in the drier regions of the Great Plains for many years, and in that region fertilizer rates are often adjusted to account for NO_3^- found in the soil prior to planting. Over the past 10 years, there has been some interest in testing corn fields for NO_3^- in the more humid regions of the eastern US and Canada, utilizing samples taken in late spring, after crop emergence, rather than before planting. This strategy, the pre-side dress nitrogen soil test (PSNT), has received a great deal of publicity and seems to provide some indication of whether additional side dressed nitrogen is needed or not. The test is still in the experimental stage in most states, and interpretations based on local research and experience are essential.

FERTILIZER PLACEMENT

Placement decisions should maximize availability of nitrogen to crops and minimize potential losses. A plant's roots usually will not grow across the root zone of another plant, so nitrogen must be placed where all plants have direct access to it. Broadcast applications accomplish this objective. Banding does also, when all crop rows are directly next to a band. For corn, banding anhydrous ammonia or UAN in alternate row middles is usually as effective as banding in each middle, because all rows have access to the fertilizer. Moist soil conditions are necessary for nutrient uptake. Placement below the soil surface can increase nitrogen availability under dry conditions because roots are more likely to find nitrogen in moist soil with such placement. Injecting side dressed UAN may produce higher corn yields than surface application in years when dry weather follows side dressing. In years when rainfall occurs shortly after application, subsurface placement is not as critical. Subsurface placement is normally used to control nitrogen losses.

Anhydrous ammonia must be placed and sealed below the surface to eliminate direct volatilization losses of the gaseous ammonia. Volatilization from urea and UAN solutions can be controlled by incorporation or injection. Incorporating of urea materials (mechanically or by rainfall shortly after application) is especially important in no-till and turfgrass situations where volatilization is aggravated by large amounts of organic material on the soil surface. Applying small amounts of "starter" nitrogen as UAN in herbicide sprays, however, is usually of little concern. Placing nitrogen with phosphorus often increases phosphorus uptake, particularly when nitrogen is in the NH_4^+ form and the crop is growing in an alkaline soil. The reasons for the effect are not completely clear, but may be due to nitrogen increasing root activity and potential for phosphorus uptake, and nitrification of NH_4^+ providing acidity which enhances phosphorus solubility.

TIMING OF NITROGEN APPLICATIONS

Timing has a major effect on the efficiency of nitrogen management systems. Nitrogen should be applied to avoid periods of significant loss and to provide adequate nitrogen when the crop needs it most. Wheat takes up most of its nitrogen in the spring and early summer, and corn absorbs most nitrogen in mid-summer, so ample availability at these times is critical.

If losses are expected to be minimal, or can be effectively controlled, applications before or immediately after planting are effective for both crops. If significant losses, particularly those due to denitrification or leaching, are anticipated, split applications, where much of the nitrogen is applied after crop emergence, can be effective in reducing losses. Fall applications for corn can be used on well-drained soils, particularly if the nitrogen is applied as anhydrous ammonia amended with N-Serve; however, fall applications should be avoided on poorly drained soils, due to an almost unavoidable potential for significant denitrification losses. When most of a crop's nitrogen supply will be applied after significant crop growth or positioned away from the seed row (anhydrous ammonia or UAN banded in row middles), applying some nitrogen easily accessible to the seedling at planting ensures that the crop will not become nitrogen deficient before gaining access to the main supply of nitrogen.

MINIMIZING NITROGEN FERTILIZER LOSSES

The major mechanisms for nitrogen fertilizer loss are denitrification, leaching, and volatilization. Denitrification and leaching occur under very wet soil conditions, while volatilization is most common when soils are only moist and are drying.

ANHYDROUS AMMONIA

Ammonia is a basic industrial material and is the material from which other nitrogen fertilizers are made. It has been manufactured on a commercial scale since 1913, and the process used to produce it, originally developed by Fritz Haber, has changed little since that time.

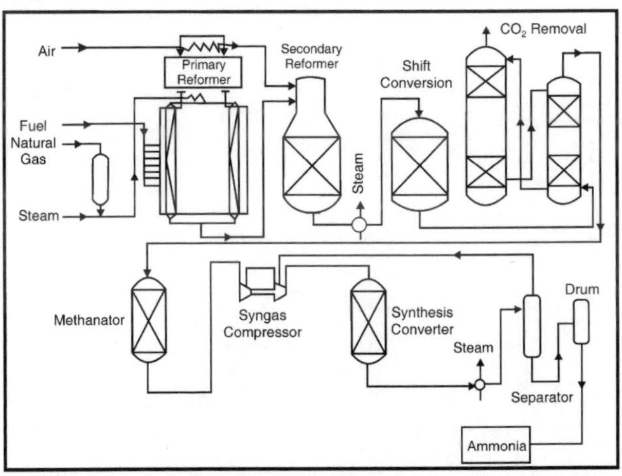

Fig. Production of Ammonia

In the Haber process, gaseous H_2 and N_2 are combined under high temperature and pressure conditions to produce ammonia. Ammonia production is an energy intensive process, not only because of its temperature requirements, but also because it requires large quantities of purified H_2 that is usually derived from natural gas, or methane (CH_4).

Naphtha and coal are also used as feed stocks in some parts of the world. Hydrogen is produced from methane (CH_4) in a multi-step process. First, a "synthesis gas" is produced by "reforming" methane and steam to hydrogen and carbon monoxide (CO) over a catalyst at 1300-1600° F. This synthesis gas is mixed with air in a second reforming reaction over another catalyst, producing a mixture of N_2, H_2, CO, CO_2 and steam.

The carbon-monoxide and steam then undergo a final "shift conversion", producing carbon dioxide and more hydrogen. This final synthesis gas is then purified to remove CO_2 that would poison the ammonia-synthesis catalyst. Any remaining traces of carbon oxides are removed by absorption into other materials or are converted back to methane over a methanation catalyst.

The purified N_2-H_2 synthesis gas is compressed to pressures of 2000-5000 psig (depending on process) and passed over an iron catalyst where ammonia is formed. The resulting gas mixture is cooled to condense and remove ammonia. Unreacted products are recycled through the synthesis converter. The resulting liquid ammonia is then stored in insulated tanks.

AMMONIUM NITRATE

Nitrate Ammonia can be oxidized in air to produce nitric acid (HNO_3). This nitric acid can then be neutralized with more ammonia to produce a solution that is typically 83 percentammonium nitrate and 17 per cent water.

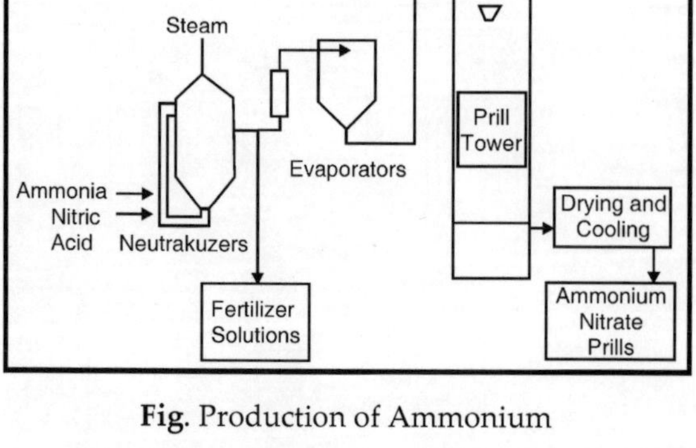

Fig. Production of Ammonium

This solution can be used to produce nitrogen fertilizer solutions or can be processed further to produce solid ammonium nitrate. The ammonium nitrate solution is concentrated to 96-99 per cent NH_4NO_3 and the "melt" transported to the top of a prilling tower, where it is sprayed into a rising air stream.

The droplets crystallize and condense into hard, spherical "prills" that are dried, cooled, and sized for shipment. Ammonium nitrate can also be granulated by several processes.

UREA

Pressurized carbon dioxide gas and heated liquid ammonia are mixed in a condenser to form a mixture containing an intermediate compound, ammonium carbamate. The mixture flows to a reactor where it is dehydrated to convert a portion of the carbamate to urea. This mixture then passes through a stripper, where a stream of carbon dioxide removes unreacted ammonia and recycles it back to the condenser. In the stripper, unreacted ammonium carbamate is decomposed back to ammonia and carbon dioxide, and is also recycled to the condenser. The remaining urea solution is removed from the stripper and purified. The solution can then be used to make nitrogen fertilizer solutions, or can be concentrated and used to produce solid urea in prills or granules.

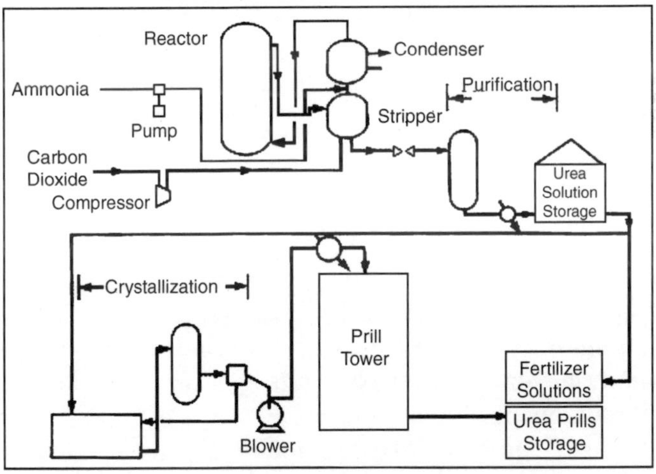

Fig. Production of Urea

UAN SOLUTIONS

Urea-ammonium-nitrate solutions are made by mixing the final solutions produce during production of urea and ammonium nitrate. Mixing in various ratios allows production of materials with different nitrogen concentrations, from 19 to 32 per cent.

AMMONIUM SULFATE

Ammonium sulfate is a byproduct of the steel and caprolactam industries, made by neutralizing waste sulfuric acid with ammonia. Solid ammonium sulfate is removed from the solution by heating the solution to boiling, evaporating water and concentrating the salt which eventually crystallizes out. The crystal slurry is dewatered in a centrifuge and the wet crystals dried in warm air.

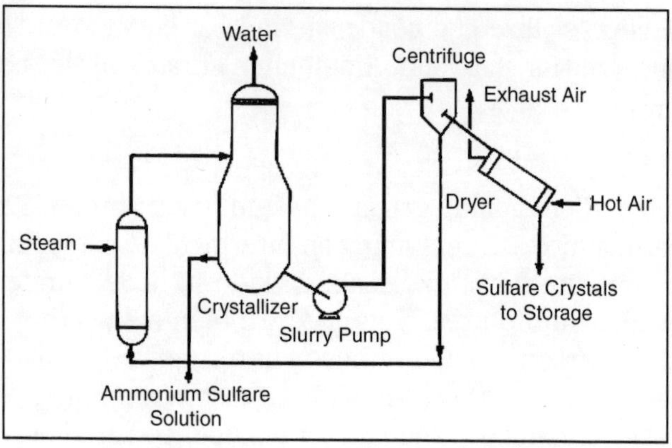

Fig. Ammonium Sulfate Crystallization

AMMONIUM PHOSPHATES

These materials are produced by neutralizing wet process phosphoric acid (H_3PO_4) with anhydrous ammonia to produce ammonium orthophosphates (diammonium phosphate (DAP), mono-ammonium phosphate (MAP)) or by neutralizing concentrated "superphosphoric" acid with ammonia to produce ammonium polyphosphates. The orthophosphates are generally granulated as dry materials, while the polyphosphates are used as fluid materials.

WATER QUALITY CONCERNS INVOLVING NITROGEN

Nitrogen fertilizer use is currently linked to some water quality problems involving surface and underground drinking water supplies. Current public health regulations in the United States and other countries require water suppliers to notify customers when NO_3^- nitrogen concentrations in their water exceed 10 micrograms per liter.

This is because ingestion of large amounts of NO_3^- by very young infants can lead to development of a potentially fatal, but treatable, condition called methemoglobinemia. Nitrate ingestion is usually not harmful to older children or adults.

Nearly all natural water contains some NO_3^- at some time during the year and in some regions water has always shown elevated NO_3^- concentrations. However, concentrations may rise temporarily in streams draining watersheds in intensive agricultural production in some years, or may become elevated in groundwater in agricultural regions. Pollution of individual wells can often be traced to improper well construction, coupled with a concentrated loading, such as feedlot run-off.

Elevated concentrations in streams or aquifers are often due to excessive nitrogen applications (often as manure) or failure of the crop to efficiently utilize the nitrogen applied. HighNO_3^- concentrations in water, logically, often occur in years following droughts. While droughts, and their effects, cannot be predicted in advance, following best management practices for nitrogen use will help ensure that most NO_3^- concerns do result from abnormal weather rather than improper management decisions.

NITROGEN FERTILIZERS OF PLANT

At one time the fertilizer industry relied on organic or direct-mined materials as sources of nitrogen. Today, virtually all nitrogen materials are manufactured, usually from ammonia. Such materials are less expensive, more concentrated, and just as plant available as the organics used in the past.

READILY SOLUBLE MATERIALS

ANHYDROUS AMMONIA (NH_3) AND AQUA AMMONIA

Anhydrous ammonia is 82 per cent nitrogen and is a gas at normal atmospheric temperatures and pressure. It is stored as a liquid under pressure and is injected several inches below the soil surface, where it vaporizes and dissolves in the soil water to form NH_4^+ ions.

Ammonia (not ammonium) is harmful or toxic to living organisms, and applications near living plants may cause temporary injury. Applications too close to seeds or seedlings may cause stand problems. Though some soil sterilization occurs in the immediate zone of ammonia application, the effect is temporary, and soil life is restored in the zone within a few weeks of application.

Aqua ammonia, a solution of ammonia in water (usually 21 per cent nitrogen), is sometimes used as a fertilizer. It is also injected underground, due to the ammonia's tendency to volatilize from the solution.

AMMONIUM NITRATE (NH_4NO_3)

Ammonium nitrate is usually used as a solid material with an analysis of up to 34 per cent nitrogen. It contains both NH_4^+ and NO_3-forms of nitrogen, and is used as a source of nitrogen in many blends of liquid and dry fertilizers, as well as being applied directly. Pure ammonium nitrate is very hygroscopic and can be explosive under certain conditions; however, present fertilizer grades of the material are specially conditioned, and when stored and handled properly, pose no problem or hazard.

UREA [$CO(NH_2)_2]_2$

Urea is a manufactured, organic compound containing 46 per cent nitrogen that is widely used in solid and liquid fertilizers. It has relatively desirable handling and storage characteristics, making it the most important solid nitrogen-fertilizer material, worldwide. It may contain small concentrations of a toxic decomposition product, biuret; however, urea manufactured using good quality control practices rarely contains enough to be of agronomic significance. Urea is converted to ammonium carbonate by an enzyme called urease when applied to soil. Ammonium carbonate is an unstable molecule that can break down into ammonia and carbon dioxide. If the ammonia is not trapped by soil water, it can escape to the atmosphere. This ammonia volatilization can cause significant losses of nitrogen from urea when the fertilizer is applied to the surface of warm, moist soils, particularly those covered with plant residues (no-till) or those drying rapidly. Relatively high surface pH also aggravates nitrogen volatilization from urea.

NON-PRESSURE NITROGEN SOLUTIONS

These materials are generally composed of urea and/or ammonium nitrate dissolved in water. The most popular are urea-ammonium-nitrate (UAN) solutions. Nitrogen concentrations generally range from 19-32 per cent. These are very versatile materials and are often used as carriers for herbicides or applied in irrigation water. They can be injected, or banded on the soil surface, to minimize potential volatilization losses of the urea component in no-till or surface application situations. Spraying over live foliage can cause severe tissue burn, but the problem can often be minimized by applying in narrow streams rather than by broadcasting. Because the solubility of the nitrogen salts in these solutions decreases as temperature declines, these materials are subject to "salting out". This limits storage and use of more concentrated formulations in cooler seasons and regions.

AMMONIUM SULFATE [$(NH_4)_2SO_4$]

Ammonium sulfate is a by-product of many industrial processes and contains 20 per cent nitrogen and 24 per cent sulfur. It is not a widely used material, due to its relatively low nitrogen content, but may become more popular as sulfur deficiencies become more wide spread. It is not as prone to nitrogen volatilization as urea, but is slightly more sothan ammonium nitrate, particularly on alkaline soils.

AMMONIUM PHOSPHATES

Ammonium phosphates are manufactured in a wide variety of forms and formulations including mono-ammonium phosphate (MAP), diammonium phosphate (DAP) and several ammonium polyphosphates. These materials are used mainly as carriers for phosphorus, but can also be significant in nitrogen management programmes, particularly when they are used as starter and row-applied fertilizers. When used in a starter programme, DAP (at any rate) is usually placed several inches from the seed to avoid damage due to any ammonia released as the material dissolves.

NITRATE SALTS

Several salts of nitrate, including sodium nitrate ($NaNO_3$, 16 per cent nitrogen), calcium nitrate ($Ca(NO_3)_2$, 15.5 per cent nitrogen), and potassium nitrate (KNO_3, 14 per cent nitrogen) are available, but are not widely used due to relatively high cost and low analysis. They are used mainly in specialty situations, such as production of horticultural crops.

SLOW-RELEASE MATERIALS

There are a number of materials available that release nitrogen slowly into the soil rather than very shortly after application as do those listed above. They are useful in situations where the producer does not want high levels of soluble nitrogen in the system at any one time, due to potentials for excessive losses or excessively rapid plant growth. Such situations occur in container-based horticultural production and inturfgrass management, among others.

Inorganic compounds, such as magnesium ammonium phosphate, and organic compounds, such as urea-formaldehyde (urea form) and isobutylidene diurea (IBDU),are often used to supply nitrogen at slow but fairly predictable rates. Nitrogen release from these compounds is slowed by very low solubility in water, and release from the organic materials is usually related to rate of microbial degradation. Simpler nitrogen compounds, such as urea, can be coated with less soluble materials, including polymers and waxes, and nitrogen release controlled by the rate at which the coating deteriorates.

Sulfur-coated urea (SCU), made by spraying urea granules with molten sulfur, is a common example of such a product, though more are becoming available. All of these products are relatively expensive sources of nitrogen, and their use is restricted to specialty markets and use on high-value crops.

NITROGEN IN PLANTS

Plants are surrounded by the nitrogen (N) in our atmosphere. Every acre of the earth's surface is covered by thousands of pounds of this essential nutrient, but because atmospheric gaseous nitrogen is present as almost inert nitrogen (N_2) molecules, this nitrogen is not directly available to the plants that need it to grow, develop and reproduce. Despite nitrogen being one of the most abundant elements on earth, nitrogen deficiency is probably the most common nutritional problem affecting plants world wide.

Healthy plants often contain 3-4 per cent nitrogen in their above ground tissues. These are much Higher concentrations than those of any other nutrient except carbon, hydrogen and oxygen, nutrients not of soil fertility management concern in most situations. Nitrogen is an important component of many important structural, genetic and metabolic compounds in plant cells. It is a major component of chlorophyll, the compound by which plants use sunlight energy to produce sugars from water and carbon dioxide (*i.e.* photosynthesis).

It is also a major component of amino acids, the building blocks of proteins. Some proteins act as structural units in plant cells while others act as enzymes, making possible many of the biochemical reactions on which life is based.

Nitrogen is a component of energy-transfer compounds, such as ATP (adenosine triphosphate) which allow cells to conserve and use the energy released in metabolism. Finally, nitrogen is a significant component of nucleic acids such as DNA, the genetic material that allows cells (and eventually whole plants) to grow and reproduce. Nitrogen plays the same roles (with the exception of photosynthesis) in animals, too. Without nitrogen, there would be no life as we know it.

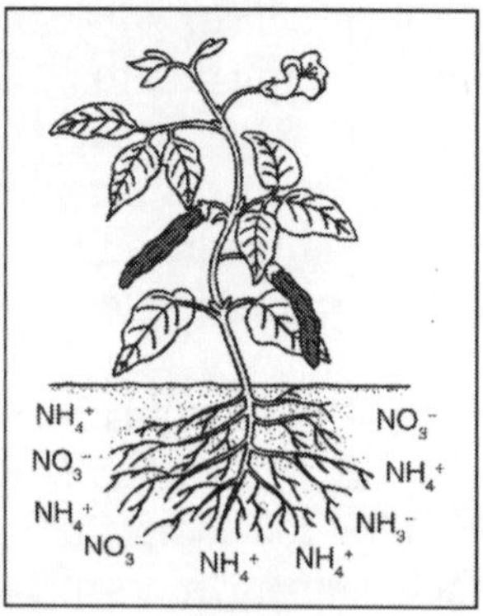

SOIL NITROGEN

Soil nitrogen exists in three general forms - organic nitrogen compounds, ammonium (NH_4^+) ions, and nitrate (NO_3^-) ions. At any given time, 95-99 per cent of the potentially available nitrogen in the soil is in organic forms, either in plant and animal residues, in the relatively stable soil organic matter or in living soil organisms, mainly microbes such as bacteria.

This nitrogen is not directly available to plants, but some can be converted to available forms by microorganisms. Avery small amount of organic nitrogen may exist in soluble organic compounds, such as urea, that may be slightly available to plants.

The majority of plant-available nitrogen is in the inorganic (sometimes called mineral nitrogen) NH_4^+ and NO_3^- forms. Ammonium ions bind to the soil's negatively-charged cation exchange complex (CEC) and behave much like other cations in the soil.

Nitrate ions do not bind to the soil solids because they carry negative charges, but exist dissolved in the soil water, or precipitated as soluble salts under dry conditions. Some NH_4^+ and NO_3^- may also exist in the crystal structure of certain soil minerals, and maybe quite available; however, such nitrogen is important in only a few soils.

NATURAL SOURCES OF SOIL NITROGEN

The nitrogen in soil that might eventually be used by plants has two sources–nitrogen containing minerals and the vast storehouse of nitrogen in the atmosphere. The nitrogen in soil minerals is released as the mineral decomposes. This process is generally quite slow and contributes only slightly to nitrogen nutrition on most soils.

On soils containing large quantities of NH_4^- rich clays (either naturally occurring or developed by fixation of NH_4 added as fertilizer), however, nitrogen supplied by the mineral fraction may be significant in some years. Atmospheric nitrogen is thought to be a major source of nitrogen in soils. In the atmosphere it exists in the very inert N_2 form and must be converted before it becomes useful in the soil. This conversion is accomplished two ways. Some N_2 is oxidized to NO_3 by lightning during thunderstorms.

The NO_3 dissolves in raindrops and falls into the soil. The quantity of nitrogen added to the soil in this manner is directly related to thunderstorm activity, but most areas probably receive no more than 20 lb nitrogen/acre per year from this source.

Some microorganisms can utilize atmospheric N_2 to manufacture nitrogenous compounds for use in their own cells. This process, called biological nitrogen fixation, requires a great deal of energy; therefore, free-living organisms that perform the reaction, such as *Azotobacter*, generally fix little nitrogen each year (usually less than 20 lb nitrogen/acre), because food energy is usually scarce. Most of this fixed nitrogen is released for use by other organisms upon death of the microorganism. Bacteria such as *Rhizobia*, that infect (nodulate) the roots of, and receive much food energy from, legume plants can fix much more nitrogen per year (some well over 100 lb nitrogen/acre). When the quantity of nitrogen fixed by *Rhizobia* exceeds that needed by the microbes themselves, it is released for use by the host legume plant. This is why well-nodulated legumes do not often respond to additions of nitrogen fertilizer. They are already receiving enough from the bacteria.

NITROGEN TRANSFORMATIONS AND LOSSES IN SOILS

Nitrogen can go through many transformations in the soil. These transformations are often grouped into a system called the "nitrogen cycle", which can be presented in varying degrees of complexity. Because microorganisms are responsible for most of these processes, they occur very slowly, if at all, when soil temperatures are below 50 °F, but their rates increase rapidly as soils become warmer. The heart of the nitrogen cycle is the conversion of inorganic to organic nitrogen, and vice-versa. As microorganisms grow, they remove NH_4^+ and NO_3^- from the soil's inorganic, available-nitrogen pool, converting it to organic nitrogen in a process called immobilization.

When these organisms die and are decomposed by others, excess NH^{4+} can be released back to the inorganic pool in a process called mineralization. Nitrogen can also be mineralized when microorganisms decompose a material containing more nitrogen than they can use at one time, materials such as legume residues or manures. Immobilization and mineralization are conducted by most microorganisms, and are most rapid when soils are warm and moist, but not saturated with water. The quantity of inorganic nitrogen available for crop use often depends on the amount of mineralization occurring, and the balance between mineralization and immobilization. Ammonium ions (NH_4^+) not immobilized or taken up quickly by higher plants are usually converted rapidly to NO_3^- ions by a process called nitrification. This is a two step process, during which bacteria called Nitrosomonas convert NH_4^+ to nitrite (NO_2^-), and then other bacteria, Nitrobacter, convert the NO_2^- to NO_3^-. This process requires a well aerated soil, and occurs rapidly enough that one usually finds mostly NO_3^- rather than NH_4^+ in soils during the growing season. The nitrogen cycle contains several routes by which plant-available nitrogen can be lost from the soil.

Nitrate-nitrogen is usually more subject to loss than is ammonium nitrogen. Significant loss mechanisms include leaching, denitrification, volatilization, and crop removal. The nitrate form of nitrogen is so soluble that it leaches easily when excess water percolates through the soil. This can be a major loss mechanism in coarse-textured soils where water percolates freely, but is less of a problem in finer-textured, more impermeable soils, where percolation is very slow.

These latter soils tend to become saturated easily, and when microorganisms exhaust the free oxygen supply in the wet soil, some obtain it by decomposing NO_3^-. In this process, called denitrification, NO_3^- isconverted to gaseous oxides of nitrogen or to N_2 gas, both unavailable to plants. Denitrification can cause major losses of nitrogen when soils are warm and remain saturated for more than a few days.

Losses of NH_4^+ nitrogen are less common and occur mainly by volatilization. Ammonium ions are basically anhydrous ammonia (NH_3) molecules with an extra hydrogen (H^+) attached. When this extra H^+ is removed from the NH_4 ion by another ion such as hydroxyl (OH^-), the resulting NH_3 molecule can evaporate, or volatilize from the soil. This mechanism is most important in high pH soils that contain large quantities of OH^- ions. Crop removal represents a loss because nitrogen in the harvested portions of the crop plant are removed from the field completely. The nitrogen in crop residues is recycled back into the system and is better thought of as immobilized rather than removed. Much is eventually mineralized and may be reutilized by a crop.

NITROGEN NEEDS OF AND UPTAKE BY PLANTS

Plants absorb nitrogen from the soil as both NH_4^+ and NO_3^- ions, but because nitrification is so pervasive in agricultural soils, most of the nitrogen is taken up as nitrate. Nitrate moves freely towards plant roots as they absorb water. Once inside the plant NO_3^- is reduced to an NH_2 form and is assimilated to produce more complex compounds. Because plants require very large quantities of nitrogen, an extensive root system is essential to allowing unrestricted uptake. Plants with roots restricted by compaction may show signs of nitrogen deficiency even when adequate nitrogen is present in the soil.

Most plants take nitrogen from the soil continuously throughout their lives and nitrogen demand usually increases as plant size increases. A plant supplied with adequate nitrogen grows rapidly and produces large amounts of succulent, green foliage. Providing adequate nitrogen allows an annual crop, such as corn, to grow to full maturity, rather than delaying it. A nitrogen-deficient plant is generally small and develops slowly because it lacks the nitrogen necessary to manufacture adequate structural and genetic materials.

It is usually pale green or yellowish, because it lacks adequate chlorophyll. Older leaves often become necrotic and die as the plant moves nitrogen from less important older tissues to more important younger ones. On the other hand, some plants may grow so rapidly when supplied with excessive nitrogen that they develop protoplasm faster than they can build sufficient supporting material in cell walls. Such plants are often rather weak and may be prone to mechanical injury. Development of weak straw and lodging of small grains is an example of such an effect.

ORGANIC NITROGEN IN SOILS

The supply of available nitrogen in soils is often supplemented by nitrogen released from soil organic matter or organic materials added to soils (manure, residues of forage legumes, etc.). The quantity of nitrogen released from any of these materials depends on the composition of the material, particularly its ratio of carbon to nitrogen, and the weather.

When microorganisms decompose materials such as manure and alfalfa residue, that have C/N ratios of 10:1 to 20:1 (lb carbon per lb nitrogen), net mineralization occurs, because the microbes are releasing nitrogen from the residue faster than they can utilize it. When they are decomposing residues such as cornstalks or wheat straw, that have much larger C/N ratios, they must take some extra mineral nitrogen from the soil to utilize all of the carbon they are digesting, so net immobilization occurs. The rate at which these processes occur depends on weather and soil conditions. When soils are warm and moist, decomposition proceeds rapidly, and nitrogen released from legume residues or manures may be significant, but when soils are cold and wet, or very dry, nitrogen release may be very much less than expected.

There are situations in which nitrogen applications can be reduced or eliminated when corn follows good stands of alfalfa or another legume. However, the nitrogen relationships involved here are so site-specific and weather-related that local recommendations should be consulted before making decisions regarding nitrogen credits. Legume species, stand density, age, soil drainage, and timing of operations are all factors to be considered.

The timing and rate of manure application should also be considered when calculating nitrogen credits, and again, due to influences of weather and soil, local guidelines should be used. Manure can be applied at rates that supply all nitrogen, phosphorus, and potassium to a crop, but such rates usually cause significant over-applications of phosphorus and potassium. Over-applications of phosphorus are important water quality concerns in many regions, so manure applications should be at rates low enough to avoid this over-application. At such low application rates, supplementary nitrogen-fertilizer applications are usually needed in crops such as corn.

CROP RESPONSE TO MICRONUTRIENTS

Inorganic sources include oxides and carbonates, and metallic salts such as sulfates, chlorides, and nitrates. The sulfates are the most common of the metallic salts and are sold in crystalline or granular form. An ammoniated $ZnSO_4$ solution also is used in polyphosphate starter fertilizers. Oxides of manganese and zinc also are commonly used, and are sold as fine powders and in granular form.Because oxides such as ZnO and MnO are water insoluble, their immediate effectiveness for crops is rather low in granular form.

Also, the available divalent form of manganese in MnO will oxidize to the unavailable tetravalent form of manganese, so there is very little residual availability of manganese fertilizers for succeeding crops. Thus, agronomic effectiveness of granular MnO may be rather low. Since manganese in MnO_2 already is in the unavailable form, it should not be used as a manganese fertilizer. Oxysulfates are oxides, usually industrial by-products, which have been partially acidulated with sulfuric acid, and generally are sold in granular form.

The percentage of water-soluble manganese or zinc in oxysulfates is directly related to the degree of acidulation by sulfuric acid. Research results have shown that about 35 to 50 per cent of the total zinc in granular zinc-oxysulfate should be in water-soluble form to be immediately effective for crops. Similar results would be expected for manganese-oxysulfate. Inorganic sources usually are the least costly sources per unit of micronutrient, but they may not always be the most effective for crops.

SYNTHETIC CHELATES

These sources are formed by combining a chelating agent with a metal through coordinate bonding. Stability of the metal-chelate bond affects availability to plants of the micronutrient metals—copper, iron, manganese, and zinc. An effective chelate is one in which the rate of substitution of the chelated micronutrient for other cations in the soil is quite low, thus maintaining the applied micronutrient in chelated form.

Relative effectiveness for crops per unit of micronutrient as soil-applied chelates may be from two to five times greater than that of inorganic sources, while chelates costs per unit of micronutrient may be five to 100 times higher. Several types of chelates are sold, so relative effectiveness values depend on the sources of chelates and inorganic products being compared.

NATURAL ORGANIC COMPLEXES

These complexes are made by reacting metallic salts with some organic by-products of the wood pulp industry or related industries. Several types of these complexes are the lignosulfonates, polyflavonoids and phenols. The types of chemical bonding of the metals to the organic components are not well understood. Some bonds may be coordinate as in the chelates, but other types of chemical bonds also may be present. While natural organic complexes are less costly per unit of micronutrient, they usually are less effective than synthetic chelates. They also are more readily decomposed by microorganisms in soil. These sources are more suitable for foliar sprays and mixing with fluid fertilizers.

FRITS

Fritted glassy products (frits) in which solubility is controlled by particle size and changes in matrix composition. Micronutrient concentrations vary from 2 to 25 per cent, and more than one micronutrient may be included in a fritted product. Fritted micronutrients generally are used only on sandy soils in regions of high rainfall were leaching occurs. This class of materials is more appropriate for maintenance programmes than for correcting severe micronutrient deficiencies. Therefore, frits only have a small share of the micronutrient market.

APPLICATION WITH MIXED FERTILIZERS

The most common method of micronutrient application for crops is soil application. Recommended application rates usually are less than 10 lb/acre (on an elemental basis), so uniform application of micronutrient sources separately in the field is difficult. Therefore, both granular and fluid NPK fertilizers are commonly used as carriers of micronutrients. Including micronutrients with mixed fertilizers is a convenient method of application and allows more uniform distribution with conventional application equipment. Costs also are reduced by eliminating a separate application.

Four methods of applying micronutrients with mixed fertilizers are:

1. Incorporation during manufacture
2. Bulk blending with granular fertilizers
3. Coating onto granular fertilizers
4. Mixing with fluid fertilizers

INCORPORATION WITH GRANULAR FERTILIZERS

Incorporation during manufacture results in uniform distribution of micronutrients throughout granular NPK fertilizers. Because the micronutrient source is in contact with the mixed fertilizer components under conditions of high temperature and moisture, the rate of chemical reactions which may reduce the plant availability of some micronutrients is increased.

For example, acid decomposition of ZnEDTA or any synthetic chelate may occur if they are mixed with phosphoric acid before ammoniation during manufacture, which results in reduced plant availability of the micronutrient. Immediate plant availability of applied zinc in granular ammoniated phosphates also decreases with the level of water-soluble zinc in these products.

BULK BLENDING WITH GRANULAR FERTILIZERS

Bulk blending of micronutrients with granular NPK fertilizers is a common practice in the U. S. The main advantage is that fertilizer grades can be produced which will provide the recommended micronutrient rates for a given field at the usual fertilizer application rates. The main disadvantage is that segregation of nutrients can occur during the blending operation and with subsequent handling. Segregation results in no uniform application, which is critical with micronutrients since their application rates are quite low. Segregation can be minimized by properly matching particle sizes of micronutrient sources with those of the NPK components of the blend. Mechanical devices to minimize coning and segregation of the materials during handling and storage are available. Blending of various sized fertilizer particles results in no uniform application because of segregation in the applicator during transport and spreading operations.

COATING GRANULAR FERTILIZERS

Coating powdered micronutrients onto granular NPK fertilizers decreases the possibility of segregation, which is the main disadvantage of bulk blending micronutrients with mixed fertilizers. Fertilizer solutions are preferred as binding agents because the fertilizer grade is not decreased so much as with use of water, oils and waxes. Some binding materials are unsatisfactory because they do not maintain the micronutrient coatings during bagging, storage, and handling. This results in segregation of the micronutrient sources from the granular NPK components. Agronomic effectiveness of micronutrients coated onto soluble granular NPK fertilizers should be similar to that with incorporation during manufacture. This method of micronutrient application is not commonly used because of the extra costs associated with coating.

FLUID FERTILIZERS

Mixing micronutrients with fluid fertilizers has become a popular method of application, especially in the U. S. Clear liquids are commonly used as starter fertilizers for row crops and some micronutrients, especially zinc sources, are easily applied with these fluids. Solubility of some micronutrient sources is higher in polyphosphate fertilizers such as 10-34-0 than in orthophosphate clear liquids. Micronutrients also may be applied with nitrogen solutions such as UAN, but solubility of many sources is rather low. Compatibility tests should be made before tank mixing operations of micronutrients with fluid fertilizers are attempted; otherwise, problems could occur when incompatible sources are mixed. Suspension fertilizers also are used as micronutrient carriers. Oxides also can be applied with suspensions since complete solution is not required.

FOLIAR SPRAYS

Foliar sprays are widely used to apply micronutrients, especially iron and manganese, for many crops. Soluble inorganic salts generally are as effective as synthetic chelates in foliar sprays, so the inorganic salts usually are chosen because of lower costs. Suspected micronutrient deficiencies may be diagnosed with foliar spray trials with one or more micronutrients. Correction of deficiency symptoms usually occurs within the first several days and then the entire field could be sprayed with the appropriate micronutrient source. Inclusion of sticker-spreader agents in the spray is suggested to improve adherence of the micronutrient source to the foliage. Caution should be used because of leaf burn due to high salt concentrations or inclusion of certain compounds in foliar sprays.

Advantages of foliar sprays are:

- Application rates are much lower than for soil application;
- A uniform application is easily obtained;
- Response to the applied nutrient is almost immediate so deficiencies can be corrected during the growing season.

Low residue foliar sprays of manganese and zinc have been used to correct deficiencies of citrus and other fruit crops, but sprays which will discolor the fruit should be avoided.

Disadvantages of foliar sprays are:

- Leaf burn may result if salt concentrations of the spray are too high;
- Nutrient demand often is high when the plants are small and leaf surface is insufficient for foliar absorption;
- Maximum yields may not be possible if spraying is delayed until deficiency symptoms appear;
- There is little residual effect from foliar sprays. Application costs will be higher if more than one spray is needed, unless they can be combined with pesticide spray applications.

MICRONUTRIENT RATES AND EXAMPLE RESULTS

BORON

Recommended application rates of boron are rather low (0.5 to 2 lb/acre), but should be carefully followed because the range between boron deficiency and toxicity in most plants is narrow. Uniform application of boron in the field is very important for the above reason. Boronated NPK fertilizers (those containing boron sources incorporated at the factory) will insure a more uniform application than most bulk blended fertilizers. Foliar sprays also insure a rather uniform application, but costs generally are higher.

In an experiment in Arkansas, cotton yields were increased by 490 and 584 lb/acre at boron rates of 0.3 and 0.5 lb/acre, respectively. Response of cotton in Tennessee to boron application is shown below. Without applied boron, cotton yields decreased with increasing soil pH. Yields were increased at all soil pH levels when boron was applied at a rate of 0.5 lb/acre. Soil tests should be included in boron fertilization programmes, first to assess the level of available boron and later to determine possible residual effects (buildup). The most common soil test for boron is the hot-watersoluble test. This test is more difficult to conduct than most other micronutrient soil tests, but most boron response data have been correlated with it.

COPPER

Recommended copper rates range from 3 to 10 lb/acre as $CuSO_4$ or finely ground CuO. Residual effects of applied copper are very marked, with responses being noted up to eight years after application.

Because of these residual effects, soil tests are essential to monitor possible copper accumulations to toxic levels in soils where copper fertilizers are being applied. Plant analyses also can be used to monitor copper levels in plant tissues. Copper applications should be decreased or discontinued when available levels increase beyond the deficiency range.

IRON

Soil applications of most iron sources generally are not effective for crops, so foliar sprays are the recommended application method. Spray applications of a 3 to 4 per cent $FeSO_4$ solution at 20 to 40 gallons/acre are used to correct iron deficiencies. The application rate should be high enough to wet the foliage. More than one foliar application may be required for correction of iron chlorosis. Inclusion of a sticker-spreader agent in the spray is suggested to improve adherence of the spray to the plant foliage for increased iron absorption by the plant.

MANGANESE

Recommended application rates range from 2 to 20 lb/acre of manganese, generally as $MnSO_4$. Application rates of MnO would be similar if applied as a fine powder or in NPK fertilizers.

Band application of manganese sources with acid-forming fertilizers results in a more efficient use of applied manganese because the rate of oxidation of applied manganese to the unavailable tetravalent form (as in MnO_2) is decreased.

There are no residual effects of applied manganese for the same reason, so annual applications are needed. Foliar spray applications of $MnSO_4$ also are used and require lower rates than soil applications.

Table. Soybean Response to Manganese

lb/acre Mn	Yield, bu/acre
0	13
2.5	24
5.0	26
10.0	33

Results given below show that soybean yields were more than doubled when 10 lb of manganese/acre was applied as $MnSO_4$ to a poorly drained sandy soil in Georgia. Cotton also responded to applied Manganese in Georgia, as shown below.

Table. Cotton Response to Manganese

lb/acre Mn	Yield, lb/acre	Increase, lb/acre
0	2,657	--
2.5	2,857	200

MOLYBDENUM

Recommended molybdenum rates are much lower than those for the other micronutrients, and uniform application is very important. Broadcast application of molybdenized phosphate fertilizers prior to planting or to pastures has been used to correct molybdenum deficiencies. Soluble molybdenum sources also can be sprayed on the soil surface before tillage to obtain a uniform application.

Table. Soybean Response to Molybdenum

	Mo Applied, oz/acre	
Soil pH	0	1
	—Yield, bu/acre—-	
5.3	25.7	45.0
5.7	34.3	45.0
6.3	41.8	46.1

Seed treatment is the most common method of molybdenum application. Molybdenum sources are coated onto the seed with a sticking agent and/or conditioner. This method insures a uniform application and sufficient amounts of molybdenum can be seed coated to provide sufficient molybdenum. Data in the following table show the effectiveness for soybean in Georgia of seed-coated molybdenum at a rate of one ounce of molybdenum/acre. Soybean yields without applied molybdenum increased with increases in soil pH, but not as high as those with seed-applied molybdenum at each soil pH level.

ZINC

Recommended rates of zinc generally range from 1 to 10 lb/acre. Band or broadcast applications are used, but foliar applications also are effective. Band applications of zinc sources with starter fertilizers is a common practice for row crops. Foliar sprays of a 0.5 per cent $ZnSO_4$ solution applied at a rate of 20 to 30 gallons/acre also will supply sufficient zinc, but several applications may be necessary.

As with copper, residual effects of applied zinc are substantial, with responses found at least 5 years after application. Because of these residual effects, soil test levels of available zinc generally increase after several applications. Many states have reduced their recommended zinc application rates because of these residual effects. Crop response to several zinc sources each banded with a 10-34-0 starter fertilizer at zinc rates up to 3 lb/acre for corn in Nebraska is shown below. Results show that ZnEDTA was much more effective at the lower zinc rates, but all zinc sources were about equally effective at the highest zinc rate.

Micronutrients are as important as the primary and secondary nutrients in plant nutrition. However, the amounts of micronutrients required for optimum nutrition are much lower. Micronutrient deficiencies are widespread because of increased nutrient demands from the more intensive cropping practices. Soil tests and plant analyses are excellent diagnostic tools to monitor the micronutrient status of soils and crops. Visual deficiency symptoms of these nutrients also are well recognized in most economic crops. Micronutrient recommendations are based on soil and plant tissue analyses, the type of crop and expected yield, management level, and research results. Numerous micro-nutrient fertilizers are on the market.

These sources are classified as inorganic, synthetic chelates, natural organic complexes, and fritted glasses. Some industrial by-products are used as micronutrient fertilizers because of their lower cost. Micronutrient sources vary considerably in physical form, chemical reactivity, cost, and relative availability to plants. While most micronutrient fertilizers are applied to soils, foliar sprays also are used on tree crops and some vegetables. Because recommended rates usually are low, most micronutrients are applied with NPK fertilizers by incorporation during manufacture or bulk blending with granular fertilizers, or by mixing with fluid fertilizers just before application to soil. Choice of micronutrient source depends on the method of application, compatibility with the NPK fertilizer, convenience of application, and the relative agronomic effectiveness and cost per unit of micronutrient.

PLANT NUTRIENTS AND SOIL COMPOSITION

Soil is made up of three main things: clay, humus and sand. There are also many small organisms that live in the soil, and most of these are useful to the plants. *Clay* Clay is made from the breakdown and recombination of silicate rocks. It is made up of alternating layers of silicon oxides then aluminium oxides, with various cations such as Ca^{2+} loosely bound in between the layers.

Anions adsorb onto the oxide surfaces, with doubly and triply charged cations sticking better than singly charged ones. Clays are the main source of nutrients in the soil. *Humus* This is any organic matter in the soil - *i.e.* the products of the decay of plants and animals. It is mostly made up of aromatic compounds. Over time it breaks down to carbon dioxide and water so it needs to be continually replaced. Humus is important in regulating the amount of water in the soil. *Sand* This is solid particles of ground up rock. A small amount of sand is necessary to ensure the correct water content in the soil.

SOURCES OF PLANT NUTRIENTS

The nutrients available in a given soil ultimately depends on the rock from which the soil was made. If the plants grown from this soil die and decay where they have grown then their nutrients are recycled. However, if the plants have been grown for agriculture then they are removed from the area in which they have grown and their nutrients cannot be recycled. So in soils that are used for cropping, essential nutrients continually have to be replaced.

PHOSPHORUS

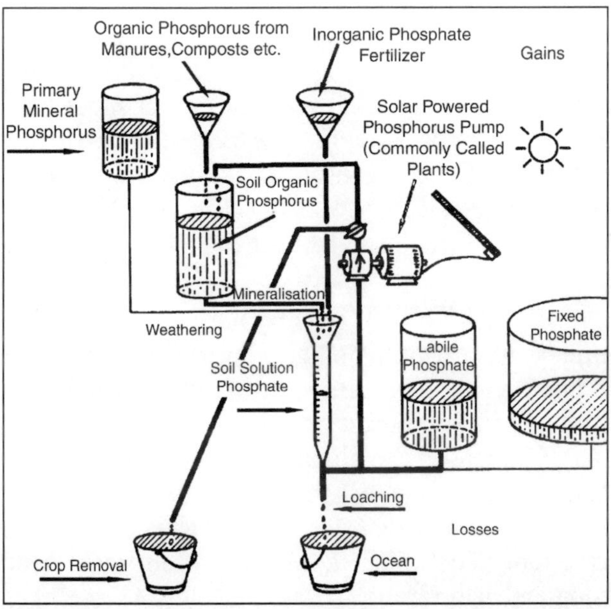

The phosphorus originally comes from the *weathering* of the parent material, which releases phosphate ions into the soil solution. This phosphate is *adsorbed* onto soil constituents, forming "reservoirs" of various types and capacities, some of which release phosphate ions into the soil solution easily, while others strongly bind phosphate. Plants absorb phosphate from the soil solution, and incorporate it into their tissues as they grow. When the plants die, decay processes slowly convert the complex phosphorus compounds in the plant tissues back to simple phosphate ions (*mineralization*) and the cycle repeats.

Under "natural" conditions the losses by leaching are to some extent made up by weathering of more parent minerals to release fresh phosphate. If a crop is removed from the area then the soil will contain less phosphate, the phosphate reservoirs will gradually become depleted and the soil will become less fertile.

The phosphate content of the soil can be increased by addition of phosphorus as "inorganic" phosphate which goes straight into the soil solution (if soluble), or dissolves slowly (if sparingly soluble). Alternatively, phosphorus can be added as complex organic phosphorus compounds, such as are present in plant or animal materials, which release their phosphorus slowly as they decay. Phosphorus is bound so firmly to soil that leaching losses into waterways are generally small, though erosion which carries phosphorus-rich soil particles into the waterways can encourage the growth of algae and weed in rivers and lakes (*eutrophication*).

NITROGEN

Nitrogen is one essential nutrient which behaves quite differently. There is virtually no nitrogen in rocks; most of the world's nitrogen is in the air, which is 80 per cent dinitrogen, N_2. However, most plants are unable to make use of dinitrogen, and can only absorb nitrogen from the soil solution in the form of nitrate or ammonium ions. The dinitrogen of the air must be converted into nitrate or ammonia by the process of *nitrogen fixation* before it becomes available to plants. This nitrogen fixation can occur by either a physical process, or a biological process. The physical process occurs whenever air is heated to a high temperature. The dinitrogen and dioxygen of the air combine to form nitric oxide

$$N_2(g) + O_2(g) \rightarrow 2NO(g).$$

This reaction occurs naturally in lightning strikes and forest fires. Human activity causes many more high temperature processes, such as in internal combustion and jet engines, thermal power plants, steel and cement works, etc, and has greatly increased the amount of nitrogen fixation.

The nitric oxide formed undergoes further reactions in the atmosphere, and is finally converted into nitric acid, which is washed out by rain and enters the soil as nitrate ions. Too much of this type of nitrogen fixation can cause atmospheric smog and acid rain.

Biological processes in the soil reduce the nitrate to ammonium ions. Biological fixation occurs in certain micro-organisms which have the ability to convert dinitrogen into ammonia, using the enzyme *nitrogenase*, for which the element molybdenum is required. Some of these micro-organisms live freely in the soil (*e.g*. the bacterium *azotobacter*). These bacteria fix the nitrogen for their own needs, but when they die and decay the nitrogen in their cells becomes available for plants. Other bacteria form an alliance (*symbiosis*) with certain plants (*e.g rhizobia* with legumes).

The plant provides a safe place for the bacteria to live (in nodules on its roots), and a food supply (carbohydrate obtained from photosynthesis). In return the bacteria provide the plant with fixed nitrogen as ammonium ions. Much of New Zealand agriculture is dependent on this relationship. Clover in the pasture has bacteria in root nodules which fix nitrogen, so the clover plant is rich in nitrogen. This is eaten by herbivores, and the excess nitrogen is excreted onto the soil in the animal's urine, when it becomes available to the grasses which form the bulk of the pasture. Farmers would have to use much more nitrogen fertilizer to achieve the same amount of pasture growth without the assistance of the nitrogen fixing bacteria in nodules on clover roots.

The life of fixed nitrogen in the soil is short. Both nitrate and ammonium ions are bound weakly by ion-exchange processes, so they are both leached out by rainwater, and finish up in ground water in wells and in rivers. Nitrate and ammonium ions encourage the growth of weed and algae in rivers and lakes, and when these are used for water supplies the toxicity of nitrate ions can cause health problem if the concentrations become too high. Nitrogen is also lost from the soil by the activities of *denitrifying bacteria*.

These convert the nitrate and ammonium ions back to dinitrogen, or more usually, to nitrous oxide (N_2O). This is a *greenhouse gas* and so increasing the amount of fixed nitrogen in the soil indirectly increases the amount of nitrous oxide in the atmosphere and potentially contributes to the raising of the temperature of the earth. Farming practices which lead to high levels of nitrogen in the soil can therefore contribute to environmental problems.

OTHER NUTRIENTS

Cycles similar to that described for phosphorus can be drawn for other major nutrients such as potassium, calcium or magnesium, and for trace nutrients, such as cobalt, molybdenum or selenium, which are derived from the soil parent material.

REPLACEMENT OF PLANT NUTRIENTS

This can be done either using manure or compost (which are relatively poor sources of nutrients) or a synthetic fertilizer. Synthetic fertilizers are usually used to replace nitrogen (in the form of urea, ammonium sulfate, ammonium nitrate or diammonium hydrogen phosphate), phosphorous, sulfur and potassium. The nitrogen is obtained from the air, the phosphorous is mined as $Ca_3(PO_4)_2$, the sulfur is mined in elemental form and reacted with oxygen to form sulfate, and the potasium is obtained from seawater. Other minor nutrients are also sometimes replaced using specialised artificial fertilizers.

COMPOSTS AND MANURES

In *organic* agriculture nutrients are replenished using only natural materials, though what is and what is not natural is often debated. The main "organic" nutrient sources are composts and manures.

COMPOSTS

Composts are formed from plant residues, and have much the same chemical composition as the plants from which they are made. Composts are *not* a rich source of plant nutrients, and need to be used in large amounts if they are the only source of nutrient replenishment.

Their main use is to boost the amount of soil organic matter (humus), and composts should be regarded primarily as *soil conditioner*s. There is also a philosophical problem that if compost is made from "imported" plant material, then the area from which the material was obtained has been depleted in nutrients.

As example, continual use of lawn mowings to make compost depletes the lawn area of nutrients, which need to be replenished to keep the lawn healthy. Therefore, while all available plant material should be converted into compost to maintain humus levels and to recycle nutrients, this does not really solve the nutrient deficit problem. An additional problem arises if a region is deficient in some particular nutrient. The plants (and animals) grown in the region will also be deficient in this nutrient, and so also will be composts made from plants grown in the area.

MANURES

Manures are usually the faeces of herbivores. In the digestion process the animal absorbs as much as it can of the nutrient elements for its own needs, so the residue finally excreted consists of plant material depleted in nutrients.

The nutrients not incorporated into the animal are excreted in the animal's urine, which is a better source of plant nutrients than the faeces. As with compost, manures are useful mainly as soil conditioners. They are not a rich source of plant nutrients, and need to be used in large amounts if they are the only source of nutrient replenishment.

The old gardening advice "to dig in six inches of well rotted stable manure" is fine in principle, but not very practicable for most modern gardeners, and there remains the problem that the area where the animal was feeding has been depleted of the nutrients.

Animal residues, such as "blood and bone" form another type of "natural" fertilizer. These can contain high levels of nutrients, which are generally released slowly into the soil solution as the complex compounds are broken down, which has advantages. Again these nutrients have been "mined" from the soil where the animals fed.

FERTILIZERS

The only answer to the "nutrient mining" problem is to bring fresh nutrients into the area from some richer source, and fortunately natural processes have concentrated the major nutrients in locations around the world. All "synthetic" nitrogen fertilizers obtain their nitrogen from the atmosphere. This is "fixed" using the *Haber process*, in which dihydrogen and dinitrogen at high temperature and pressure are combined over a catalyst, which speeds up the reaction forming ammonia:

$$N_2(g) + 3H_2(g) \rightarrow 2NH_3(g).$$

The hydrogen required for the Haber process is normally made using methane separated from natural gas. Methane is reacted with water and air at high temperatures over catalysts to form a mixture of dihydrogen, dinitrogen (from the air) and carbon dioxide.

Carbon dioxide and water are removed (the carbon dioxide is used later to make urea), leaving dihydrogen and dinitrogen in the correct proportion to combine to form ammonia (unbalanced equation)

$$CH_4(g) + H_2O(g) + [O_2(g) + N_2(g)] \rightarrow [N_2(g) + 3H_2(g)] + CO_2(g).$$

The ammonia obtained from the Haber process is converted into a form more useful as fertilizer. It can be combined with carbon dioxide to form urea, which is chemically identical to the urea excreted in the urine of animal:

$$2NH_3(g) + CO_2(g) \rightarrow O{=}C(NH_2)_2(s) + H_2O(l)$$

or combined with sulfuric, nitric or phosphoric acid to form ammonium sulfate, ammonium nitrate or diammonium hydrogen phosphate $(NH_4)_2(HPO_4)$. The nitric acid is formed by oxidizing ammonia over a catalyst

$$NH_3(g) + 2O_2(g) \rightarrow HNO_3(g) + H_2O(l)$$

PHOSPHORUS

Phosphorus is a relatively abundant element in the earth's crust, and occurs naturally in marine deposits of *phosphate rock*, as calcium phosphate, $Ca_3(PO_4)_2$, or the minerals *apatite*, $CaF_2.3Ca_3(PO_4)_2$, or *hydroxy apatite*, $Ca(OH)_2.3Ca_3(PO_4)_2$ (hydroxy apatite is the mineral that forms bones and teeth). The phosphate rock from the Pacific region is found on coral islands where flocks of sea birds roost.

It is thought that the phosphate in their excreta, derived from their fishy diet, combines over time with coral to form the phosphate rock. Some phosphate rock is sufficiently soluble so that the phosphorus from the finely ground rock is available to plants. Other phosphate rock is too insoluble for direct use, and is commonly treated with sulfuric acid to make a mixture of calcium sulfate and the more soluble calcium hydrogen phosphate, called *superphosphate*.

$$Ca(OH)_2.3Ca_3(PO_4)_2 + H_2SO_4 \rightarrow CaSO_4 + Ca(HPO_4)_2$$

Phosphate rock can also be reacted with excess sulfuric acid to make phosphoric acid, H_3PO_4, which can be converted into ammonium hydrogen phosphate, $(NH_4)_2(HPO_4)$ for use in soluble fertilizers, or into sparingly soluble magnesium ammonium phosphate $Mg(NH_4)(PO_4)$ (magamp) for use as a slow-release fertilizer.

SULFUR

Sulfur is also an essential nutrient, and some areas of New Zealand are sulfate deficient. Use of superphosphate, or ammonium, potassium or calcium sulfate (gypsum) as fertilizer supplies the needed sulfate. The sulfuric acid needed to make superphosphate is obtained by burning sulfur to form sulfur dioxide

$$S(s) + O_2(g) \rightarrow SO_2(g)$$

which is oxidised over a catalyst to form sulfur trioxide,

$$2SO_2(g) + O_2(g) \rightarrow SO_3(g)$$

which reacts with water to form sulfuric acid,

$$SO_3(g) + H_2O(l) \rightarrow H_2SO_4(aq).$$

The sulfur is obtained from deposits of elementary sulfur found in volcanic areas of the world, or is extracted from natural gas, which in some places contains hydrogen sulphide. Sulfur dioxide is released into the atmosphere from volcanoes. This is ultimately converted into sulfuric acid in the atmosphere; this is washed out by rain, so providing a "natural" source of sulphur.

The 1995 eruption of Ruapehu "topdressed" much of the central North Island with economically useful amounts of sulphate. Sulfur dioxide is also released into the atmosphere when coal or oil containing sulfur is burned in power stations, and the resulting sulfuric acid in the atmosphere causes "acid rain" in some industrialized areas.

POTASSIUM

Potassium is the other main nutrient needed, and this is obtained from sea water, directly or indirectly. When sea water evaporates sodium chloride (common salt) crystallises first. If the liquid remaining (bittens) is evaporated potassium chloride crystallises. In some places this occurred over geological times when oceans were isolated by geological processes and then evaporated, leaving large deposits of salt (which crystallised first) and of potassium chloride (which crystallised later).

The potassium chloride can be used directly as a fertilizer, and is satisfactory for use at low rates, but this has the disadvantage that chloride is also added, and many plants are not very tolerant of high levels of chloride. It is therefore usual to convert the potassium chloride to potassium sulfate (which also occurs in natural deposits) for use as a fertilizer.

-The so-called "un-natural" inorganic fertilizers are thus derived from the very natural sources of the air (plus natural gas as a hydrogen source) for nitrogen, coral rock plus sea-bird droppings (plus sulfur from volcanoes or "sour" natural gas to make superphosphate) for phosphorus, and the ocean for potassium.

SOIL ANALYSIS

If soil has an inappropriate composition for a given plant species then that plant will not thrive (or maybe will not grow at all) in that soil. To prevent this being a problem, samples of soil are often analysed so that nutrients can be added before seeds are planted. Soil is commonly analysed for potassium, magnesium, calcium, nitrogen and phosphorous (the exact choice of elements to monitor depends on which plants are to be grown) and also for pH.

Some of these tests are so simple they can be carried out at home and others require specialised equipment. Plants grow in the thin upper layer of the Earth's crust known as soil. Soil is formed over very long times from igneous or sedimentary rock, volcanic ash, sand or peat. With the exception of peat, these are all silicate minerals, so soil is largely made up of silicates and organic material known as humus. Plants use this soil for anchorage and as a source of nutrients and water.

WHY PLANTS NEED SOIL

Green plants are very effective photochemical factories: they use the energy of light to combine carbon dioxide from the air with water from the soil to make carbohydrates:

$$6CO_2(g) + 6H_2O(l) + \text{energy} \rightarrow C_6H_{12}O_6(aq) + 6O_2(g).$$

This reaction occurs in the chloroplasts of all green plants, using the green pigment chlorophyll to trap the energy from sunlight. The basic building blocks of glucose are then used to build up more complex carbohydrates such as cellulose. However, plants need more than just carbon dioxide, water and sunlight to keep alive.

The elemental composition of a 'typical' plant. It can be seen that the composition of plants is not much different from that of animals, and includes significant amounts of nitrogen and phosphorous, as well as trace amounts of many other elements. Although there is enormous variation between the abundance of the most common element, hydrogen, and the least common, molybdenum, both are very important. For example, plants can only use nitrogen in the form of the ammonium ion. Nitrates are converted to ammonium ions by the enzyme *nitrate reductase*, and as this enzyme cannot function without molybdenum, the presence of molybdenum is essential if the plant is to receive nitrogen in a form that it can use. These trace elements, generally called plant nutrients, normally come from the soil in which the plant is growing.

The plants absorb their nutrients from the soil solution (the film of water which covers all the particles in moist soil), usually in very simple forms. A typical composition of the main constituents of the soil solution from which plants must obtain all their requirements for growth. It is possible to grow plants without soil, using *hydroponics*. The plant is grown in a solution which contains all the required plant nutrients, in the correct proportions.

It is possible to determine just how much of any particular nutrient a plant needs by starting with a balanced nutrient solution, and then varying the amount of each nutrient present, one nutrient at a time. When insufficient of the nutrient is present the plant will not thrive.

As the amount of nutrient is increased the growth improves, until all that the plant can utilize is present, when adding more of that nutrient produces no additional growth. If the amount of the nutrient is increased still further, the excess may eventually cause toxic effects, and growth is retarded. For some nutrients, *e.g.* potassium, plants can tolerate a large excess before ill effects occur; for others, *e.g.* boron, the optimum range is narrow.

THE COMPOSITION OF SOIL

Apart from peat, which is derived from plant material, soil is composed of silicate minerals of various types, composed largely of oxygen and silicon, with smaller amounts of aluminium, iron, sodium, potassium, calcium and magnesium, and variable and much smaller amounts of all the other elements.

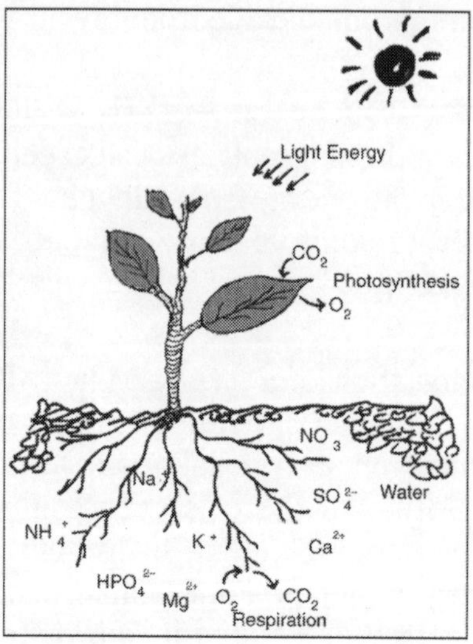

Fig. The Plant Environment

As rainwater saturated with carbon dioxide percolates through silicate minerals it dissolves out the simple soluble cations, such as Na^+, K^+ and Ca^{2+}, and anions such as Cl^-, in the process of *weathering*. The residue consists largely of oxides of silicon, aluminium and iron.

Table. Typical Composition of Soil Solution

Cation	Conc./mmol L^{-1}	Anion	Conc./mmol L^{-1}
Ca^{2+}	10	NO_3^-	5
Mg^{2+}	3	SO_4^{2-}	4
K^+	1	Cl^-	2
Na^+	1	HCO_3^-	2
NH_4^+	0.5	HPO_4^{2-}, H_2PO_4	0.01

Over hundreds to thousands of years this recrystallises as *clay minerals*. These are *aluminosilicate* compounds, composed largely of silicon, aluminium and oxygen, which have *layer structures* built up like stacks of sandwiches.

Silicon atoms, each surrounded by a tetrahedron of oxygen atoms, are linked into sheets by sharing oxygen atoms. Aluminium atoms, each surrounded by an octahedron of oxygen atoms (since aluminium is larger than silicon), similarly form sheets. These sheets combine to form layers by sharing oxygen atoms. In the *kaolin* type of clay minerals one silicon-oxygen sheet is combined with one aluminium-oxygen sheet to form each layer. These layers are then loosely stacked together like stacks of bread-and butter (—bread-butter—bread-butter—-). In another type (*smectite* type) a central silicon-oxygen sheet has an aluminium-oxygen sheet on each side; these are then loosely stacked like a bread and butter sandwich (—bread-butter-bread—bread-butter-bread—-).

The crystals of the clay minerals are very small, typically less than 4 ìm in size (such a particle would contain about 5,000 layers), and so have large total surface areas.The attractive forces between the layers are weak, and for some clay minerals water can diffuse between the layers, making the effective surface area even larger.

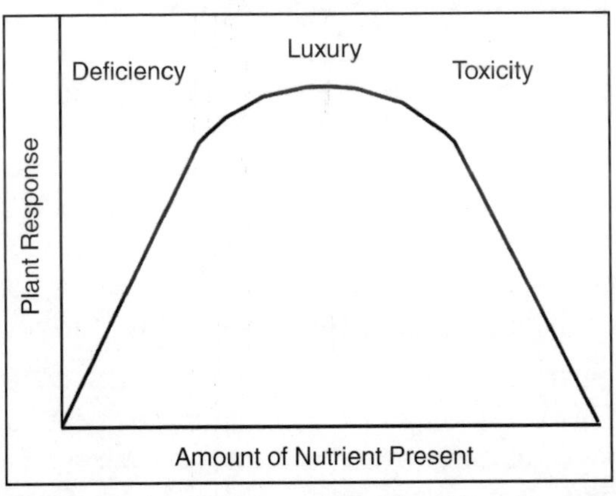

Fig. Response of Plant to Amount of One Nutrient

The clay structures are generally not perfect, and "wrong" atoms are frequently incorporated into the layers. Such mistakes can upset the charge balance in the layers, and lead to unbalanced negative charges appearing in the oxygen sheets.

These charges are balanced either by incorporating a hydrogen ion (converting an -O- to -O-H), or by having a cation, such as Ca^{2+}, somewhere nearby between the layers. The edges of the layers also have excess negative charges, which have to be similarly balanced. Figure represents a typical clay mineral, with loosely bound cations between the layers and at the surface of the particles. The places where hydrogen or cations are present are called *exchangeable cation sites*.

The cations bound by the clay particle can interchange with cations in solution. Clay minerals thus have an ability to loosely bind cations and *exchange* them with other cations in the soil solution. This *cation exchange capacity* of clays is an important property of soils. Figure shows the way in which the cations loosely held in soils can be exchanged for others from the soil solution.

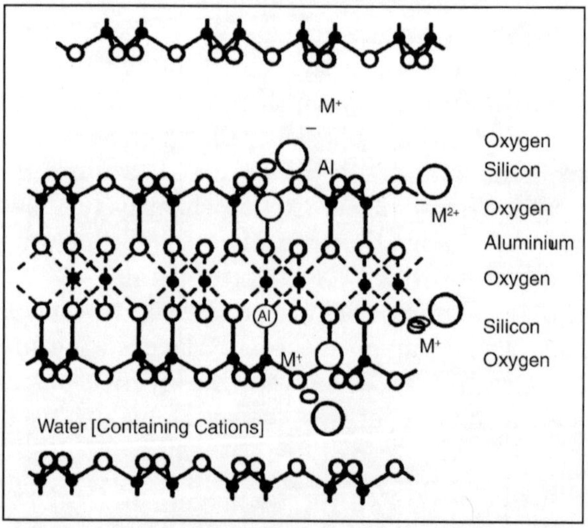

Fig. The Structure of a Clay Mineral, Showing Cation Exchangeable Sites

The clay minerals, and also tiny particles of the oxides of iron and aluminium present in soil, have another important property; they adsorb anions on their surfaces. The strength of this anion binding increases with charge, so that singly charged anions, such as Cl^- or NO_3^- are poorly retained by soil, doubly charged SO_4^{2-} is more strongly bound, and triply charged PO_4^{3-} is very strongly bound.

Some New Zealand soils formed from volcanic ash contain a clay mineral called *allophane*. This has a very imperfect aluminosilicate structure in which the layers are thought to curl around to form tiny porous balls about 5 nm in diameter, rather like plastic practice golf balls. These form soils with a very large total surface area—about a rugby-field of surface for every 10 grams of allophane. These soils have the ability to bind phosphate so strongly that it becomes unavailable to plants, and is said to be "fixed". Soils also contain organic matter, called *humus*.

This is the ultimate product of the decay of plant and animal material, and is intimately mixed in with the clay by the actions of the soil's population of organisms. Humus is largely derived from the most resistant of plant material, the lignin from the cell walls.

This contains much aromatic material (six-carbon benzene rings) with attached hydroxy (-OH) and carboxylic acid (-COOH) groups. These can also ionize to form -O- and -COO- groups, and these charges have to be balanced by a cation in the vicinity, so humus also contributes to the ion exchange capacity of the soil. The humus is slowly, continuously, being broken down to simpler compounds, and ultimately to CO_2 and H_2O and must continuously be replenished with new plant residues to maintain the soil's "condition".

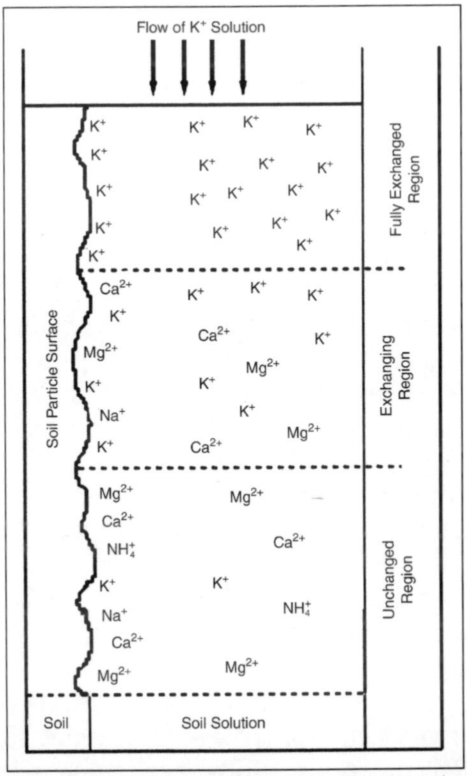

Fig. Exchange of Soil-bound Cations by Solution Containing Potassium Ions. The Anions Necessary for Charge Neutralization are Omitted for Clarity

Without the ability of soils to bind both cations and anions the essential plant nutrients would be rapidly washed out of the soil, and plant life would probably not have developed in the forms we know. The physical properties of a soil are very important in determining its usefulness for gardening or agriculture.

A fertile soil will generally have a balance of clay minerals and organic material, and probably also coarser particles of "sand" as silica or other resistant minerals of the original rock. The humus present improves the physical properties of the soil, and gives it the ability to drain excess water yet retain sufficient moisture for plant growth.

The balance of these constituents determines the soil's properties; how well it holds water, and drains, whether it is "sticky" or friable, how well it stands up to repeated cultivation, etc, and so determine its usefulness for agriculture. Soils have a thriving living population, ranging in size from earthworms to single celled organisms, living on plant and animal residues, which are partly responsible for the creation of fertile soil from weathering rock. Most soil organisms are beneficial, though some eat plant roots or cause or transmit plant diseases.

SYSTEMS OF SOIL CLASSIFICATION

The systems of soil classification developed particularly in relation to highly differentiated climatic conditions, as for example by K. Glinka, soon led to the recognition of climatic soil types but also the appreciation that under less extreme conditions, such as may obtain in western Europe and especially in Britain, the climatic conditions can cease to be the main factor in soil genesis and the nature of the substrate from which the soil is derived becomes of great significance.

In Britain the criss-cross, of changing climatic chara-cteristics from East to West and North to South, superimposed upon the rapidly changing geological strata from the more recent towards the South and East to the older in the West and North, has produced in the British Isles a peculiarly intricate mosaic of soil conditions that has been still further complicated by the glaciations and the post-glacial loess that now appears to have been even more widespread than had hitherto been envisaged as shown by the investigations of Perrin, Coombe and others.

These various soils may enhance or ameliorate the effect of present-day local climate and this has enabled a greater diversity of immigrant species to persist in British plant communities than might otherwise be able to survive. For a long time the significance of soil aeration was ignored and Clements in 1921 did a service to this aspect of ecology by bringing together much of the relevant literature. Recent work would seem to emphasize its importance not merely by reason of its influence on the extent of the root development but also because the actual intake of the nutrients appears to be directly related to efficient aeration of the adsorbing cells that can provide the energy for intake against an adverse gradient.

From the data furnished by Hopkins and his associates it would appear that the intake, of plant nutrients generally, decreases with the reduction of the oxygen content of the atmosphere that surrounds the root-system. The effect upon the intake of the trace elements appears to be less than for the major nutrients whilst sodium intake would seem actually to augment with diminished oxygen supply, a feature that may be of considerable import for vegetation growing on soils liable to flooding by sea-water.

Both the oxygen content of the soil-atmosphere and its carbon-dioxide content are important and depend upon the activity of soil organisms and the freedom of diffusion with the air above, which latter is facilitated by volumetric changes due to temperature fluctuations. So it is that under turf or woodland litter high carbon-dioxide concentrations may obtain, in illustration of which we may note that Russell and Appleyard recorded over 3 per cent under grassland and Harley and Brierley under beech-leaf litter from three to eleven times the amount in the atmosphere above.

A high organic content is often associated with a large microbial population so that the production of carbon dioxide is copious and the surface litter tends to retard diffusion as well as insulating the surface soil from rapid temperature fluctuations. In assessing the climatic and edaphic factors as they affect plant life we have in the past been perhaps too apt to regard climate and soil conditions as separate abstractions so that a true soil ecology has been distorted by the partialities that are only valid in isolation. The investigations of plant physiologists during recent decades have contributed to our appreciation of the direct influence of environment upon vegetation, notably in relation to two types of factors, the climatic and the edaphic.

It is now over thirty years since Garner and Allard published the results of their classical experiments demonstrating the influence of the relative duration of daylight and darkness upon growth and reproduction. Since that time the vast volume of literature on the photoperiodic response has revealed that, for a large number of species of flowering plants, flower formation is dependent either on the stimulus of a short period of darkness and a long day or the stimulus of a longer period of darkness and a shorter day. Variations in the relative duration of day and night could moreover retard or inhibit reproduction by seed and also influence the degree and character of vegetative growth.

For annual species whose growth rhythm is adjusted to such stimuli the latitudinal changes in photoperiodism may clearly operate as a limiting factor to their distribution, but for spreading perennials the substitution of vegetative multiplication for the production of fruit may well extend their geographical range. *Mercurialis perennis,* for example, so abundant in our basic woodlands, achieves success by means of its rhizomatous aggression although the reproductive capacity from seed towards its northern limits is negligible.

The specialization, which response to such stimuli implies, may therefore be advantageous, or the reverse, when the species concerned extends beyond the climatic conditions of its original home. It is significant therefore to note that some species have not developed any response relation to photoperiodic stimulation, as a time keeper of their growth rhythms, and that these include some of our commonest weeds such as the Chickweed *(Stellaria media),* the Shepherd's Purse *(Capsella spp.),* and Poa annua which flower and fruit at all seasons when temperature conditions permit and are well-nigh cosmopolitan in their distribution.

For these and other such species the lack of photoperiodic response is manifestly a factor in their success and determines in part the role they play in vegetation. It is also to be noted that photoperiodic response can be markedly influenced by temperature and humidity whilst within one and the same species long-day and short-day strains can be evolved. Probably the chief contribution of physiology to an understanding of the habitat's direct influence in recent years was the discovery of what have been termed the micronutrients or trace elements. That extremely small quantities of certain elements, mostly only requisite in parts per million, were nevertheless essential for plant life is due to their role in forming catalytic agents of various metabolic activities. Such micronutrients now regarded as generally necessary for the higher plants are boron, copper, zinc, manganese, and molybdenum.

But in addition to these, various other elements would appear to be at least beneficial, if not actually essential, to certain plants, as for example sodium and iodine, whilst though there is some experimental evidence for stimulation of growth by minute traces of, for example, titanium and arsenic, it seems unlikely that these play any significant part in determining the composition of plant communities or the degree of dominance of species. It should, however, always be borne in mind that owing to the "compound interest law" minute initial advantages may lead to significant final effects.

QUANTITATIVE ESTIMATION OF AMOUNTS

The quantitative estimation of amounts of such diminitude as that in which the trace elements occur, with an accuracy of within 10 per cent or less, is one of the triumphs of modern spectroscopic technology and these determinations have shown that deficiencies may often occur either through their inadequacy in the soil, through their absence, or through being rendered non-available.

The last condition is exemplified in respect to manganese, zinc, and boron in markedly alkaline soils whilst copper deficiencies or manganese deficiencies can be brought about by microbiological activity. The observation that certain characteristic appearances are associated with deficiencies of each of the various nutrients, both macro- and micro- alike, made practicable the survey of such occurrences over a far wider area than would have been possible if analyses alone could have sufficed. The outcome has been the recognition that the availability of most of the essential elements for the nutrition of the higher plants is very variable. In natural soils rainwater charged with CO_2 readily brings the calcium bicarbonate into solution and by base exchange with potassium and magnesium these also are readily leached from the soil so that deficiencies of all three are by no means uncommon.

Calcium is never wholly absent from soils that can support higher plants since it is an essential constituent of their cell walls but the great range of concentration, from the so-called non-calcareous soils, devoid of free carbonates, to the highly calcareous soils of downs and dunes, is accompanied by a corresponding change in the character of the vegetation which may involve the replacement of a calcifuge by a calcicole species occupying a similar niche or the change from acid moorland to calcareous grassland.

A recent review of the role of magnesium as a plant nutrient shows that many natural soils in Europe may contain very small amounts, sometimes less than 0.01 per cent and magnesium deficiency is, in fact, frequent. The large areas in South Australia that have been rendered fertile, by amending deficiencies of molybdenum, zinc, and copper, is sufficient evidence for the ecological significance of the distribution and availability of trace elements. Though this is admittedly an extreme example the areas where deficiencies of one type or another have been recognized are continually being added to.

It is of considerable import that evidence has been furnished that some species are much more susceptible to such deficiencies than others whilst some plants can accumulate trace elements in a marked degree so that, when these are essential nutrients, such capacity must be an important asset in competition. On seleniferous soils grass herbage may be quite innocuous to stock whilst species of *Astragalus* if present can accumulate selenium to such an extent as to render the herbage lethal.

So, too, clovers will accumulate molybdenum and the Sun Spurge, *Euphorbia helioscopia,* is known to be able to accumulate boron, facts which suggest that competition for the micronutrients may, because they are frequently marginal in concentration, be more severe than for the macronutrients which are more commonly considered from this standpoint. Of these latter, deficiency of nitrogen is common in acid soils and those containing little organic material, whilst phosphate deficiency also characterizes acid soils in areas of high rainfall but especially when these are rich in organic matter since phosphorous in organic combination is even less readily available to the plant than from inorganic compounds such as iron-phosphate. Potassium deficiency is common on chalk, sandy soils, and peat.

Although much is known as to the respective demands that crop plants make upon the various nutrients and of their susceptibility to toxic concentrations of manganese or aluminium, our knowledge of wild species in these respects and of their powers of accumulation of nutrient ions is as yet fragmentary although these might well furnish the clue to many apparent idiosyncrasies of presence and absence in communities.

That the precise form in which an element is present may be far more significant than its amount is manifest from the efficiency of chelated compounds of iron and manganese for relieving deficiencies of these elements where other compounds are wholly ineffectual. For the ecologist, as for the agricultural chemist, the problem is not the accurate determination of the total amounts of the various nutrients present but the estimation of the proportion of each which is in a form that the plant can assimilate.

A recent investigation of New Zealand soils by D. E. Hogg, with respect to potassium, is illuminating in this connection. Utilizing a number of standard methods of assessment he found that determinations of the exchangeable potassium corresponded closely to the response made by plants when potassium was supplied but only in the regions of high rainfall. In the areas of low rainfall assessment based upon nitric acid extraction showed a closer relationship with crop response.

It is manifest that the varying availability of the macro- and micro-nutrients may have a direct effect upon the distribution of species and communities but the possibilities of indirect effects must not be ignored. We have already referred to the differential absorption of such potentially lethal elements as selenium and molybdenum which can be economically serious in relation to agriculture, but under natural conditions such differences might materially influence the biotic pressure exerted by the grazing animal.

So obvious is it that a plant community is a social aggregate of its constituent species that one is apt to forget its implication, that all these must have similar environmental preferences or tolerances, except in so far as the presence of one or more species ameliorate, or exacerbate, the conditions for others. The obvious manifestation of this is the creation of shelter by taller species but there may be many other influences more subtle, as for instance the capacity of some species to tolerate and accumulate ions of alumina and manganese which may conceivably render conditions acceptable to other plants more susceptible to their toxic action.

It may be urged with some justice that many of the features we have mentioned as affecting distribution are mostly as yet only known to operate locally and determine the success of individual species rather than communities. That individual species can bring about vegetation changes of a widespread character is, however, witnessed by the disastrous extension of *Opuntia inermis* in Queensland or the spectacular spread of RosebayWillow-herb in Britain. It would appear to be the initial establishment and the build up of the population of a species to the requisite "infection pressure" that presents the major obstacles to marked extension and it is not least in this context that one should perhaps take particular cognizance of such discriminating causes.

So far we have made no reference to one class of biotic influences, namely the effects of animal life upon the nature of the plant community. No doubt in some degree it is true that the vigour of growth of vegetation and its rate of reproduction is adjusted to the hazards of the predators upon it, but it has been shown that Herbert Spencer's facile generalization is invalid and even were it true the imposed burden for survival is a serious handicap in competition. Recently C. B. Huffaker has expressed the view that "changes in plant populations and insect populations are often reciprocally related phenomena". The profound effect of a high rabbit population in Australia and in Britain and its drastic diminution by myxomatosis has recently brought into prominence the differential influence that an animal population can impose. Only in recent years have population studies shown what high numbers vegetable feeders such as field-voles can attain and whilst the importance of animal life, both large and small, is doubtless profound, yet the assessment of its differential effects on plant communities still remains largely the subject-matter for future research.

Competition between species and its severity in a plant community is dependent upon the requirements of the constituents both as to what demands each makes upon the habitat and the degree of intensity of these demands with respect to time. All flowering plants require water and mineral nutrients and most require light but the claims made upon these may vary greatly with the species concerned.

Competition ensues only when the supply is inadequate to meet the full requirements of all, a condition which may be attained with respect to light in open communities and with respect to water in aquatic ones. Competition for light we have already briefly referred to. It is largely dependent upon potential height and leaf area on the one hand and on the other upon the specific compensation point. Competition for water and nutrients is clearly dependent upon the differing capacity of a species to obtain these and to develop a root system, in the presence of others, adequate to its demands with sufficient rapidity and this can only be accomplished if, *inter alia,* the requisite photosynthetic products are available to effect this growth.

In assessing the adequacy of supplies of essentials it is necessary to distinguish, as in the human counterpart, between mere subsistence and what is requisite for vigourous growth. It has been asserted that there is no competition between the herbaceous layer of a forest and the dominant trees but, as we have pointed out, the former may deprive the deeper-rooted trees of their water supply although in general it is between the species of the same stratum that competition is most severe. Competition is perhaps most acute in the early stages of growth when mortality is highest.

Between individuals of the same species an equilibrium may be established so that over a wide range of population density the total productivity is very similar as has been shown in numerous crop spacing experiments. Between different species the reactions vary greatly. Some, which are closely allied and make similar demands upon the habitat (*e.g. Cerastium semidecandrum and C. tetrandrum*) may attain a measure of equilibrium but with open communities priority of establishment is significant and hence the mode and times of germination and the processes of dispersal play their part.

In perennials the mode of growth and rate of potential spread can play significant roles. Plant succession can be thought of as in large part a competitive drift in which at each phase until the climax the constituent species render the habitat more favourable to their successors than to themselves. From this aspect the few species that appear to be able to establish an almost static condition, gain a special significance. The maritime communities dominated by *Halimione portulacoides and Spartina townsendii* may be cited as possible examples though their seeming stability may be but an illusion due to the slowness of their change.

The importance of the competition aspect of succession is that it emphasizes the slow rate at which competitive influences may operate, masking the almost imperceptible decline or augmentation of individual species despite the vagaries of seasonal fluctuations. One of the desiderata is for more actual data as to the time scale involved. We know, for instance, that weeds of arable land that reverts may survive for perhaps half a century, that some pioneer dune species may persist nearly a century, but the rate of a decline under various degrees of competitive pressure for the characteristic perennial species of communities could materially assist in estimating the relative importance of reproduction, vegetative spread, and life span as factors in aggressive capacity.

From this brief survey it is apparent that future progress demands a far more detailed analysis of the habitat factors alike in space and time and parallel with this a more meticulous assessment of the constituent organisms amongst which the importance of sub-specific taxa may far transcend the specific aggregates in ecological significance, alike with respect to their biological behaviour and their physiological attributes.

RETENTION OF CATIONS IN SOILS

Cations are held more strongly (less reversibly) when pH increases from 5 to 7. Cu, Zn, Ni, Cd and other metals become signicantly less soluble and less exchangeable when pH increases from 5 to 7.

Retention of metals in soil can occur through several processes:

- Cation exchange (non-specic adsorption),
- Specic adsorption,
- Organic complexation
- Co-precipitation. In a given situation most, if not all, of these processes contribute to metal retention in soils.

In order to maintain electroneutrality, negative charges on solid particles (soil colloids) are balanced by an equal amount of cations; an exchange refers to the exchange between counter-ions balancing the surface negative charge on the soil colloids and ions in soil solution. Such an exchange is reversible, stoichiometric and diffusion controlled.

Moreover, there is a certain degree of selectivity of the adsorbent. The higher the valency of an ion, the greater its replacing power (H^+ behaves like a polyvalent cation). By contrast, with greater degrees of hydration, a given ion will exhibit a lower replacing power. Adsorption by cation exchange represents electrostatic binding through the formation of outer-sphere complexes with the surface functional groups. An outer-sphere complex means that at least one molecule of a solvent comes between the functional group and the ion.

Specic adsorption is pH dependent and related to the hydrolysis of the heavy-metal ion. In specic adsorption partly covalent bonds are formed with the lattice ions. Partly covalent bonds are inherently stronger than electrostatic binding involved in the non-specic cation exchange (*e.g.* Zn can be adsorbed on Fe and Al oxides 7 and 26 times more strongly than their corresponding cation exchange capacity at pH 7.6 would imply).

Metals most able to form hydroxy complexes are specically adsorbed to the greatest extent:

$$Hg > Pb > Cu >> Zn > Co > Ni > Cd$$

Speciûc adsorption may also include diffusion of metals into mineral interlayer spaces and their xation there. Such diffusion increases with an increase in pH. Organic matter may either increase or decrease availabil-ity of micronutrients, Al and heavy metals.

Reduced availability is due to complexation with humic acids, lignin and other organic compounds of high molecular weight (insoluble precipitates are thus formed). Conversely, increased availability may result from solubilisation and thus mobilisation of metals by low molecular weight organic ligands (*e.g.* short-chain organic acids, amino acids and other organic compounds). Stability constants of chelates with several metals occur with increasing order as:

$$Cu > Fe = Al > Mn = Co > Zn$$

Co-precipitation represents formation of mixed solids by simultaneous precipitation as occurs with Fe and Mn oxides.

SOIL–PLANT PH

Increasing acidity or alkalinity correspond to logarithmic increase in concentration of H^+ and OH^- respectively (vertical bars). Horizontal bars represent relative availability (or toxicity) at any particular pH. Most agricultural soils will be slightly acid (pH around 5.5 to 6.5) and essential nutrients are all readily available within that range. Of particular note, highly acid soils are conducive to both Al and Mn toxicity and to Mo deficiency. Highly alkaline soils are conducive to B toxicity but to Fe, Zn and Mn deficiencies.

The pH value most relevant to soil and plant chemical processes is pH of the soil solution. A soil is acidic if the pH of its aqueous solution phase is <7 and alkaline if that pH exceeds 7. Nutrient element availability varies accordingly, and beyond the range of pH 4–8 plant growth becomes a function of pH *per se*, plus pH effects on nutrient ion availability. In chemical terms, pH represents a measure of H^+ activity in a soil solution which is in a dynamic equilibrium with a negatively charged solid phase. H^+ ions are strongly attracted to these negative sites and have sufcient power to replace other cations from them. A diffuse layer in the vicinity of a negatively charged surface has higher H^+ activity than the bulk soil solution.

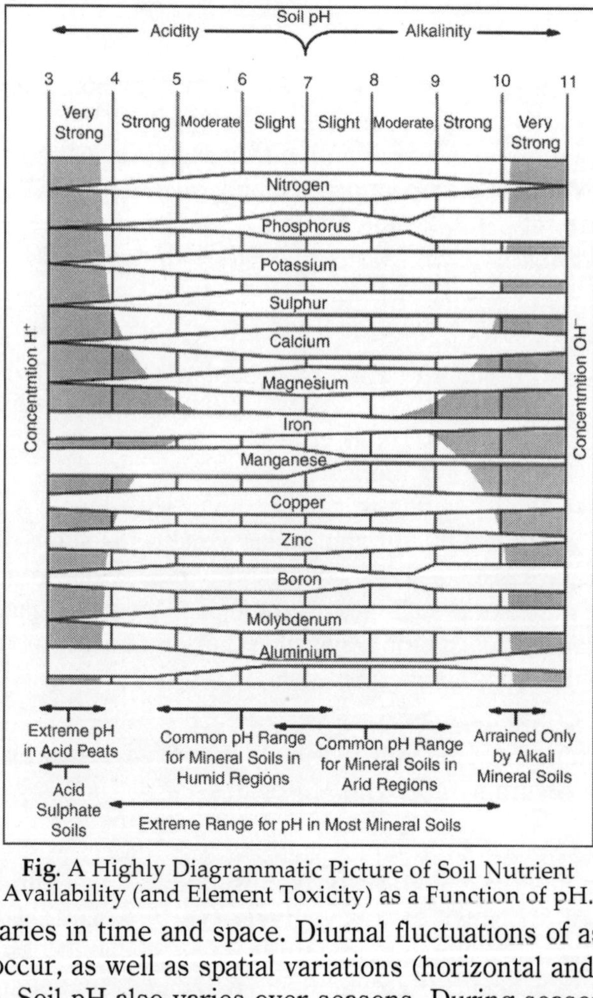

Fig. A Highly Diagrammatic Picture of Soil Nutrient Availability (and Element Toxicity) as a Function of pH.

Soil pH varies in time and space. Diurnal fluctuations of as much as one pH unit may occur, as well as spatial variations (horizontal and vertical down the soil prole). Soil pH also varies over seasons. During seasons with low to moderate rainfall when evapotranspiration greatly exceeds precipitation, salts are not being removed by deep percolation and increased salts tend to reduce pH by forcing more of the exchangeable H^+ ions into the soil solution. Conversely, during wet seasons, salts are removed from the topsoil and pH goes up. This season to season fluctuation in total salt content should not be confused with long-term soil acidication.

RELATIONSHIPS BETWEEN PH AND ION TOXICITY

Soil pH is a dominant influence on solubility and therefore availability and potential phytotoxicity of metals. Low pH favours free metal cations and protonated anions, higher pH favours carbonate or hydroxyl complexes. There-fore, availability of micronutrient and toxic ions (which are present in soil solution as cations) increases with increasing soil acidity. By contrast, availability of those present as anions (MoO_4^{2-}, CrO_4^{2-}, SeO_4^-, SeO_3^- and $B(OH)_4^-$) increases with increasing alkalinity.

RHIZOSPHERE

Plant growth is dependent on availability of water and nutrients in the rhizosphere, the soil–root interface consisting of a soil layer varying in thickness between 0.1 mm and up to a few millimetres depending on the length of root hairs. Availability of nutrients in the rhizosphere is controlled by the combined effects of soil properties and interactions between plant roots and adjacent microorganisms in the surrounding soil.

Chemical conditions in the rhizosphere are usually very much different from those in the bulk soil further away from roots. Root-induced changes in the rhizosphere pH are a result of the balance between H^+ and HCO_3^- excretion, evolution of CO_2 by respiration and loss of various organic compounds known collectively as root exudates. The balance between H^+ and HCO_3^- excretion depends upon the cation/anion uptake ratio.

Greater excretion of H^+ accompanies a greater absorption of cations than anions and results in rhizosphere acidication. Conversely, when uptake of anions exceeds uptake of cations, excretion of HCO_3^- exceeds that of H^+. The chemical form of soil N (ammonium v. nitrate) is an influential factor for the cation/anion ratio. Ammonium-fed plants take up more cations than anions, and they usually have a more acidic rhizosphere than bulk soil, while nitrate-fed plants take up more anions than cations and show the opposite relationship between rhizosphere and bulk soil pH.

Plant effects on rhizosphere pH also vary with genotype, which can in turn influence nutrient ion availability. Overall, plants and soils must be regarded as interacting components in any ecosystem, and because plants take up more basic than acidic components, any net increase in ecosystem biomass will result in some degree of soil acidication.

SOIL ACIDITY AND TOXICITIES

Elements essential for plant growth generally occur at soil concentrations that range between decient or adequate for plant growth. Instances do occur, however, where there are excess concentrations of either essential or non-essential elements. These toxicities may occur naturally, as saline soils, in soils developed from parent materials high in heavy metals, or through soil acidication. Human action can exacerbate natural processes. This has occurred with salinisation of the landscape as a result of clearing native vegetation or through irrigation. Heavy metals have accumulated in the landscape through industrial pollution, through the use of fungicides that contain Cu or the use of superphosphate which may contain Cd. Accelerated acidication has occurred through certain farming practices and is a particularly serious problem on sandy soils with low buffer capacity. Application of N fertilizers and inclusion of pasture legumes in rotations increase acidication. Also, drainage of some estuarine soils for farming or urban uses, in which S is present in the reduced state (*e.g.* as iron pyrite) under waterlogged conditions (*i.e.* acid sulphate soils), results in oxidation of S and reduction in soil pH.

Certain mine spoils are also acid, particularly those high in iron pyrite, which on oxidation result in acidication of spoil *in situ* or of run-off waters. Acid rain is a further problem, and in industrialised nations, especially in Europe and northeastern USA, there is considerable atmospheric pollution which produces acid rain containing dissolved oxides of N and S. This acid rain may fall long distances away from the source of pollution, and has led to international disputes.

SOIL ACIDICATION

As soils acidify, plant growth generally declines and there is a decrease in the range of plant species that may be grown. The activity of soil fauna and flora decreases also. The extent of acid soils and the degree to which plant growth is reduced by acid soil factors make these soils of considerable ecological and economic importance.

Soil pH (negative logarithm of the molar activity of hydrogen ions) has been used to dene acid soils. Since pH is measured on a logarithmic scale, it is important to realise that a soil with pH 4 is 10 times more acid than a soil with pH 5! Acid soils may be classied as those soils with a pH of less than 7 (*i.e.* neutral pH).

However, this has little value in practice, since most plant species will only show reduced growth at substantially lower pH. Thus, soils with a pH of less than 5.5 may be regarded as acid (*i.e.* those in which growth of many plant species is adversely affected). Acid soils occupy about 30 per cent of global land surfaces, and predominate in two major regions of the world: humid temperate forests and humid tropics and subtropics. However, many soils in other areas are acid also. Tropical acid soils alone comprise approximately two billion hectares, or 14 per cent, of the total ice-free area of the world. Approximately 40 per cent of the world's arable soils (*i.e.* those on which crops may be grown) are acid.

Eight regions of Australia have been identied in which agricultural production is reduced by acid soils. In southern Australia, these include the wheat and sheep belt of Western Australia, the southern portion of South Australia, cropping and grazing lands of southeastern South Australia along with the western, central and northeastern districts of Victoria. There are also about one million hectares of highly acidic soils in Tasmania.

Further north in New South Wales, the cropping and grazing regions of the Riverina and southwestern slopes contain many acid soils, as do the tablelands and wet sub-tropical coastal farming regions of New South Wales and Queensland. Soils of the tropical wet coast of Queensland and adjacent intensively cropped and grazed areas are also acid to a considerable extent.

Overall, acid soils are most common where rainfall exceeds 450 mm year^{-1} and more than 80 million hectares of the most productive agricultural land in Australia are acidic, with more than 40 per cent of this land being highly acidic. Acid sulphate soils of coastal areas may also acidify when drained for crop production or urban use, with consequent effects on estuarine aquatic life. This most noticeably results in sh kills, often following rains after a dry period during which oxidation of soil S has occurred.

Acid soil infertility in Australia is now recognised as an insidious and invisible form of land degradation. Worldwide, soil nutrient decline and acidication are often the most easily visible forms of land degradation. A landmark Australian study showed that soils under subterranean clover pasture acidify, with a decrease from a soil pH of around 6.0 in virgin soils to pH 5.2 where pasture had been grown for more than 30 years.

Subsequent studies have shown that the acid addition rate over extensive areas of New South Wales ranges from near zero to 3–5 kmol H^+ ha^{-1} year^{-1}, with up to 20 kmol H^+ ha^{-1} year^{-1} in some exploitative systems. With an acid addition rate of 4 kmol H^+ ha^{-1} year^{-1}, there would be a decrease of about 1 pH unit in the surface 30 cm of a sandy loam within 30 years. This would take about 120 years in more strongly buffered clay soils.

ALUMINIUM AND MANGANESE TOXICITIES

Soil pH has a marked effect on nutrient availability by affecting both solubility of soil minerals present and the ability of plant roots to absorb nutrients. Generally, a decrease in soil pH increases availability of cations, especially essential micronutrients, Fe, Mn, Cu and Zn. In contrast, availability of Mo decreases, often to such an extent that acid soils are decient in plant-available Mo. Availability of P also decreases at low pH via xation with hydrous oxides of Fe and Al, although reactions with Ca can also reduce P availablility around pH 8.5.

ALUMINIUM TOXICITY

Al toxicity has been recognised for over 60 years. Shortly after the turn of the twentieth century, soil scientists found Al present in solution leached from soils. However, later attention was given to hydrogen clays, and following development of glass electrodes research emphasis was placed on soil pH for many years—an example of technology driving science! Despite publication of research results between the late 1920s and late 1960s, deleterious effects of Al on plant growth did not attract interest again until the early 1970s. Some of those ndings are outlined here.

Table. Many froms of Al are Potentially Present in Soil Solutions of Acid Soils

From	Examples
Micro-crystals	Kaolinite, Gibbsite
Amorphous Precipitates	Al_2SiO_5, $Al(OH)_3$, $AlPO_4$
Inorgaic Polycations	$Al_2(OH)_2^{4+}$, $Al_3(OH)_3^{5+}$
	$AlO_4Al_{12}(OH)_{24}(H_2O)_{24}^{?+}$
Organic Complexes	Al citrate, Al oxalate
	Al fulvate, Al humate
Inorganic Complexes	AlF^{2+}, AlF_2^+, $AlSO_4^+$
Inorganic Monomers	Al^{3+}, $AlOH^{2+}$, $Al(OH)_2^+$

Al is the most common metal in the earth's crust (8 per cent of dry mass) and comprises some 7 per cent of soils, where it has a complex chemistry. Al is an important component of aluminosilicate compounds, including clay particles. As soils acidify, Al dissolves from these solid forms, and enters solution with the potential to then become toxic. Not all Al in the soil solution is toxic, however, providing challenges to biologists to identify those forms that are. Those shown to be toxic to plants include the inorganic Al monomers (Al^{3+}, $AlOH^{2+}$, $Al(OH)_2^+$) and the inorganic poly-cationAlO_4Al_{12} $(OH)_{24}$ $(H_2O)_{24}^{7+}$ (known as Al_{13}). The concentration of Al in the soil solution is often low (<50 μM), and activities of monomeric Al in the soil solution that reduce root growth have been found to range from 4 to 15 μM.

A further challenge to analytical chemists is to develop techniques able to discriminate between toxic and non-toxic forms of Al, made all the more difcult because of the low concentration of toxic Al. This concentration has been calculated to be about 0.1 g Al in 1 m^3 of soil (*c.* 12 μM Al or one-third of one part per million in the soil solution).

The same 1 m^3 of acid soil would contain a total of 70 kg Al. Solubility of Al decreases dramatically with an increase in soil pH, and the concentration of inorganic monomeric Al in solution is also affected by organic and inorganic anions. Complexes of Al with organic ligands of low relative molecular mass, especially citric and malic acids, are also considerably less toxic. Tolerant plants benet from such chelating mechanisms.

In addition, organic matter increases the concentration of fulvic and humic acids, resulting in the formation of Al fulvate and Al humate. These complexes are considerably less toxic than the inorganic monomeric species.

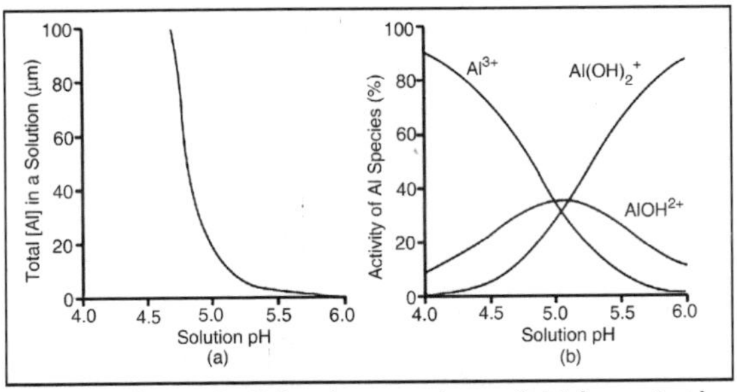

(a) Total concentration of all ionic species of Al in solution as a function of pH. In (b) a solution that initially contains 1000 μM $CaCl_2$ and 100 μM $AlCl_3$ (used to approximate a soil solution) shows a marked decrease in concentration of Al^{3+} from pH 4.0 to pH 6.0, and a change in relative activities of different inorganic monomeric Al species.

Underlying equilibria responsible for such pH effects are summarised below. Values for pK coincide with the pH of a solution where the reaction mixture would be 50 per cent dissociated. Increasing H+ ion concentration due to decreasing pH forces such equilibria towards Al^{3+}, thus inducing Al toxicity.

$$Al^{3+} + H_2O \Leftrightarrow AlOH^{2+} + H^+ \quad pK = 5.00$$

$$Al^{3+} + 2H_2O \Leftrightarrow AlOH^{2+} + 2H^+ \quad pK = 10.10$$

Biological effects of ions are often better related to activ-ity in solution rather than to concentration. In addition to soil pH effects on Al solubility, speciation of Al in solution changes with pH. At pH 4.0, most of the inorganic monomeric Al is present as Al^{3+}; this decreases with an increase in pH while the activities of the hydroxy Al species ($AlOH^{2+}$, $Al(OH)_2^+$) increase. At alkaline pH, aluminate ions (*e.g.* $Al(OH)_4^-$, $Al(OH)_5^{2-}$) enter into solution but are not toxic to plant roots.

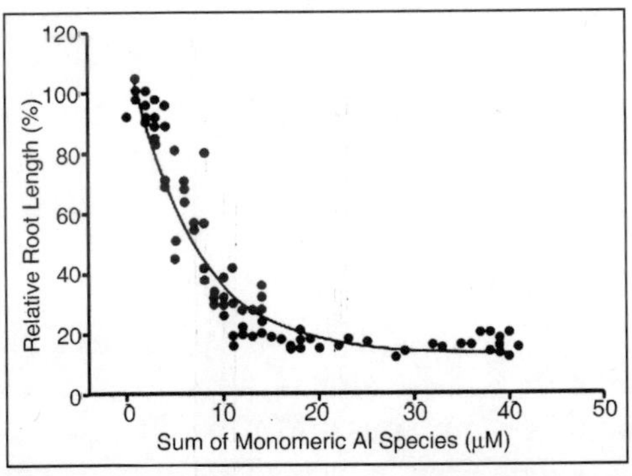

Fig. Soybean Root Growth Decreases Markedly as the Sum of Activities of Monomeric Al Species in Solution Increases.

Soluble Al in solution has a rapid effect on root growth, with the effects often visible within 2 d. Microscopically, decreased root growth may be evident within a few hours of exposure to Al, and changes in Golgi apparatus activity have been documented to occur within 5 h of placing roots in a solution containing Al. To be toxic, the root tip must be exposed to Al.

While the toxic effect of Al has been known for more than 60 years, the biochemical basis of Al toxicity has not been claried. However, Al is known to exert its primary toxic effect on roots, initial effects including a reduction in root length (through reduced cell elongation) and damage to cortical cells of the root epidermis near the root tip.

Proliferation of root hairs, important for uptake of water, essential nutrients and for infection by N-xing rhizobia in legumes, is severely reduced by Al at concentrations lower than those that reduce root growth. Proposals for the primary toxic effect have included reactions of Al with components of the cell wall, plasma membranes, cytoplasm and nucleus.

Visible effects of soluble Al on plant tops are considered to be a secondary effect through reduced nutrient uptake. Symptoms include those similar to deciencies of Ca, Mg, Fe, and P, probably as a result of decreased root proliferation and of reduced root activity. Further secondary effects include a reduction in uptake of water, increased sensitivity to drought and decreased N_2 xation by nodulated legumes.

Plant species differ markedly in tolerance to soluble Al in the soil solution. As with toxicity, the biochemical basis of genetic tolerance of Al toxicity is not clearly understood. However, malic acid excretion by root apices of Al-tolerant wheat is known to be ve- to ten-fold greater than that excreted by Al-sensitive wheat. Malic acid presumably complexes soluble Al, making it less toxic, and that capacity for excretion co-segregates with Al tolerance in progeny from crosses between near-isogenic lines. By implication, a single major gene is probably responsible for an Al tolerance that is functionally linked to Al stimulation of malic acid excretion.

MANGANESE TOXICITY

Mn, an essential element for plants, is about the tenth most abundant element in the earth's crust. Like Al, Mn chemistry in acid soils is complex, affected by both soil pH and soil redox potential. Mn occurs in rocks mostly in co-ordination with O^{2+} and as the divalent Mn^{2+} in soil solution and in natural waters.

Biotic and abiotic oxidation occurs readily so that Mn does not generally occur to excess except in acid soils or in waterlogged soils (assuming there are sufcient Mn-containing minerals). Mn toxicity may occur following soil sterilisation due to loss of microorganisms that normally oxidise Mn^{2+}. Al is a root toxin, whereas high Mn is mainly toxic to shoots (at very high Mn concentration, root growth may be affected directly). Thus, unlike Al, which often has an immediate effect on root proliferation, Mn must accumulate in shoots to become toxic. Three types of symptoms result from Mn accumulation: dark-brown, necrotic spots on lower leaves, distortion of expanding leaves (possibly an induced Ca deciency) and chlorosis of young leaves. Such chlorosis is often interpreted as an Fe deciency due to reduced Fe uptake by roots due to excess Mn. Considerable genetic differences in tolerance to high concentrations of plant-available Mn exist among plant species, and even among lines within a species. Such variation may occur via differences in root exclusion of excess Mn, complexation of Mn within roots, or shoot tolerance of high Mn. Sunflower, watermelon and cucumber all show shoot tolerance and even excrete Mn in a biologically inactive form around trichomes (hairs) on leaves and stems.

External Mn concentrations at which plant growth is reduced also vary greatly among plant species. In carefully controlled solution culture, the critical external concentration (*i.e.* the concentration required for 10 per cent reduction in plant dry mass) in the two most sensitive species, maize and wheat, was 1.4 μM Mn. In contrast, sunflower growth was only reduced with 65 μM Mn in solution. Likewise, the critical internal Mn concentrations varied also, from 200 mg kg^{-1} in maize to 5300 mg kg^{-1} in sunflower. Although cowpea and bean had a similar external critical concentration, cowpea was able to tolerate a much higher tissue Mn. A higher retention of Mn in roots of soybean enabled this species to tolerate a higher external Mn concentration than cotton.

COUNTERING ADVERSE EFFECTS OF ACID SOILS

Acid soil problems have been traditionally corrected via application of agricultural lime ($CaCO_3$) or dolomite ($CaCO_3$ + $MgCO_3$). Such amelioration raises soil pH, reducing Al and Mn concentrations in the soil solution, as well as adding the essential nutrients Ca and Mg. Lime or dolomite is generally required to raise soil pH to a level at which Al or Mn toxic-ities no longer affect plant growth (often above pH 5.5).

Rates required vary considerably depending on soil type, with clay soils requiring higher rates to increase soil pH (greater buffering capacity referred to earlier in connection with acidication). Moreover, lime requirements for sustainable agriculture vary with farming systems.

For example, lime required (kg $CaCO_3$ ha^{-1} $year^{-1}$) to balance net acid accumulation is about: 0.8 for wool, 6.0 for lamb and 7–20 for cereal enterprises. Even though cereal production is not greatly acidifying, Australia's average wheat crop of 15 million t would have removed alkalinity equivalent to 135000 t $CaCO_3$ as well as 6000 t Ca, 18 000 t Mg and 66 000 t K (along with 330 000 t N, 42 000 t P and 26 000 t S) from the land. Most of the crop was used for human consumption either in Australian cities or overseas. There would be little return of these nutrients to their original elds. Application of lime in Australia falls far below that required to balance losses of alkalinity. Annually, *c*. 0.5 million t of lime is used in Australia, but more like 2.25 million t are needed annually as prophylactic dressings. Overliming is a potential problem, especially on sandy soils, due to reduced availability of essential micronutrients, especially Zn, while root diseases such as 'Take all' of cereals are exacerbated.

Amelioration of some acid soils is possible via application of gypsum ($CaSO_4.2H_2O$) which provides a readily soluble form of Ca. Leaching soluble Ca applied as gypsum may improve root growth in acid subsoil horizons. Application of Mo is also benecial where this essential micronutrient is in short supply due to low pH. Finally, breeding and selection for crop tolerance to acid soil limitations has immediate application in agriculture, but an understanding of factors involved is crucial because tolerance to one factor limiting growth on acid soils (Al toxicity) does not imply tolerance to another such as Mn excess or Mo deciency.

CONCLUDING REMARKS

Acid soil infertility is a major limitation to plant growth worldwide, especially in the humid tropics, subtropics and temperate zones. Infertility results primarily from toxicities of Al or Mn and from deciencies of Ca, Mg, K, P and Mo. While there are many soils that are naturally acid, farming practices and industrial processes contribute to increased soil acidication. These include use of N fertilizers, especially where they exacerbate the loss of basic cations through leaching and crop removal, and growth of legumes, which increase soil N status. Acid rain is important in industrialised regions.

Two complementary approaches may be used to overcome the adverse effects of soil acidity. The rst is amelioration of acid soils through application of lime or gypsum to arrest degradation occasioned by a loss of basic cations. Unless these basic cations are replaced, land degradation through acidication and nutrient decline is inevitable. The second is a quest for genotypes better adapted to acid soil conditions. Genetic ap-proaches alone simply buy time and will result in further soil acidication, but do offer a benecial trade-off: tolerant plants would draw water and nutrients from acid subsoils which are difcult, if not impossible, to ameliorate any other way.

9

Species in Forest Ecosystems

The diversity and abundance of invertebrate species in forest ecosystems are determined by the interaction of complex ecological and environmental factors. Research into these factors is gradually providing data that can be incorporated into models, both descriptive and mathematical, of the processes involved.

At the core of such models is the description of the availability of resources that enables a particular invertebrate species to reproduce, whereas the rate of reproduction is dependent on multiple factors, including temperature, competition, etc., that determine utilization of available resources. In relation to exploitation of resources within a forest ecosystem, the primary focus for forest managers is the potential of invertebrates to cause damage and to reduce yields of the trees themselves.

This is certainly the case in managed forests where the primary aim is to grow trees for commercial gain, often with associated value from recreational, landscape or environmental considerations. Any invertebrates that feed directly on the trees, or have an influence of how the trees grow, may be classified as pests and hence may require some form of pest management to reduce the threat. But it is pertinent to consider how ecological theory and practice can be used both to understand why a pest problem has arisen and to provide guidance on how to combat that problem.

The same principles can be applied to understanding and enhancing the biodiversity of forests, including conservation of rare and beneficial invertebrates. Indeed, the two extremes of pest outbreak and invertebrate rarity are part of the same continuum of interaction between invertebrates and their host trees. In general, considering that trees are known to harbour the greatest numbers of insect species (diversity), it is surprising that serious pest status is reached only relatively infrequently.

For example, although oak *(Quercus* spp.) is known to support at least 450 species of insect in Britain, Bevan (1987) lists only 35 species as pests, of which five were classified as important or severe (XXX-XXXXX on Bevan's scale).

Wider listing of pest species has been noted for forest insects in continental Europe, although even here the number of consistently serious pests is small relative to the total diversity of insects present (Klimetzek 1993).

Naturally, the severity of a pest outbreak will be classified differently depending on the purposes of the forest manager, and thus it is not always possible to classify insects consistently as pests across geographical ranges or, indeed, between years.

Further, an insect may normally be classified as uncommon or rare but, given an abundance of resources, may be described later as a pest. This was certainly the case for oak pinhole borer, *Platypus cylindrus* (Coleoptera: Platypodidae), which prior to the 1987 great storm in southern England was listed in the British Red Data Book as rare.

Winter (1988) warned that the great increase in availability of freshly dead oak following the storm could result in an increase in numbers of *P. cylindrus* and, subsequently, it has been noted that the insect is now causing damage to high-value oak logs as a result of its habit of boring directly into the heartwood. Therefore, consideration of why invertebrate abundance might increase above a given threshold level can provide a more consistent appraisal of status and also potential approaches for reducing populations below that threshold or indicating measures that can be taken to encourage rare or endangered species.

Among the many interacting processes that determine insect diversity and abundance, a few attributes that are particularly well linked to the environmental status of trees can be distinguished. These provide a means to describe intrinsic processes but also allow comparison with other plants, thus offering the potential for a more consistent approach to descriptive population ecology of invertebrates irrespective of the host plant being considered.

LONGEVITY

Trees are the longest-lived components of an ecosystem and, in the majority of natural situations, approach or represent the climax stage of vegetation succession. This provides opportunities for continuity of insect population growth and over time this will give a particular set of insect associates for a given tree species. Continuing the theme of resource availability, the fact that trees may be present in the same location for tens or hundreds of years ensures that both migrant and, particularly, relatively sedentary invertebrates can exploit them at some stage during their growth.

However, longevity *per se* is not a quantifiable attribute but does provide the means for the other components, to influence invertebrate diversity and abundance. Increasing stability in the environment, as represented under forest conditions, leads to stability in both microclimate and resource availability, particularly favouring exploitation by specialist insect feeders and their associates.

BIOGEOGRAPHY

Tree species richness in a given region or locality varies considerably with geographical location and thus the climax community structure and longevity of individual species may vary accordingly. Although subject to uncertainty and lack of unequivocal evidence of the underlying processes, there is certainly a gradient of species richness of plant cover, especially trees, from the tropics to the poles.

Not surprisingly, the same applies to a wide range of animals and plants; as an extreme example, the species richness of orchids ranges from 15 to 2500 in a latitudinal range from 0 to 66° S. Further examples are provided by Brown (1988).

More specifically, in an analysis of species richness of deciduous forests both in the tropics and, latitudinally, within North America, Brown and Gibson (1983) contrasted the high diversity of tree species in the tropics (up to 100 ha^{-1}) with the low diversity from north to south within North America (1–30ha^{-1}). However, even within such a generalization there are exceptions, so that for example conifers are more diverse in temperate latitudes.

GEOLOGICAL TIME

The ecological structure of forests and of their associated invertebrate faunas is determined by both present and recent characteristics, particularly area, and by their history in geological time. The latter characteristic is driven as much by *time per se* as by the history of major disturbances (glaciations, volcanic eruptions, land shifts, etc.) within geological time.

Thus, present flora and fauna may well be determined by recolonization since the last major disturbance, the diversity of species being a reflection of both colonization and speciation relative to more recent ecological conditions. The gradient of species from polar to tropical regions could therefore be explained by the high level of disturbance closer to the polar masses compared with the relative stability of climate in the tropics. Evidence from habitats that have become isolated since the last major disturbance indicates that current species diversity may reflect the historical link to the original major habitat.

For example, islands that were once joined by land bridges to the larger land masses of New Guinea have species diversities of birds greater than expected compared with other islands that were never connected. However, although local ecological communities can be explained by historical patterns, the gradients over wide geographical scales are not so easily attributed to time as the major determinant.

The concept that major disturbances drive change undoubtedly carries weight, but closer analysis of tropical rain forests reveals that they too have been subject to disturbances of various sorts and are, in many cases, much changed over geological time. Based on this knowledge Haffer proposed a theory that disturbances in the tropics during the Pleistocene (alternating wet and dry periods giving rise to forest blocks surrounded by grassland) gave rise to refugia in which high levels of speciation took place, ultimately leading to the current high diversity characteristic of the tropics. While this theory is plausible and has been invoked by other authors, tests of the hypothesis have cast doubt over its wide applicability. Irrespective of the mechanistic bases for variation in species richness over wide latitudinal ranges, time is a significant determinant at a more local scale. Birks (1980) analysed the insect faunas of trees in Britain during the period since the last glaciation 13000 years ago and concluded that there was a significant correlation with the time that a given tree species had been continuously present in Britain during that period. Taking this further, Kennedy and Southwood (1984) included Birks' radiocarbon estimates of time in Britain as a variable in a multiple regression analysis of insects on British trees.

They showed that time was a significant predictor, both in its own right and as a co-predictor with log abundance (area). This is logical considering that over time there will be a tendency for well-established tree genera to increase their abundance, despite the recent proliferation in areas arising from commercial afforestation.

SIZE

A consequence of basic structure and longevity is that trees tend to be the largest components of the phylloplane, thus providing a much wider number and range of ecological niches compared with herb and shrub layer plants. Such attributes enable more insect species to colonize the plants without facing competition, both interspecific and intraspecific.

Size has many interrelated characteristics that help to explain the presence of both herbivores and their natural enemies:

- A larger physical presence is more easily detected and colonized by invertebrates. This concept, termed 'apparency', reflects a higher visual profile for colonizing adults or passively migrating immature life stages (*e.g.* mean height was a good predictor in models of the distribution of Scolytidae in Finland; Heliovaara & Vaisanen 1995) and a greater production of olfactory cues.
- Complex plant architecture is a consequence of greater size and for trees is particularly linked to the provision of many niches capable of supporting invertebrates with a wide range of feeding strategies (guilds). Although the concept of the guild is not always clearly defined, it is nevertheless a useful means of describing the trophic interactions of invertebrates and trees without having to consider their precise taxonomic status. Thus, seed, bud, leaf, bark and wood feeders can be

distinguished, each occupying a different part of the tree and therefore potentially avoiding competition.

- Increased biomass provides greater food resources per individual, thus encouraging growth and breeding success whilst reducing potential intraspecific competition. Price (1992) has reviewed the role of plant resources in insect population dynamics and concludes that, in part, the carrying capacity of an ecosystem depends on plant succession in time and space. Carrying capacity is therefore greater for trees, which generally appear late in succession, compared with herbs or shrubs. Thus biomass is available for longer and in greater quantities on trees compared with other plants.
- The concept of enemy-free space applies to the likelihood of herbivores being overlooked by searching parasitoids or predators. One aspect of size is that there is more opportunity for herbivores to occupy a part of the plant that offers protection against natural enemies (and competition from other herbivores). The complexity and number of niches present on a plant (plant architecture; Lawton 1983) therefore has an effect on enemy-free space. The greater the architectural complexity, the greater the time that natural enemies may have to spend in searching for prey, thus reducing their overall effectiveness and enhancing the survival probability of herbivores.

AREA

Forests tend to occupy greater areas than other plants. There is evidence to support the view that the greater the area occupied by the host plant, the greater the number of insect species associated with that host. This is called the species-area relationship and was originally linked to the island biogeography theories of species isolation and colonization.

Species-area relationships have been demonstrated for many plants and animals, including both highly mobile and sedentary species. With regard to the numbers of insect species on trees, data on the British insect fauna, one of the best characterized in the world, have been analysed by a number of authors.

Arising from earlier tentative analyses, Kennedy and Southwood (1984) analysed the relationships between insect species and characteristics of tree species in Britain. They concluded that there was a significant relationship between the area occupied by British tree species and the numbers of insect species associated with them. The area occupancy data were based on county and subcounty information of plant records and were thus potentially overestimates of actual areas occupied.

Claridge and Evans (1990) reanalysed the species-area relationships using the same insect data as Kennedy and Southwood but employing actual area information derived from Forestry Commission census data. They concluded that area was not a good descriptor for insect species on deciduous trees but was just significant for conifers. However, further unpublished analysis of the data broken down into numbers and areas of habitat patches occupied by the tree genera indicated that whereas total area was not a good descriptor, the degree of fragmentation of that area provided a better explanation for the observed insect species numbers. A similar conclusion, concentrating on Lepidoptera that have moved hosts to include conifers, was reached by Fraser and Lawton (1994) in an analysis of new associations on British trees. The overall finding that area provides a broad descriptor of insect diversity remains valid, although this does not explain the mechanisms that lead to these associations.

TAXONOMIC RELATEDNESS

In a given ecosystem, the closer the taxonomic relatedness of host plants, the more similar the numbers of insect species associated with those plants. This has been documented for tree species that have a long history in a particular location and thus provide rich sources of invertebrate species that may subsequently colonize other tree species, especially if they are taxonomically related. A particularly good example of the process in action is the colonization of *Nothofagus,* a member of the Fagaceae newly introduced to Britain, which has been colonized rapidly by insects associated with closely related members of the native Fagaceae.

Geographical isolation as a critical factor in determining invertebrate diversity and implications for international movement of pest organisms. Geographic isolation drives the acceleration of speciation and provides the means to derive principles on the processes involved. In addition, the fact of isolation means that some invertebrates abundant in one region will be absent elsewhere, despite conditions being suitable for their establishment.

In such situations, it is pertinent to examine the international plant health implications of geographical isolation and how modern patterns of trade influence the degree of isolation and later colonization by new insect species.

GEOGRAPHICAL ISOLATION AS A FACTOR IN DETERMINING INSECT DIVERSITY

In general, the basic principles apply across ecosystem boundaries and geographical zones extending from the poles to the tropics. In addition, the isolation afforded by relatively small islands provides further evidence of the principles of species colonization over space and time. Great Britain provides a particularly good example, because of its island position in the Atlantic Offshore climatic zone and the fact that it has been subjected to glaciation in recent geological time. Among the ecological characteristics that arise from this set of circumstances are a relatively depauperate mix of both animal and plant species that have arisen from the combination of losses resulting from the last ice age and the slow rate of recolonization arising from island status. British flora and fauna are therefore impoverished relative to continental Europe. Further influences arise from the preponderance of exotic tree species in commercial forestry in Britain, which although dominated by exotic conifer species also has increasing numbers of exotic broadleaved species.

Expanding on the principle that colonization of tree species is determined by the features, the ranges of numbers of insect species associated with trees in Britain reflect both the area (with provisos) and length of time that trees have been continuously present in Britain. However, there are exceptions, so that Sitka spruce in particular would appear to support far fewer species than predicted on the basis of area, although this probably also reflects the relatively short history of the species and recent expansions in areas planted in Britain.

IMPLICATIONS FOR INTERNATIONAL MOVEMENT OF PEST INSECTS

Trading patterns between countries have had a considerable influence on both the flora and fauna of most of the countries in the world. This reflects deliberate introductions of plants and animals for agricultural, forestry and amenity purposes and also accidental introductions by association with other commodities. In the majority of cases, the introduced species have had a neutral to beneficial effect, reflecting their 'domesticity' within managed crop or animal husbandry systems.

However, there are many exceptions to this generalization and in most cases the results of introduction of new species were not foreseen, either through lack of knowledge of the potential impacts or because the introduction was accidental and undetected. The literature on international biological control is dominated by examples where attempts have been made to rectify the effects of new pest introductions (both animal and plant).

A classic example is the control of the woodwasp *Sirex noctilio* (Hymenoptera: Siricidae) in exotic pine forests in Australia with the introduced nematode *Deladenus siricidicola* (Nematoda: Neo-tylenchidae) isolated from *Sirex* in New Zealand. This nematode has a life cycle that includes a free-living form that feeds on the fungus *Amylostereum areolatum* and a parasitic form that invades *S. noctilio* larvae and subsequently migrates to the reproductive organs of the female wasp, causing sterility. In this situation, all elements of the interaction are exotic to the new location, with little prospect of natural control being realized.

FOREST ECOLOGY AND CONFLICTS

The crisis in tropical forest resources is today recognised as the world's most severe ecological and economic crisis. This is primarily because of the linkages between the genetic resources and diversity of tropical forests and food security of the entire world, and the linkages between the ecological stability of tropical forests and the economic well-being of the majority of the world's people who live in the tropics, in what is called the Third World.

These Third World countries, the erstwhile colonies of the industrialized coun tries, provided natural resources on which the industrialization of the latter was based. Both industrialisation and economic growth in the colonial and post-colonial periods have been based on the reckless exploitation of tropical forests. Today, the cumulative impact of this over-exploitation has led to critical and almost irreversible ecological degradation.

The famine in Africa and other arid regions has shifted the world's attention to the high ecological and social costs of tropical deforestation. It has become the central concern of governments, development agencies and ecology movements. Yet the focus on tropical deforestation and its reversal is not automatically translated into the protection of tropical forests and those who depend on them for survival. As long as misconceptions about the nature of tropical forest ecosystems prevail, as long as conflicting demands are ignored and as long as the causes of tropical deforestation are inaccurately located, this degradation cannot be arrested.

As a result, tropical ecosystems will continue to be degraded, the survival of people of the Third World will continue to be threatened, and forest conflicts will continue to grow. Normally, forests are identified only with the economy associated with the commercial-industrial use of forests, however, the crisis of tropical forestry needs to be understood in the light of the conflicting demands for forest biomass, by the three fundamental economies associated with forests:

1. Nature's economy of essential ecological processes.
2. The survival economy of basic needs satisfaction of the people.
3. The market economy of industrial-commercial demands.

Nature's economy of essential ecological processes generates a demand on the products of the forest in terms of the maintenance of the stability of soil systems and the hydrological balance of the forest ecosystems. In ecologically sensitive ecosystems like upland watersheds, this economy becomes the most crucial one and should be given the necessary attention in forest management. Neglecting this economy in upland watersheds will imply tremendous negative externalities to the national exchequer as relief for regular floods and drought, which are easily described as loss due to nature's fury.

The survival economy of basic needs satisfaction reflects the requirement of forest biomass of the people living in and in the vicinity of forests in terms of fuelwood, fodder, fruits, nuts, green manure, small timber, etc. In forest areas where human settlements have either existed or are in the vicinity, the requirements of the survival economy have been satisfied all along without any major ecological damage. However, under certain situations, the pressure of the survival economy can be substantial and its neglect can lead to the unexpected and rapid degradation of forest resources.'

The market economy of industrial-commercial demands coastitutes the forest biomass demand of the total market system in the formal market economy. It includes the demand for pulpwood, plywood, furniture, house construction, etc., as well as fuelwood for the urban market. The demand for biomass from the urban industrial sector as well as the survival requirements of people have increased dramatically in the last century in India. There are many examples which illustrate that the growth of forest based industries is disproportionately beyond the ecological limit of the renewable productivity of nature under the present system of management. The growth in population adds another significant demand for biomass for domestic purposes.

It has, however, not been recognised, quantified and internalized in formal forest management. In the perspective of the increasing demand from survival and market economies, the biomass requirement for nature's economy are systematically sacrificed and nature's needs remain totally unfulfilled. In due course, this leads to the ecological destabilisation of the forest ecosystems.

Again, the conflict between survival and market economies assumes large proportions when most of the biomass produced is cornered by the economically powerful groups through the formal official mechanisms while the basic needs of fuel, fodder and small timber of the economically weak remain unsatisfied due to their weak political and economic status. Obviously, these three diverse economies push for the satisfaction of their demands either silently or loudly, resulting in both overt and covert conflicts over forest resources.

The tacit and invisible nature of these conflicts combined with a mistaken or incomplete understanding of these diverse economies can result in the lack of perception of conflicts between them. The failure to perceive conflicts in forest use can only aggravate the forestry crisis. It is thus absolutely essential to understand and perceive the covert as well as explicit demands on forest resources to evolve a forest policy that is ecologically sensitive and socially just, so that our forest and land resources can be used for the overall satisfaction of the needs of the nation and the people in an equitable and sustainable manner.

The conceptual framework based on the three economies adopted here differs from the conventional categories related to forest resources. We do not use the dichotomy between quantity/quality as categories of forest use that Hallsworth, for example, has used. According to him, Human demands on the forest fall into two categories. First a demand for forest quantities; for the actual things a forest can produce: for timber, for food and for space for cultivation or for grazing; secondly, a demand for forest qualities; for the effects that forests have on the environment of man-to protect supplies of water, provide havens for wild-life and maintain the pool of genetic resources, to protect the soil against erosion, and to provide space for recreations.

The dichotomy between quantity and quality, as between tangible and intangible, fails to capture the reality of those village women in Garhwal who launched the Chipko movement because for them water from the forest is a more significant product than wood and timber. For them water is not an 'intangible' produce or a mere quality. It is more tangible and basic in their sustenance economy than the commercial wood extracted and exported from their forests.

For similar reasons we have not adopted the local-national dichotomy between local and national interests to understand conflicts over natural resources, because the local subsumes the national interest. The national interest is or should be the integration of all local interests, not just another interest acting over and against local interests.

The demands of nature's economy and survival economy are national demands, seen throughout the country, and are not peculiar to a particular region. What has usually passed as the 'national interest' are the large scale demands of the commercial industrial sector, operating through the government. The emergence of ecology movements at the national level are indicative of the fact that these are demands of a socially narrow sector which come into conflict with socially broader sets of demands that are more appropriately termed the national interest.

We have also avoided the categories of private versus public demands because they fail to show adequately that what are called individual or private needs at the local level are located in the public context of a village community, and what usually passes as public policy in forestry is a government policy which is totally subservient to commercial/industrial interests. The 'privatisation' inherent in a commercially-oriented forest policy is What has been challenged by movements like Chipko which demand a more effective social and public control over forest resources.

The official management of forest resources in the name of 'national' or 'public' interest has, in reality, privatised people's 'common resources. This privatization trend which was rooted in the colonial period has survived to the present day. As Chhattarpati Singh observes, 'The class or "public" which more often than not benefits from such acquisitions is the rich. This is patently true in the acquisition of common land, especially forests. The public of the "public purpose" for which forests have been acquired constitutes all but the forest dwellers.

We have therefore preferred to distinguish human demands on the basis of use for survival and use for commerce and industry, and have included nature's demands to avoid an anthropocentric bias as well as to have a context to locate forest movements like Chipko which struggle for forest conservation to respect nature's rights, and not just to assert their own rights to forest resources.

ECOLOGY MOVEMENTS AND CONFLICTS OVER NATURAL RESOURCES

The recent period in human history contrasts with all the earlier ones in its strikingly high rate of resource utilization. Ever expanding and intensifying industrial and agricultural production has generated increasing demands on the world's total stock and flow of resources.

These demands are mostly generated from the industrially advanced countries of the North and the industrial enclaves in the underdeveloped countries of the South. Paradoxically, the increasing dependence of the industrialised societies on natural resources, through the rapid spread of energy and resource-intensive production technologies, has been accompanied by the spread of the myth that increased dependence on modern technologies implies a decreased dependence on nature and natural resources This myth is supported by the introduction of a long and indirect chain of resource utilisation which leaves invisible the real material resource demands of the industrial processes.

Through this combination of resource intensity at the material level and resource indifference at the conceptual and political levels, conflicts over natural resources generated by the new pattern of resource utilisation are generally shrouded and overlooked. These conflicts become visible when resource and energy-intensive industrial technologies are challenged by communities whose survival depends on the conservation of resources threatened by destruction and overexploitation, or when the devastatingly destructive potential of some industrial technologies is demonstrated as in the Bhopal disaster.

For centuries, vital natural resources like land, water and forests had been controlled and used collectively by village communities thus ensuring a sustainable use of these renewable resources. The first radical change in resource control and the emergence of major conflicts over natural resources induced by non-local factors was associated with colonial domination of this part of the world.

Colonial domination systematically transformed the common vital resources into commodities for generating profits and growth of revenues. The first industrial revolution was to a large extent supported by this transformation of commons into commodities which permitted European industries access to the resources of South Asia.

With the collapse of the international colonial structure and the establishment of sovereign countries in the region, this international conflict over natural resources was expected to be reduced and replaced by resource policies guided by comprehensive national interests. However, resource use policies continued along the colonial pattern and, in the recent past, a second drastic change in resource use has been initiated to meet the international requirements and the demands of the elites in the Third World, leading to yet another acute conflict among the diverse interests.

The most seriously threatened interest, in this conflict, appears to be that of the politically weak and socially disorganised group whose resource requirements are minimal and whose survival is primarily dependent directly on the products of nature outside the market system. Recent changes in resource utilisation have almost wholly by-passed the survival needs of these groups. These changes are primarily guided by the requirements of the countries of the North and of the elites of the South.

This book analyses environmental conflicts in contemporary human society. In general it relates to societies all over the world, but in particular it addresses the most intense and emerging social contradictions in India related to conflicts over natural resources. Science and technology are central to these conflicts because while scientific knowledge has been used by contemporary societies to considerably enlarge man's access to natural resources, it has also allowed the utilisation natural resources at extremely high rates.

The contemporary period is characterised by the emergence of ecology movements in all parts of the world which are attempting to redesign the pattern and extent of natural resource utilisation to ensure social equality and ecological sustainability. Ecology movements emerging from conflicts over natural resources and the people's right to survival are spreading in regions like the Indian subcontinent where most natural resources are already being utilised to fulfil the basic survival needs of a large majority of people.

The introduction of resource and energy-intensive production technologies under such conditions leads to economic growth for a small minority while, at the same time, undermines the material basis for the survival of the large majority. In this way, ecology movements have questioned the validity of the dominant concepts and indicators of economic development. The ideology of economic development, which remained almost monolithic in the post World War II period, is thus faced with a major foundational challenge.

In this chapter an attempt has been made to provide a systematic conceptual framework for analysing the processes and structures of modern economic development from an ecological -perspective. It attempts to analyse the relationship between economic development and conflicts over natural resources to trace the roots of ecological movements. Further, in the light of the ecological perspective, it examines the fundamental assumptions and categories of modern development economics that are used to determine the objectives of economic development as well as the criteria for the choice of technologies that are used to achieve these objectives.

ECONOMIC DEVELOPMENT AND ENVIRONMENTAL CONFLICTS IN INDIA

A characteristic of Indian civilization has been its sensitivity to natural ecosystems, vital renewable natural resources like vegetation, soil and water were managed and utilised according to well defined social norms that respected the known ecological processes. The indigenous modes of natural resources utilisation were sensitive to the limits to which these resources could be used It is said that the codes of visiting important pilgrim centres Badrinath in the sensitive Himalayan ecosystem, included a maximum stay of one night so that the temple area would not put excess pressure on the local natural resources base.

In the pre-colonial indigenous economic processes, the levels of utilisation of natural resources were not significant enough to result in drastic environmental problems. There were useful social norms for environmentally safe resource utilisation and people protested against the destructive use of resources even by kings. A major change in the utilisation of natural resources of India was introduced by the British who linked the resources of this country with the direct and large non-local demands of Western Europe.

Natural resource utilisation by the East India Company, and later by the colonial rulers, replaced the indigenous organizations for the utilisation of natural resources, like water, forest and minerals, that were mainly managed as commons. With the establishment of British colonial rule in India, the ever increasing resource demands of-the industrial revolution in England were largely met from colonies like India.

Forced cultivation of indigo in Bengal and Bihar, cultivation of cotton in Gujarat and the Deccan led to large-scale commitment of land for the supply of raw materials for the British textile industry, the flagbearer of the industrial revolution. Forests in the sensitive mountain ecosystems like the Western Ghats or the Himalayas were felled to build battleships, or to meet the requirements of the expanding railway network.

Forests of the Bengal-Bihar-Orissa region were used for running wood fuel locomotives in the early stages of railway expansion. The latter stages of colonial resource utilisation and control included the monopolization of water rights as in the Sambhar Lake of Rajasthan or the Damoda' Canal in Bengal. Colonial intervention in natural resource management in India led to conflicts over vital renewable natural resources like water or forests and induced new forms of poverty and deprivation.

Changes in resource endowments and entitlements introduced by the British came into conflict with the local people's age old rights and practices related to natural resource utilisation As a result local responses were generated through which people tried to regain and retain control over local natural resources. The indigo Movement in Eastern India, the Deccan Movement for land rights or the forest movement in all forest areas of the country, the Western Ghats, the Central Indian Hills or the Himalayas, were obvious expressions of protest generated by these newly created conflict's. Conflicts generated by the colonial modes of natural resource exploitation could not, however, grow with a local identity. With the progress of the anti-colonial people's movement at the national level, these local protests merged with the national struggle for independence. With the collapse of colonial rule internationally, and the emergence of sovereign independent countries in the Third World like India, resolution of these conflicts at the local level became a possibility.

While political independence vested the control over natural resources with the Indian state, the colonial institutional framework for natural resource management did not change in essence. Where colonialism collapsed, the slogan of economic development stepped in. There was unfortunately no alternative institutional mechanism other than that of the classical model of development left by the British, with which the newly formed Indian state could respond to the accentuated aspirations of the Indian people for a better life.

The same institutions and concepts, nurtured and developed by the colonial rulers were applied to objectives which were exactly opposite to those of the colonial period. Concepts and categories relating to economic development and natural resource utilisation that had emerged in the specific context of capitalist growth and industrialization in the centres of colonial power were raised to the level of universal assumptions and applicability. The processes which led to deprivation were now entrusted with the responsibility of basic needs satisfaction.

No serious thought was given to the fact that the historical specificity of early industrial development in Western Europe necessitated the permanent occupation of the colonies and the undermining of the local 'natural economy'.' This inexorable logic of resource exploitation, exhaustion and alienation integral to the classical model of economic development based on resource intensive technologies led Gandhi to seek an alternate path of development for India when he wrote:

God forbid that India should ever take to industrialism after the manner of the West. The economic imperialism of a single tiny island kingdom (England) is today keeping the world in chains. If an entire nation of 300 million took to similar economic exploitation, it would strip the world bare like locusts.

While Gandhi's critique was a forewarning against the problems likely to arise by following the classical path of resource-intensive development, at the time of India's independence, there was no clear and comprehensive work plan to realise the Gandhian dream of alternate development that would be resource prudent and would satisfy basic needs.

The issues of resource constraints of economic development were, therefore, not highlighted at the theoretical level, partly due to the tremendous pressure of the enhanced developmental aspirations of a newly independent nation, and partly due to the lack of internalization of natural resource parameters within the framework of economics. As the scale of economic development activities escalated from one Five Year Plan to another, the disruption of ecological processes that maintain the productivity of the natural resource base started becoming increasingly apparent.

The classical model of economic development in the case of the newly independent nations resulted in the growth of urban-industrial enclaves where commodity production was concentrated, as well as rapid exhaustion of the internal colonies whose resources supported the enhanced demands of these enclaves. In the absence of ecologically enlightened resource management methods, the pressure of poverty enhanced the pace of economic development activities in the hope of a quick improvement in the standard of living for all, as in the case of Western Europe.

For example, commercial forestry earned more revenue by making increasing amount of timber and pulpwood available in the market but in the process reduced the multipurpose biomass productivity or damaged the hydrology of the forests. People dependent on non-timber biomass outputs of forests like leaves, twigs, fruits, nuts, medicines and oils were unable to sustain themselves, in the face of the commercial exploitation of forests. The changed hydrological character of the forests affected both the micro-climate and the stream flows, disturbing the hydrological stability and affecting agricultural production. There are similar examples from all parts of the country, related to almost all massive developmental interventions in India's natural resource system. Ecological degradation and economic deprivation generated by the resource insensitivity and intensity of the classical model of development have resulted in environmental conflicts, an understanding of which is imperative for the reorientation of our current development priorities and concepts. It is becoming in creasingly clear that these classical concepts and priorities are being used as an alibi to direct 'development' at the national level, while the educated minority elite is the main beneficiary of these 'development' processes.

The ecology movements that have emerged as major social movements in many parts of India are making visible many invisible externalities and pressing for their internalisation in the economic evaluation of the elite-oriented development process. In the context of a limited resource base and unlimited development aspirations, ecology movements have initiated a new political struggle for safeguarding the interests and survival of the poor, the marginalised, including women, tribals and poor peasants.

ECOLOGY MOVEMENTS AND SURVIVAL

The intensity and range of ecology movements in independent India have continuously widened as predatory exploitation of natural resources to feed the process of development has increased in extent and intensity.

This process has been characterised by the massive expansion of energy and resource-intensive industrial activity and major development projects like large dams, forest exploitation, mining and energy-intensive agriculture. The resource demand of development has led to the narrowing of the natural resource base for the survival of the economically poor and powerless, either by direct transfer of resources away from basic needs or by destruction of the essential ecological process that ensure renewability of the life-supporting natural resources.

In the light of this background, ecology movements emerged as the people's response to this new threat to their survival and as a demand for the ecological conservation of vital life-support systems. The most significant life-support systems in addition to clean air are the common property resources of water, forests and land on which the majority of the poor people of India depend for survival. It is the threat to these resources that has been the focus of ecology movements in the last few decades. Among the various ecology movements in India, the Chipko movement (embrace the trees to oppose fellings) is the most well known.

It began as a movement of the hill people in the state of Uttar Pradesh to save the forest resources from exploitation by contractors from outside.' It later evolved into an ecological movement that was aimed at the maintenance of the ecological stability of the major upland watersheds in India. Spontaneous people's response to save vital forest resources was seen in Jharkhand area in Bihar-Orissa border region as well as in Bastar area of Madhya Pradesh where there were attempts to convert the mixed natural forests into plantations of commercial tree species, to the complete detriment of the tribal people.

In the southern part of India the Appiko movement, which was inspired by the success of the Chipko movement in the Himalayas, is actively involved in stopping illegal over-felling of forests and in replanting forest lands with multipurpose broad leaved tree species. In Himachal Pradesh the Chipko activists have intensified their opposition to the expansion of monoculture plantation of the commercial Chir Pine (Pinus roxburghii). In the Aravalli Hills of Rajasthan there has been a massive programme of tree planting to give employment to those hands which were hitherto engaged in felling of trees.

The exploitation of mineral resources, in particular the opencast mining in the sensitive watersheds of the Himalayas, the Western Ghats and Central India have also resulted in a great deal of environmental damage. As a consequence, environmental movements have come up in these regions to oppose the reckless mining operations. Most successful among them is the movement against limestone quarrying in the Doon Valley. Here, volunteers of the Chipko movement have led thousands of villagers, in peaceful resistance, to oppose the reckless functioning of limestone quarries that is seen by the people as a direct threat to their economic and physical survival.'

While the Doon Valley instance has a long history of popular opposition to the quarrying of limestone and a Supreme Court order has restricted the area of quarrying to a minimum, examples of such success' of ecology movements are rare People's ecology movements against mineral exploitation in the neighbouring areas of Almora and Pithoragarh still seem to be ignored, probably due to the relative isolation of these interior areas. Beyond the Himalayas, the ecology movement in the Gandhamardan Hills in Orissa against the ecological havoc of bauxite mining has gained momentum and it draws inspiration from the Chipko movement.

The mining project of the Bharat Aluminium Company (BALCO) in the Gandhamardan Hills is being opposed by local youth organisations and tribal people whose survival is directly under threat. The peaceful demonstrators have claimed that the project could be only continued 'over our dead bodies. The situation is more or less the same in large parts of Orissa-Madhya Pradesh region where rich mineral and coal deposits are being opened up for exploitation and thousands of people in these interior areas are being pushed to deprivation and destitution. This is also true of the coal mining areas around the energy capital of the country in Singrauli. In these interior areas of Central India, movements against both mining and forestry are becoming increasingly volatile and people's resistance is growing.

Large river valley projects, which are coming up in India at a very rapid pace, is another group of development projects against which people have organised ecology movements. The large-scale submersion of forest and agricultural lands, a prerequisite for the large river valley projects, always takes a heavy toll of dense forests and the best food growing lands. These have usually been the material basis for the survival of a large number of people in India, specially tribal people.

The Silent Valley project in Kerala was opposed by the ecology movement on the ground of its being a threat, not to the survival of the people directly, but to the gene pool of the Tropical Rainforests threatened by submersion. The ecological movement against the Tehri high dam in the UP Himalaya exposes the possible threat to people living both above and below the dam site through large-scale destabilization of land by seepage and strong seismic movements that could be induced by impoundment.

The Tehri Dam Opposition Committee has appealed to the Supreme Court against the proposed dam by identifying it as a threat to the survival of all people living near the river Ganga up to West Bengal. Most notable among the people's movements against dams on the issue of direct threat to survival from submersion are Bedthi lcchampalli, Bhopalpatnam, Narmada Sagar, Koel-Karo, Bodhghat, etc.

In the context of the already overutilised land resources, the proper rehabilitation on a land-to-land basis of millions of people displaced through the construction of dams seems impossible. The cash compensation given instead is inadequate in all respects for providing an alternate livelihood for the majority of the displaced. Destitution is thus the first and foremost precondition for initiating large dam projects

While the process of construction of dams itself invites opposition from ecology movements, the functioning of water projects dependent on the constructed dams results in further ecological disasters and movements. People's movements against widespread water-logging, salinisation and the resulting desertification in the command areas of many dams have been registered.

Among them are instances of protests against the Tawa, Kosi, Gandak, Tungabhadra, Malaprabha, Ghatprabha projects and the canal irrigated areas of Punjab and Haryana. While excess water led to ecological destruction in these cases, improper and unsustainable use of water in the arid and semi-arid regions generated ecology movements in a different way. The anti-drought and desertification movement is gaining momentum in the dry areas of Maharashtra, Karnataka, Rajasthan, Orissa, etc. Ecological water use for survival is being advocated by water based movements like Pani Chetana, Pani Panchayat, and Mukti Sangharsh.

Another major movement originating from the ecological destruction of resources by growth based development is spreading all along the 7,000 km long coastline of India. It is the movement of the small fishing communities against the ecological destruction caused by mechanised fishing whose instant profit motive is destroying the coastal ecology and its long-term biological productivity in a big way. No amount of threat to survival in India from environmental hazards can be complete without a reference to the Bhopal tragedy on 2 December 1984, in which several thousand people died and several lakhs faced serious health hazards following the leakage of poisonous Methyl Iso Cyanate from a pesticide plant of Union Carbide (India) Limited. People's movements for clean air and water are growing in ail parts of the country just as ecologically irresponsible industrialization is moving deeper into the hinterland in search of new resources.

Bibliography

A K Roy: *Rhizosphere Biotechnology : Plant Growth - Retrospect and Prospect*, Scientific Publication, Jaipur, 2007.

A.K. Agarwal: *Plant Sciences and Genetics in Agriculture*, Monalisa Enterprises, Mumbai, 2011.

A.K. Sarkar, R.P. Singh and P. Mahapatra: *Soil, Water and Plant Nutrient Management*, Agrotech Publication, Delhi, 2011.

Amarjit S. Basra: *Plant Growth Regulators in Agriculture and Horticulture : Their Role and Commercial Uses*, International Book, Delhi, 2004.

Anita Chatterjee: *Academic Dictionary of Medicinal Plant*, Isha Books, Delhi, 2005.

B.K. Garg: *Saline Wastelands, Environment and Plant Growth*, Scientific Publication, Jaipur, 1997.

Bidhan Roy and Asit Kumar Basu: *Abiotic Stress Tolerance in Crop Plants : Breeding and Biotechnology*, New India Publication, Delhi, 2009.

C.K. Naidu: *Plant Growth and Development*, Pointer Publication, New Delhi, 2004.

Carlepon Ellis, M.W. Swaney and Tom Eastwood: *Soilless Growth of Plants*, Agrobios Publication, Jaipur, 2002.

Dharamvir Hota: *Synthetic Plant Growth Regulators*, Gene Tech Books, Delhi, 2007.

E John Russell: *Soil Conditions and Plant Growth*, Biotech Books, New Delhi, 2001.

H.C. Lakshman: *Bioinoculants for Integrated Plant Growth*, MD Publication, Delhi, 2010.

M. Ashraf and P.J.C. Harris: *Abiotic Stresses Plant Resistance Through Breeding and Molecular Approaches*, International Book

Distributing Co., Delhi, 2006.

M. S. Reddy, S. Desai, R. Z. Sayyed, V. K. Rao, Y. R. Sarma, B. C. Reddy, K. R. K. Reddy, A. R. Podile and J. W. Kloepper: *Plant Growth Promotion by Rhizobacteria for Sustainable Agriculture*, Scientific Publication, Jaipur, 2010.

Madhusudan Mishra and Prabhakar Sharma: *A Textbook of Plant Genetics*, Wisdom Press, Delhi, 2012.

Mukta Bhargava: *A Textbook of Plant Diseases and Their Control (2 Vols-Set)*, Dominant Publication, Delhi, 2007.

Mushtaq A. Wani: *Soil, Plant and Water Analysis Manual*, Agrotech Publishing Academy, Delhi, 2011.

N.S. Kute and A.R. Aher: *Principles of Plant Breeding*, Agri-Biovet Press, Delhi, 2013.

N.S. Subba Rao: *Soil Microorganisms and Plant Growth*, Oxford & IBH Publishing House, Delhi, 1995.

Nafees A. Khan and Sarvajeet Singh: *Abiotic Stress and Plant Responses*, I.K. International Publication, Delhi, 2008.

P. Vidhyasekaran: *Principles of Plant Pathology*, CBS Publication, Delhi, 2010.

P.C. Ram and G.S. Chaturvedi: *Abiotic Stresses and Plant Productivity*, Aavishkar Publication, Delhi, 2010.

P.C. Trivedi: *Plant Physiology in Agriculture and Forestry*, Aavishkar Publication, Delhi, 2009.

P.K. Gupta: *Soil, Plant, Water and Fertilizer Analysis*, Agrobios Publication, Jaipur, 2009.

Piya Chatterjee: *A Time for Tea : Women, Labor and Post-Colonial Politics on an Indian Plantation*, Zubaan, 2003.

Ravinder Choudhary: *Agriculture Growth in India*, Educational Publication, Delhi, 2011.

Robert W. Allard: *Principles of Plant Breeding*, Wiley India, Delhi, 2010.

S.S. Chahal, I.B. Parashar, H.S. Randhawa and S. Arya: *Achievements and Prospects in Mycology and Plant Pathology*, International Publication, Delhi, 1997.

Seshadri Kannan: *Advances in Plant Physiology : From the Green to the Grain*, APH Publication, Delhi, 2007.

Shagufta: *Soil, Plant, Water and Fertilizer Analysis*, A P H Publication, Delhi, 2011.

Shilpa Shyam, H.N. Verma and S.K. Bhargava: *Air Pollution and It's Impact on Plant Growth*, New India Publication, Delhi, 2006.

Susheela M. Das: *A Textbook of Plant Physiology: Theory and Objectives*, Wisdom Press, Delhi, 2011.

Susheela M. Das: *A Textbook of Plant Taxonomy : Theory and Objectives*, Wisdom Press, Delhi, 2013.

Trevor A. Thorpe: *Plant Tissue Cutlure : Methods and Applications in Agriculture*, Academic Press in imprint of Elsevier, 2013.

U.N. Bhale and V.S. Sawant: *Plant Science for Sustainable Agriculture and Rural Development*, Daya Publication, Delhi, 2012.

William E. Fry: *Principles of Plant Disease Management*, Academic Press an imprint of Elsevier, 2013.

Index